W9-CPB-645

Liberal Arts Chemistry
Worktext and Laboratory Manual
Fourth Edition

Otis S. Rothenberger
James W. Webb

Illinois State University
WWW Home Page Information
chemagic.com

KENDALL/HUNT PUBLISHING COMPANY
4050 Westmark Drive   Dubuque, Iowa 52002

Copyright © 1984, 1988, 1994, 1998 by Kendall/Hunt Publishing Company

ISBN 0-7872-9986-3

All rights reserved. No part of this publication may be reproduced,
stored in a retrieval system, or transmitted, in any form, or by any
means, electronic, mechanical, photocopying, recording, or otherwise,
without the prior written permission of the copyright owner.

Printed in the United States of America
10 9 8 7 6 5 4

"Sir! I don't teach history. I teach essence!"

Robert L. Crouch
Professor of Political Science

"How do I get there?"

Dorothy, A Little Girl
L. Frank Baum, American Journalist

*"You must walk. It is a long journey through a country
that is sometimes pleasant and sometimes dark and terrible.
However I will use all the magic arts I know of to keep you from harm."*

Good Witch of the North, A Good Witch
L. Frank Baum, American Journalist

Contents

Preface

This is a book about chemistry. It is designed for people who are not entirely sure they know what chemistry is. It is also designed to deal with some very specific objectives, which are stated below in the form of five questions:

1. What is chemistry?
2. How is chemistry done?
3. What kind of people do chemistry?
4. What questions do chemists ask and attempt to answer?
5. What services do chemists offer society?

This book attempts to answer these questions.

This is not a book about lofty "relevant" contemporary issues. The relevance of this book is limited to its stated objectives. It is written in the belief that there is an essence of chemistry that the nonscientist can readily understand. There is a proper order to all things, and the consideration of lofty "relevant" contemporary issues must follow an appreciation of essence. For example, the two molecular models on the front cover represent a useful theoretical code. Understanding this code is prerequisite to understanding the relevance of the two chemical compounds represented by this code. The topmost model (1,1,1-trichloro-2,2,2-trifluoroethane) represents a compound that can be used as a refrigerant. The bottom-most model (hydantoin) represents a compound that is used by the pharmaceutical industry. This textbook will help elucidate these seemingly mysterious codes. Following this elucidation, the related relevance can be approached with a more complete understanding.

The teaching vehicle of this book is the nineteenth century. The book, however, is not about the history of chemistry. It is about contemporary chemistry. There are two reasons that the nineteenth century is a useful vehicle for teaching the essence of chemistry. First, there are many aspects of modern chemistry that are products of nineteenth century thinking. For example, the chemical calculations introduced in Chapter 5 are still an important aspect of the modern chemist's daily routine. The second reason is even more important than the first. Examples of nineteenth century chemical reasoning can be appreciated in their entirety, even by nonscientists. This includes, of course, both the qualitative and the quantitative reasoning modes that are appropriate to the study of chemistry. To appreciate the full "flavor" of the essence of chemistry, it is necessary to experience chemical thinking. To this end, the nineteenth century is a particularly appropriate teaching device.

There are two major changes in this edition of the textbook. First, the chapter end problems have been expanded to include household chemistry experiments, library research problems, and formal laboratory exercises. Taken together, these research approaches emphasize the mechanism by which modern chemists continue to learn. A more complete discussion of these extensions to the chapter end problems can be found in the Chapter 1 problem section. It should, however, be stated in this Preface that the formal laboratory exercises do not represent an attempt at discovery learning. These laboratory exercises are designed to supplement textbook chapter content and to give students an opportunity to experience some of the things that chemists actually do. In many of the experiments, "unknown" identities are revealed in the introduction to the laboratory exercise. At Illinois State University, these formal laboratory exercises are an important part of the course, but laboratory instructors allow students to work on an experimental procedure until they get satisfactory results. Since "unknown" identity is not an issue in this approach, Appendix III of this textbook is a *brief* instructors manual for the formal laboratory exercises. This appendix lists all the materials and chemicals that are required for the formal laboratory exercises. The solution concentrations are also listed in this appendix.

The second major change in this edition of the textbook is the inclusion of World Wide Web (WWW) support. Textbook chapters are strongly supported by interactive World Wide Web resources designed specifically for students using this textbook. These resources include interactive review problems, animations of textbook figures, calculation tools, interactive three dimensional molecular models, and reviews of recent literature articles related to textbook topics. All of these resources can be freely accessed from the "Liberal Arts Chemistry: Worktext" home page. The URL for this home page is chemagic.com.

Most courses in chemistry designed for nonscience majors follow the development of chemistry into the twentieth century. This is true of the nonmajors course taught at Illinois State University. The material contained in this textbook serves as a foundation for understanding these twentieth century chemical developments.

The title of this book represents an attempt to identify an audience and a philosophy. To the extent that a book title should identify content, two alternative titles are suggested:

"The Molecule—The Evolution of A Human Idea"
"The Essence of Chemistry"

Bon Voyage
Otis S. Rothenberger
James W. Webb
Normal, Illinois

Chapter One

Chemistry ? _____

1.1 Introduction

We are living in an age when chemistry touches almost every facet of our daily lives, and yet many educated people have only a cursory understanding of this important discipline. A close look at the ingredients listed on the label of any processed food or medicine shows one example of the daily influence of chemistry. Additive lists such as the one shown in Figure 1.1 also raise some very interesting questions. What is the function of each additive? Are the additives totally safe? Who decides that a particular additive is to be allowed? In a free society, who should decide that a particular additive is to be allowed? Understanding the answers to all these questions demands more than a cursory knowledge of chemistry, and additives represent only a tiny percentage of our daily interface with chemistry.

There is a modern trend in the teaching of nonscience major chemistry to attack these "relevant" questions directly. This approach makes about as much sense as trying to teach a person to swim without the use of water. As outlined in the preface, one of the purposes of this book is to tell you something about the basic nature of chemistry. That something will be the raw material out of which you will fabricate your own answers to the "relevant" questions. In this spirit, we begin our study of chemistry with a very fundamental question. What is chemistry?

1.2 The Science of Chemistry

The division of human knowledge into academic disciplines can be a very artificial exercise. There is one body of human knowledge, and chemistry is part of that body. The artificial nature of disciplinary divisions can be illustrated by considering the similarities in the following two quotations:

> "To tell us that every species of things is endowed with an occult specific quality by which it acts and produces manifest effects, is to tell us nothing. But to derive two or three general principles of motion from phenomena, and afterwards to tell us how the properties and actions of all corporeal things follow from those manifest principles, would be a very great step in Philosophy . . ." – Sir Isaac Newton, *Opticks* (1730)

Active Ingredients: Special Vicks Medication (menthol, thymol, eucalyptus oil, camphor, tolu balsam, benzyl alcohol) in a soothing Vicks' sugars base.
Vicks is reg. U.S. Pat. & T.M. Off.

Figure 1.1 Vicks Medicated Cough Drops

"We hold these truths to be self evident, that all men are created equal, that they are endowed by their Creator with certain unalienable Rights, that among these are Life, Liberty, and the pursuit of Happiness." – Thomas Jefferson, *Declaration of Independence* (1776).

Although these quotations are apparently different, both reflect the same philosophical revolution. Both reflect a belief in fundamental principles of determinism. But academic disciplines are useful. Few people have the intellect to deal with human knowledge *en masse*, and so, we learn about the work of Newton in physics class and the work of Jefferson in political science class.[1] We also learn about the discipline called chemistry in chemistry class.

In order to understand what chemistry is, you must understand that chemistry is a **natural science**, and as such it is based on certain philosophical assumptions that the natural sciences hold in common:

Basic Assumptions of the Natural Sciences

1. The real world is composed of matter.
2. Matter is in a constant state of change.
3. This change is measurable.
4. By the process of rational thought, this change is understandable.

You may be surprised that these statements are called assumptions, but in the context of total human knowledge, they are not necessarily true. What this means is that the scientific concept of truth is tainted by these assumptions. This is no trivial point, and we shall see more of it later.

Well then, what is chemistry? At this point, it is easy to define. Of all the changes that matter undergoes, perhaps the most mystifying is its ability to change identity. Interest in this type of material change dates back to humans in their most primitive form, and the origin of chemistry is that old. **Chemistry** is the natural science that attempts to understand the ability of matter to change its identity. The study of chemistry is represented schematically in Figure 1.2.

1.3 The Nature of Scientific Thought

At this point in your study of chemistry, it is important that you understand the nature of scientific thought. The objective of this section is to investigate this topic from a general philosophic point of view. In the next section of this chapter, we will deal with the specifics of scientific reasoning. For convenience, illustrative examples in both sections are drawn from natural sciences other than chemistry. Although the traditional core natural sciences include biology, chemistry, and physics, these core disciplines form a tree with many branches. Medicine, psychology, astronomy, geology, and many more are all branches of the natural scientific tree. For the sake of convenience, the various natural sciences are often simply referred to as **sciences**.

Matter I → Matter II

Figure 1.2 The Study of Chemistry

---

1.   The late Jacob Bronowski was a notable exception to this rule of intellect, and his *The Ascent of Man* is a beautiful example of knowledge without barriers. J. Bronowski, *The Ascent of Man*, Little, Brown and Co., Boston (1974)

Throughout history, science has been assaulted by nonscientific thought processes alien to its basic nature. Even today, the fads associated with modern pop science have risen almost to the level of a scientific counter culture. The issue in controversies between science and nonscience has always involved a basic question: What is science? Perhaps the easiest way to answer this question is with a historical example.

In the second century A.D., the Greek astronomer Claudius Ptolemy summarized many centuries of astronomical observation with the publication of the *Almagest*. In this work, the geocentric model (theory) of the solar system was chosen as a convenient calculating mechanism for predicting astronomical events (Fig. 1.3A). According to this view, the apparent motions of the sun, moon, and planets relative to the fixed stars from night to night was explained by placing the earth at the center of the planetary system. The nightly progression of the sun, moon, and planets could then be explained by the orbits (cycles) and orbits within orbits (epicycles) of these heavenly objects. The epicycles were necessary in order to explain the occasional retrograde motion of the planets. The usual night-to-night progress of a planet relative to the fixed stars is from west to east. Occasionally, planets reverse this apparent motion and move east to west relative to the fixed stars. The planetary epicycles explained this retrograde motion.

The Ptolemaic model was a thing of beauty. Many students of modern science are taught to view this Earth centered model as a product of narrow thinking, but they are not taught this view by scientists. The Ptolemaic model was a masterpiece. What made it a masterpiece was that it was simple, and it worked. It did predict astronomical events. The geocentric model was useful and intellectually satisfying, but these qualities did not imply ultimate truth.[2] In this respect, the Ptolemaic model was good science.

We have arrived at a point of major importance. What is a **scientific model**? A scientific model is a way of rationalizing the behavior of nature. Because a scientific model rationalizes the behavior of nature, it becomes a tool for predicting the behavior of nature. In addition to being a useful predictor of nature, a scientific model is also one of the most efficient means of communication ever devised by human kind. With a scientific model, centuries of sensory experience can be conveyed to young minds in a very short time period. Minds thus relieved of sensory minutiae can go off in new practical and creative directions. The Ptolemaic model served this function for fourteen centuries! But what of the Copernican revolution and the "correct" model (Fig. 1.3B)?

Many scientific models share a common fate. As the body of human knowledge grows, they become more complex. At some point, they become so complex that they are no longer useful as prediction or communication devices. This is exactly what happened to the Ptolemaic model at the end of the fifteenth century when the Polish mathematician Nicolaus Copernicus developed the heliocentric theory. This theory, which was published in *De Revolutionibus* following the death of Copernicus in 1543, placed the sun at the center of the planetary system. According to the Copernican model, the retrograde motion of the planets was a natural consequence of the earth's noncentral position. The swing toward the Copernican model was not a swing from a false idea to ultimate truth, but rather it was a swing from a useless idea to useful idea.

It is important to note a few characteristics of scientific models. First, they rarely explain all the facts perfectly. They are, therefore, constantly tested and modified. This evolutionary process is probably related to the fact that the universe is infinitely complex, but the human mind is finite. Second, their utility as prediction and communication devices is much more important than their ultimate truth. Ultimate truth involves the infinite. It is the subject matter of theology, and science can't touch it. Finally, their utility as prediction and communication devices is so important that they are dogmatically retained. Science must be dogmatic. Frequent revolution is not compatible with the kind of activity demanded in science.

Well then, what is science? Science is a rational attempt to communicate sensory perceptions from one generation to the next by the simplest possible predictive mechanism or model. An essential feature of science is that scientific models provide a highly useful language that can be used to express real problems in such a way that they are rendered solvable. The utility of science in general, and chemistry in particular, is a recurrent theme in this text.

---

2.   The term *ultimate truth* is not being used in this discussion merely to imply correctness. For a further explanation, please continue reading.

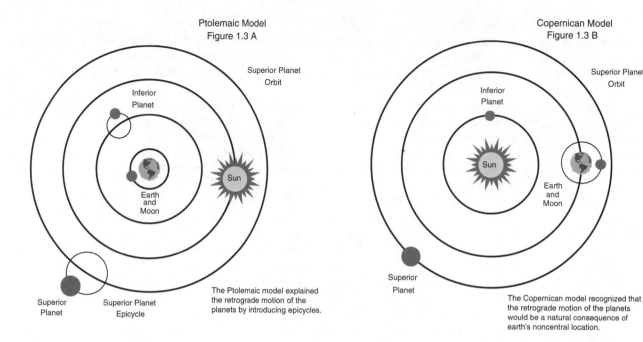

Figure 1.3 A Scientific Revolution

1.4 The Scientific Method

A key phrase that was used in the definition of chemistry is *rational thought process*. The purpose of this section is to examine the nature of the thought processes that are used in chemistry and the other natural sciences. An important point should be made at the onset of this discussion. In order to determine if a particular investigation falls into the discipline of chemistry, how the investigation is carried out is equally important as what is being investigated.

Human beings reach conclusions by relationships called **arguments**. Consider the following two examples:

Argument I
If that object is a bird, then it has wings.
That object has wings.
Therefore, that object is a bird.

Argument II
If that object is a bird, then it has wings.
That object is a bird.
Therefore, that object has wings.

If the letter A is allowed to represent the statement "that object is a bird," and if the letter B is allowed to represent the statement "that object has wings," then the two argument forms reduce to the following symbolic forms:

Argument I	**Argument II**
If A, then B	If A, then B
B	A
Therefore, A	Therefore, B

If it seems to you that there is something disquieting about argument I, you are correct. Any argument fitting the form of argument I is said to be **invalid**. Conversely, an argument fitting the form of argument II is said to be **valid**. Technically, argument II is referred to as *modus ponens* (Latin *ponere*—to affirm), and it is an example of a valid mixed hypothetical **syllogism**. Argument I is referred to as the fallacy of affirming the consequent. It is important to note that it is the logical form of these arguments that determines whether they are valid.

Although argument I is invalid, it is useful in the process of reasoning from individual facts to general principles—reasoning by **induction**. One simple use of induction, but not the only use in science, is to derive useful summaries of many factual observations. In science, a statement that summarizes many facts is called a **law**. Argument form II is useful in reasoning from general principles to individual facts—reasoning by **deduction**. For example, deduction is used in mathematics to draw conclusions from general mathematical principles.

EXAMPLE I

Analysis of an Argument Form

It is the logical form of an argument that determines the validity of an argument. Is the following argument form valid or invalid?

> If A, then B
> Not A
> Therefore, not B

SOLUTION: invalid

Replace *A* and *B* with statements such that the first line of the argument is a self-evident proposition. A proposition is self-evident if its denial results in a contradiction. Thus, the proposition "All birds have wings" is self-evident because the proposition "It is not the case that all birds have wings" is a contradiction. By definition, birds are feathered vertebrates with modified forelimbs called wings.

After establishing the first statement in the argument as a self-evident proposition, use intuitive logic to assess the entire argument. Does the argument reach a conclusion that has contradictions, or does the argument lead to a conclusion with no contradictions? This approach might outrage a professional philosopher or logician, but it allows the consideration of a topic essential to understanding science. Rigorous analysis of validity would require an entire course in logic![3]

> If that thing is a bird, then it has wings.
> That thing is not a bird.
> Therefore, it does not have wings.

By intuitive logic, the argument form is invalid. What about airplanes? Technically, this argument form is referred to as the fallacy of denying the antecedent.

□
---

3.   For example, a logician would demand a much more rigorous definition of a self-evident proposition than the one being used in this discussion. The logician would also demand a much more rigorous description of intuitive logic.

EXAMPLE II

Analysis of a Real Life Argument

Which of the following argument forms is used when the conclusion is made that a person is guilty based on circumstantial evidence?

Argument I	**Argument II**
If A, then B	If A, then B
B	A
Therefore, A	Therefore, B

SOLUTION: argument I

Consider, for example, fingerprints. If Moriarty committed the murder, then his fingerprints will be on the weapon. Moriarty fingerprints are on the weapon. Therefore, Moriarty committed the murder.

This is invalid argument I. Confidence in the conclusions drawn by this method can be increased by pursuing a large number of hypothetical predictions.

EXAMPLE III

Analysis of a Scientific Law

A scientific law is a statement based on the observation of repeated events. A trivial example of a scientific law is the statement "All crows are black." This generalization is based on many observations of crows. Create an if/then argument to show that this statement is an example of inductive reasoning.

SOLUTION:

If all crows are black, then all observed crows are black.
All observed crows are black.
Therefore, all crows are black.

This is an example of the fallacy of affirming the consequent. Of course, confidence in scientific laws is bolstered by basing them on a huge number of observations. In the world of everyday events, we place great trust in scientific laws. Still, it is important to appreciate that a scientific law is the product of inductive reasoning. For example, in the modern science of quantum mechanics, the possibility of the failure of the next induction is not a trivial matter.

Now by historical example, let us investigate further how these argument forms come into play in the scientific thought process.

In the 1850's, the process of fermentation became the subject of an intensive scientific investigation. This investigation was the outgrowth of a revolutionary research program undertaken by the Dean of the Faculty of Sciences at the University of Lille in France. Under this program, research efforts would be directed toward the practical problems of local industry! Since the University of Lille was in an area of France that contained a significant fermentation industry, it was not surprising that the problems of this industry soon became the problems of the university and its dean, the chemist, Louis Pasteur.[4]

Pasteur's researches into industrial fermentation eventually led to consulting work with the French wine industry. In order to understand the nature of one of the wine industry's problems investigated by Pasteur, it is necessary to understand some of the basics involved in the wine making process. In manufacturing wine, the first step is to prepare a "mash" or crushed grape mixture. If the "mash" is allowed to stand at an appropriate temperature, natural yeast will convert the grape sugar into alcohol. The yeast cells, which are unicellular fungi, are carrying out a chemical transformation that is known as fermentation. During fermentation, the yeast cells are carrying out normal life processes, and they are using the grape sugar as food, converting it into alcohol as a waste product. The industrialization of this natural process to produce consumable alcoholic beverages (beer, wine, mead, etc.) represents an example of chemical technology (Fig. 1.4). The origin of this technology is almost as old as human civilization. Laws governing the production and sale of beer can be found in the code of King Hammurabi (*circa* 1900 B.C.).

In the production of wine, the mixture that results after fermentation will produce the desired effect if consumed, but as every wine connoisseur knows, the true qualities of wine require aging. Hence the wine ferment is allowed to stand in the bottle for long periods of time (years!), and it is during this aging process that the wine develops flavor, aroma (bouquet), and other delicate qualities. It is important to realize that at the time of Pasteur the role of yeast as a living organism was not understood. In fact, although the microscope was a common laboratory tool at the time, the general presence of microscopic life was not recognized.

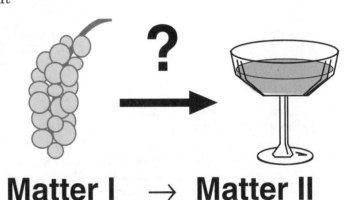

Figure 1.4   Early Industrial Chemistry

The nature of one important wine industry problem to be investigated by Louis Pasteur is now easily stated. A considerable quantity of French wine was turning sour during the aging process. This represented a serious economic problem. Sour wine meant loss of revenue, and the loss of revenue meant an increase in human suffering. It was to such practical problems that Pasteur wished to dedicate his science, and so began the "Studies on Fermentation." It should also be noted that Pasteur helped to usher in a new academic era. The ivory tower of the university would never be the same. The practical problems of society had assaulted the ivory tower, and they were there to stay.

Pasteur's approach to this problem was straightforward. His observations were simple, and his conclusions were elegant. Since we are interested in the method by which these conclusions were reached, we will briefly review Pasteur's approach to the problem, as well as his observations and conclusions.

The approach that Pasteur took in dealing with this problem was to compare the observable properties of sour (diseased) wine with the observable properties of normal (healthy) wine. Among other comparisons, Pasteur was eventually led to a microscopic observation of the two products. Pasteur noted that the sediment of diseased wine always had a larger variety of microscopic shapes. Present in both wine sediments were plate-like microscopic shapes, but the diseased wine also contained a variety of rod-like shapes. Could it be, Pasteur hypothesized, that the microscopic shapes were really living organisms that fed on the grape sugar, one producing alcohol during fermentation, the other producing acid (sour) during aging? This hypothesis could be tested by a simple experiment. Allow fermentation to

---

4.   In the following discussion, it is important to realize that the process of fermentation was not understood until after the work of Pasteur. See for example: L. Pasteur, "Studies on Fermentation: The Diseases of Beer," Trans. F. Faulkner and D. Robb, Macmillan and Co. London (1879)

take place, and then, by some method, kill all microscopic life before aging. For this microscopic massacre, Pasteur settled on gently heating the fermented product, and, sure enough, gently heated (Pasteurized!) wine ferment always produced healthy wine. The hypothesis had become a conclusion.

Pasteur went on to even greater fame. From sick French wine caused by microscopic life (germs), Pasteur went on to solve the problem of sick English beer. Pasteurized English beer no longer arrived at British colonies in a soured state. Next there was the problem of the French silkworms that were producing an unusable quality of silk fiber. Could it be that the now infamous germs infect not only wine and beer, but also silkworms? Indeed, Pasteur solved the silkworm problem by applying the germ hypothesis, which had become a conclusion. Finally, in perhaps the greatest triumph of modern medicine, the germ hypothesis was extended by Pasteur to the sickness of all living things, and the germ theory of disease was born. All this from a young chemist with a simple goal, the application of his craft to the practical problems of society.

Well, enough hero worship. Let us return to the central question. By which of the two argument forms did Pasteur convert his hypothesis into his conclusion? Return to the italicized paragraph that describes Pasteur's reasoning. Identify the observed facts (F), the hypothesis (H), the testing experiment(s) (E), and the conclusion. Now notice that the entire paragraph can be represented by the following scheme: 1) The facts suggest a hypothesis. 2) If the hypothesis is correct, then certain experiments follow. 3) These experiments work. 4) Therefore, the hypothesis is correct. This scheme is shown in symbolic form below:

$$F \text{ suggests } H$$
$$\text{If } H, \text{ then } E$$
$$E$$
$$\text{Therefore, } H$$

This scheme is the essence of the **scientific method**, and it is the "how" of chemical investigations. *F suggests H* is a complex process that involves both argument forms I and II, induction and deduction. This process also involves at least one other argument form that is discussed in problem six at the end of this chapter. By another name, *F suggests H* is called human insight, and you understand it intuitively because you possess it. Not all humans possess the same insights, of course, but we share a common bond as humanity. We all reason the same way, and thus we can share in each other's insights. We can all appreciate a Louis Pasteur or a Marie Curie.

The final point is a most important one. It is a fact about the scientific method that is misunderstood by many nonscientists, and this misunderstanding leads to many unnecessary conflicts between science and nonscience. Observe how Pasteur's hypothesis becomes his conclusion:

$$\text{If } H, \text{ then } E$$
$$E$$
$$\text{Therefore, } H$$

This is argument I. In becoming a conclusion, the hypothesis has passed through argument form I. To state that a scientific **hypothesis** is an attempt to explain fact, and that a scientific **theory** is a generally accepted hypothesis, is only partially correct. The process by which these scientific conclusions are reached is an important part of their definition. Although confidence in the conclusion can be increased by pursuing a larger number of hypothetical predictions (experiments), it is important to note how modern science uses this invalid argument form. The process of formulating a hypothesis and using this hypothesis to suggest testing experiments is called the hypothetico-deductive method. Although this technical term is useful because it emphasizes the importance of induction and deduction in scientific reasoning, the hypothetico-deductive method is more simply called the scientific method. At issue in this process of arguing from experience is not so much the ultimate truth of its conclusions (hypotheses and theories), but rather their utility. It is perhaps a weakness of our language that truth has but one meaning:

"There are trivial truths and the great truths. The opposite of a trivial truth is plainly false. The opposite of a great truth is also true." – Niels Bohr, Danish Physicist

1.5 An Important Note about Review Problems

At the end of each chapter in this book, there is a list of review problems. These problems are designed to reinforce important concepts contained within the chapter. The problems should be considered a logical extension of reading the chapter, and very often they will form the basis of class discussion. Answers to many of the end-of-chapter problems are located at the back of this text.

In addition to understanding the concepts that are presented in each chapter, testing success in chemistry demands the development of some very specific basic skills. Ideally, course examination structure should not be a major issue. However, it is through the examination process that students demonstrate their ability to think logically using the basic concepts of chemistry. During a chemistry examination, students must think like chemists. The important issue is the logical solution of chemical problems, not the mere memorization of facts.

In an attempt to facilitate the learning of basic skills, a list of performance objectives is presented at the end of each chapter. The first performance objective in each list reviews the boldfaced terms presented in the chapter. For some of the boldfaced notations in the chapter, formal definitions are given, but usually, the words are simply noted in context. It is not as important to memorize definitions for these terms as it is to understand their usage. The remaining performance objectives, which represent skills that must be developed to ensure testing success, should be mastered before attempting the problems at the end of the chapter.

1.6 An Important Note about Internet Support

"Liberal Arts Chemistry: Worktext" is strongly supported by interactive World Wide Web (WWW) resources designed specifically for students using this textbook. These resources include interactive review problems, animations of textbook figures, calculation tools, interactive three dimensional molecular models, and reviews of recent literature articles related to textbook topics. All of these resources can be freely accessed from the "Liberal Arts Chemistry: Worktext" home page. The current URL for this home page can be obtained by sending an e-mail request to the authors at the address listed on the title page.

Textbook chapter end problems address a wide range of of learning issues, and it is sometimes difficult for students to focus on specific testing skills in such a varied problem list. The "Liberal Arts Chemistry: Worktext" home page menu is link indexed according to textbook chapter. The interactive problems associated with each of these chapter links are updated each teaching semester using the Illinois State University Liberal Arts Chemistry examinations from the previous semester. Although examination philosophy may vary from one institution to another, the WWW interactive problems are an excellent starting point for homework assignments. These WWW interactive problems are actual examination questions structured to provide instructor (textbook author) feedback for every homework answer submitted by students.

Chapter One
Performance Objectives

P.O. 1.0

Review all of the boldfaced terminology in this chapter, and make certain that you understand the use of each term.

argument	chemistry	deduction
hypothesis	induction	invalid
law	natural science	science
scientific method	scientific model	syllogism
theory	valid	

P.O. 1.1

You must be able to assess the validity or invalidity of a symbolic argument form. The approach used to assess validity or invalidity is intuitive.

EXAMPLE:
Is the following argument form valid or invalid?

> If A, then B
> B
> Therefore, A

SOLUTION: invalid

Textbook Reference: Section 1.4

ADDITIONAL EXAMPLE:
The following mixed hypothetical syllogism is referred to as *modus tollens* (Latin *tollere*—to deny). It is used in the process of disproving scientific theories. Is the *modus tollens* a valid or invalid argument form?

> If A, then B
> Not B
> Therefore, not A

ANSWER: valid

P.O. 1.2

You must be able to determine the symbolic argument form being used in a real life situation.

EXAMPLE:

There is a fundamental principle in mathematics that states that two quantities that are equal to the same quantity are also equal to each other. If this is true, consider the following conclusion regarding the quantity X and the quantity Z:

$$X = Y$$
$$Z = Y$$
$$\text{Therefore, } X = Z$$

Which of the following argument forms was used in arriving at this conclusion?

Argument I	**Argument II**
If A, then B	If A, then B
B	A
Therefore, A	Therefore, B

SOLUTION: argument II

Textbook Reference: Section 1.4

ADDITIONAL EXAMPLE:

Each morning a man walks to work by an identical route. When he arrives at his office, he must reach the conclusion that this is, indeed, his office. He reasons as follows:

"I know that if I follow a certain route, then I will come to a building which is my office. Today I followed that route, therefore this building must be my office."

Is the man's argument valid or invalid?

ANSWER: valid

Chapter One
Problems

General Questions

1. Both astronomy and astrology can be defined as a study of heavenly objects. Of the two, however, only astronomy is a natural science. Explain how two disciplines devoted to identical studies can differ.

   STUDENT SOLUTION:

2. One of the important objectives of this chapter was to explain the basic nature of chemistry. Using your own words, define the academic discipline called chemistry.

   STUDENT SOLUTION:

3. Occasionally, scientific models become accepted as fact. Can you think of any examples of scientific models that have successfully made the transition from theory to fact?

   STUDENT SOLUTION:

4. The figure printed below (Fig. 1.5) represents the total estimated fertility rates in the United States from 1800 to 1976. Can you formulate hypotheses to explain the trends in U.S. fertility rates?

   STUDENT SOLUTION:

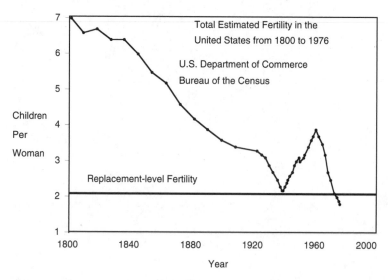

Figure 1.5   Fertility of U.S. Women 1800–1976

Argument Forms and the Scientific Method

5. Since it is the logical form of an argument that determines whether or not it is valid, the argument form is a useful tool in the study of logical relations. By making appropriate substitutions in the following argument forms, determine whether or not they are valid.

invalid   *vg*   *invalid*   *invalid*

Argument I	**Argument II**	**Argument III**	**Argument IV**
If A, then B	If A, then B	If A, then B	If A, then B
B	A	Not A	Not B
Therefore, A	Therefore, B	Therefore, not B	Therefore, not A

STUDENT SOLUTION:

6. As indicated in the text, the argument form by which scientific theories are verified is invalid. If *A* in argument IV above represents a hypothesis and *B* represents a testing experiment, then this argument form becomes the one by which scientific hypotheses are rejected. Prove this to yourself by placing appropriate statements into this argument form. Compare the logic of hypothesis verification and rejection.

STUDENT SOLUTION:

7. Consider the following paragraph from Arthur Conan Doyle's "A Study in Scarlet." In this paragraph, Sherlock Holmes explains how he reached a certain conclusion about Dr. Watson. Which of the argument forms presented in question one is Holmes using to reach this conclusion?

    "Nothing of the sort. I knew you came from Afghanistan. From long habit the train of thoughts ran so swiftly through my mind, that I arrived at the conclusion without being conscious of intermediate steps. There were such steps however. The train of reasoning ran, 'Here is a gentleman of a medical type, but with the air of a military man. Clearly an army doctor, then. He has just come from the tropics, for his face is dark, and that is not the natural tint of his skin, for his wrists are fair. He has undergone hardship and sickness, as his haggard face says clearly. His left arm has been injured. He holds it in a stiff and unnatural manner. Where in the tropics could an English army doctor have seen much hardship and got his arm wounded. Clearly in Afghanistan.' The whole train of thought did not occupy a second. I then remarked that you came from Afghanistan, and you were astonished."

STUDENT SOLUTION:

8. Consider the following quote:

"It is wise to remember that a law obtained by the process of induction may at any time be found to have limited validity. Conclusions that are reached from such a law by the process of deduction should be recognized as having a probability of being correct that is determined by the probability that the original law is correct." – Linus Pauling, American Chemist

Paraphrase this quote by making use of the following argument forms:

If A, then B	If A, then B	If A, then B
B	A	Not B
Therefore, A	Therefore, B	Therefore, not A

STUDENT SOLUTION:

9. Prior to 1970, the composition of the surface of the moon had to be inferred by observations made at a distance of 240,000 miles. Consider the following inferences:

Inference A
The moon is yellow. Cheese is yellow. Therefore, the moon is made of cheese.

Inference B
The moon is yellow and pot-marked. Swiss Cheese is yellow and pot-marked. Therefore, the moon is made of Swiss Cheese.

Identify the argument form that is being used in each of these inferences. Is the argument form valid or invalid?

STUDENT SOLUTION:

10. The properties of sunlight reflected from the moon are not the same as the properties of sunlight reflected from Swiss Cheese. How does this fact affect the validity of either inference in question 9? Recall that the validity of an argument is determined by its form. How does this fact affect the scientific acceptability of either inference according to the scientific method?

STUDENT SOLUTION:

11. Suppose that the properties of sunlight reflected from the moon *were* the same as the properties of sunlight reflected from Swiss Cheese. How would this fact affect the validity and scientific acceptability of the inferences in question 9?

STUDENT SOLUTION:

Problems Involving Household Chemistry and Science
Library Problems
Formal Laboratory Exercises

The authors of this text discovered chemistry at an early age by mixing household staples at a kitchen sink. During these early years, there was a sense of wonder associated with the color changes and the other effects of this mixing. Tragically, passage into the adult years brings a level of sophistication that can totally stifle childhood wonder. Although the indiscriminate mixing of household chemicals can be dangerous, the authors feel an obligation to end each chapter in this text with examples of reasonably safe household chemistry problems. These problems are dedicated to those adult students who recognize the importance of keeping childhood wonder alive.

Although the experiments described are reasonably safe, they should be performed with great regard for safety. Warning labels on all household chemical containers should be read before doing experiments. It is also essential to wear eye protection. This latter point is particularly important because most people do not normally wear safety glasses during casual exposure to household chemicals. Performing experiments with these chemicals, however, does not represent casual exposure. For example, the use of tincture of iodine in the experiment at the end of Chapter 5 is not the same as a casual use of tincture of iodine as an antiseptic. Readers of this text who do any work in a home workshop are already familiar with the concept of basic eye protection, and inexpensive plastic safety glasses can be purchased in any home workshop supply store.

Following the problems involving household chemistry and science, two library problems involving the *Journal of Chemical Education* will be presented. The *Journal of Chemical Education*, which was founded in 1924, continues to be an excellent source of chemical information for chemistry teachers and chemistry students. Library problems are a logical extension of the household chemical investigations. Together, these research approaches emphasize the mechanism by which modern chemists continue to learn.

Finally, each chapter will conclude with a formal laboratory exercise. Each of these exercises explores a laboratory oriented problem related to the associated chapter. The problems are stated in a more formal manner than the household chemistry problems, and the experiments associated with these problems are designed to be done in a supervised chemical laboratory. Safety issues associated with the equipment and chemicals used in these experiments require proper pre-laboratory instruction. An essential prerequisite to this pre-laboratory instruction is the complete reading of the laboratory exercise. Appendix III lists all of the materials and chemicals required for the formal laboratory exercises. All of the solution concentrations are also listed in this appendix.

Problems Involving Household Chemistry and Science

12. Most ten-year-old "chemists" working at the kitchen sink eventually discover the wonders associated with a bottle of vinegar. Add a tablespoon of vinegar to ⅛ teaspoon of each of the following chemicals found in the kitchen: corn starch, baking soda, and table salt. Which of these solid interactions with vinegar would most ten-year-old "chemists" mark for further investigation?

Corn Starch + Vinegar: ?
Baking Soda + Vinegar: ?
Table Salt + Vinegar: ?

Baking *powder* is a multi-component kitchen chemical. Repeat the above experiment with baking *powder* and vinegar. Try the experiment with baking *powder* and water. One of the important topics discussed in the first chapter was the scientific process of formulating and testing hypotheses. Based on the observations made during this experiment, formulate a hypothesis about the possible composition of baking *powder*. What other tests could be performed to help verify this hypothesis?

If a fresh egg in small glass is covered with vinegar . . . But that's another story.

STUDENT SOLUTION:

13. Align five pennies in a row so that each penny is touching. Move a penny on one end about 5 cm from the row and rapidly push the penny into the other four as indicated in Figure 1.6. Repeat this experiment by pushing two pennies into the remaining three. Based on the results of this experiment, what would happen if three pennies were pushed into the remaining two? Test your prediction. What type of reasoning was involved in your prediction? The behavior of the pennies in this experiment is controlled by a natural law called the law of conservation of momentum.

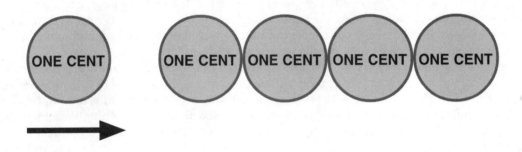

Figure 1.6   The Penny Experiment

STUDENT SOLUTION:

Chapter Two

Measurement and Arithmetic Techniques _____

2.1 Introduction

The discipline of chemistry is almost as old as humankind, and a study of the history of chemistry is filled with all of the excitement and romance that humans are capable of generating. Although this book is not intended as a book on the history of chemistry, history can provide the student with a convenient framework on which to build an understanding of chemistry. History can also give us a more complete picture of the present. For example, one of the largest chemical companies in the United States can trace its development directly to a political execution in eighteenth century France.[1]

Many historians of chemistry view the birth of modern chemistry as a product of the eighteenth century. Although many significant discoveries were made prior to the eighteenth century, there is no doubt that the study of chemistry changed drastically during this century. The purpose of this chapter is to review the measurement and arithmetic techniques that were introduced into the study of chemistry during the eighteenth and nineteenth centuries. These techniques will be the tools for the study of modern chemistry.

It should be stated at the beginning of this chapter that the review material in the chapter is presented at a very basic level. Not all students will find the material in this chapter equally useful. After many years of teaching nonscience major chemistry at the college level, however, the authors of this text have come to realize that the mathematical skills of a significant number of college students are really quite rusty. The authors have, therefore, opted to write this chapter assuming considerable student mathematical naiveté. Students who feel that they do not need this type of review are encouraged to skim this chapter, and simply verify that they can do the problems at the end of the chapter.

2.2 The Metric System of Measurement

In 1786, at the suggestion of Thomas Jefferson, the infant United States adopted a decimal currency. American children no longer learned that 4 farthings make 1 penny; 12 pennies make 1 shilling; and 20 shillings make one pound. Rather, making full use of their ten finger decimal instincts, Americans learned that 10 mills make 1 cent; 10 cents make 1 dime; 10 dimes make 1 dollar; and 10 dollars make 1 eagle (sawbuck). In 1799, when the rest of the world was adopting a decimal system for all measurements, the United States joined with Great Britain in refusing to adopt this new system, the **metric system**. Since the advantages of a decimal system were obvious to natural scientists, the metric system soon became the measurement system of science, even in the United States. And so, for most of our history, we have been a nation with two systems of measurement.

The metric system evolved in revolutionary France, and the original reasons for nonadoption may well have been political. For example, the meter was originally defined as one ten-millionth of the earth's quadrant. By using the earth as a natural universal unit, the creators of the metric system hoped to avoid arbitrary nationalistic standards that were common in older systems of measurement. For technical reasons, however, the length of the meter was originally set by measuring the distance

---

1.   See Chapter 4 for details of this story.

between Dunkerque in France and Barcelona in Spain. This arc on the earth's surface lies almost totally in France, and Thomas Jefferson felt that constituted an arbitrary national standard. Whatever the history of this situation, the bottom line is this: American chemistry students must learn the metric system; chemistry students in most other countries already know it.

Most students rebel at having to learn the metric system for three reasons:

1. Conversion of metric units to English units and vise versa is confusing.
2. A large number of new metric units need to be committed to memory.
3. A large number of decimal fraction and multiple prefixes need to be committed to memory.

In this book, an attempt will be made to deal with each of these problems. First, whenever possible we will avoid metric/English conversions. Instead, real world reference measurements will be used. For example, the fact that a United States nickel has a mass of about 5 grams is much more meaningful than being able to compute that a 5 gram mass weighs 0.176 ounces. To be sure, there is nothing difficult about conversion factors, and we will use them if their use helps to clarify a point, but in general, we will not allow conversion factors to cloud the basic logic of the metric system. Second, metric units will only be introduced as they are needed. This is a chemistry book of limited scope, and only those metric units required for understanding its content will be considered relevant. Finally, of the large number of decimal fraction and multiple prefixes that are part of the metric system, only three find common usage in routine chemistry. The student will be asked to learn only these three. With these points in mind, let us consider those aspects of the metric system that are of immediate concern to our chemical objectives.

2.3 Length Measurement

The basic unit of length in the metric system is the **meter** (abbreviated m). Although it might seem contradictory to the above paragraph, the easiest way for American students to deal with the meter is to realize that it is almost equal to one yard. What is important here is not remembering that a meter is 1.09 yards, but that a meter is about a yard (Fig. 2.1). If someone asks you to tell them how long a foot is, you probably will raise your hands and separate them by about a foot. You will not tell them, "Oh, a foot is 0.305 meters." You have a common sense of foot, and in the same way, you can have a common sense of meter. How long is a meter? I hope you are answering with a gesture.

2.4 Volume Measurement

All matter occupies space, and the amount of space that matter occupies is termed **volume**. In any system of measurement, volume can be measured in cubic linear units. Hence a cube one foot on an edge is said to occupy one cubic foot (abbreviated ft^3). In a similar manner, a unit of volume in the metric system is the cubic meter (abbreviated m^3). For the purpose of dealing with fluids (liquids and gases), it is also convenient to have a separate unit of volume or holding capacity. For example, the English system measures volume in cubic feet and gallons. The metric system also has a basic unit of holding capacity. It is the **liter** (abbreviated L), and it is defined in such a way that conversion between cubic linear units and capacity units is ridiculously simple.[2] This point will be explained later.

If you have a common sense of meter, you obviously have a common sense of cubic meter (Fig. 2.1).

---

2. Abbreviations for metric units are represented by lower case letters. If, however, the unit is named in honor of a person, then upper case abbreviations are used. The liter appears to be an exception to this rule. The word liter comes from an old French word *litron*, an obsolete volume measurement.

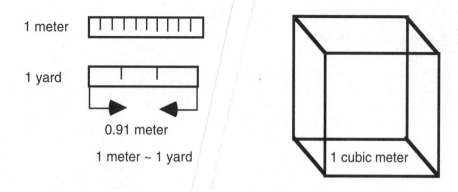

Figure 2.1 The Metric Unit of Length and Volume

1 meter

1 yard

0.91 meter

1 meter ~ 1 yard

1 cubic meter

There are several ways to relate to the liter in a common sense manner. One way is to appreciate that one standard kitchen measuring cup is almost exactly ¼ of a liter, and that if the liquid in such a cup were to be poured into a medium sized paper cup, it would fill the paper cup all the way to the top. If medium sized paper cup is too vague, try this comparison. A liter is equal to about a quart. Again, this is not meant as a conversion factor, but rather, as a touchstone to common sense.

It may seem that a system with two units for measuring volume is unnecessarily complex. In fact, from a theoretical point of view, such a system is a bit awkward, but natural science is more than just theory. An important part of any natural science is experimentation, and a dual system for measuring volume is experimentally quite useful, cubic meters for regular solids and liters for fluids. In order to deal with the theoretical problems of multiple units, not only of volume, but of other units also, world scientists agreed in 1960 to adopt an extension of the metric system called the **Système International** (SI). This system fixes one unit for each of seven different fundamental physical quantities. The units selected are selected for their theoretical utility, and all other units of measurement can be derived from these seven. In the case of volume, the derived SI unit is the cubic meter. The seven fundamental units of the *Système International* are shown in Figure 2.2.

Although the *Système International* has great scientific utility, many other units of measurement are used in common commerce. For example, the derived SI unit of energy is the kilojoule (Chapter 7), but most Americans understand nutritional energy in terms of kilocalories. Even in the scientific community, other units of measurement are common. A review of annual industrial productivity reports in the American Chemical Society's *Chemical and Engineering News* illustrates the importance of measurement units used in commerce. For this reason, some of the measurements in this textbook will be expressed as fundamental and derived *Système International* units, but the textbook will also make use of measurement units used in the market place.

Measurement	Unit Name	Abbreviation
Mass	kilogram	kg
Length	meter	m
Time	second	s
Temperature	kelvin	K
Electric Current	ampere	A
Substance Amount	mole	mol
Luminous Intensity	candela	cd

Figure 2.2 The Seven Fundamental SI Units

2.5 Mass Measurement

As we have seen, volume is the measure of how much space matter occupies. You will come to realize that a point of equal importance to the study of chemistry is the amount of matter contained in a given volume. Is volume the best measure of how much matter an object contains? If you consider a liter of balsa wood and a liter of iron, the answer to this question is obviously "no." The amount of matter contained in a given object is the **mass** of that object, and the unit of mass in the metric system is the **gram** (abbreviated g).

As mentioned earlier, the best common sense touchstone for the size of the gram is an appreciation that a United States nickel has a mass of about 5 grams. This means that the gram is a very small unit of mass.

One problem that arises in dealing with mass is the concept of weight. Chemists are particularly guilty of using the concepts of mass and weight interchangeably, and this is technically not correct. Mass is a measure of how much matter an object contains. In the metric system this is measured in grams, and for a given object in a given reference frame it is a constant. **Weight**, on the other hand, is the pull (force) that an object feels due to gravity, and it is anything but constant. An object on Mars does not weigh the same as an identical object on the Earth, although both objects have identical masses. On Earth, however, an object of fixed mass also has essentially a fixed weight, hence the interchangeable word usage. In fact, on the planet Earth, the pull of gravity is actually used to measure an object's mass (Fig. 2.3).

The distinction between weight and mass is further confused by a quirk of the English language. In the English language, there is no accepted verb form of the noun "mass." Hence it is common for English speaking chemists to use the verb "weigh" to mean "determine the mass." For the time being, we will be doing ourselves a big favor if we deal exclusively with the concept of mass.

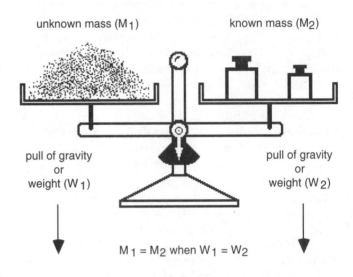

Figure 2.3 The Measurement of Mass

2.6 Time Measurement

The basic unit of time in the metric system is the **second** (abbreviated s). Since the second is also the basic unit of time in the English system, this unit should be easy for you to deal with.

2.7 The Prefixes

As in any system of measurement, the units of the metric system can be divided into smaller units or increased by multiples to larger units. A particularly useful feature of the metric system is the fact that all multiples of units are decimal (based on ten) in nature. These decimal fractions and multiples are indicated by prefixes. In this book, only three common **metric prefixes** need to be committed to memory:

Common Metric Prefixes

$1000 \times$	represented by kilo (abbreviated k)
$\dfrac{1}{100} \times$	represented by centi (abbreviated c)
$\dfrac{1}{1000} \times$	represented by milli (abbreviated m)

The following less commonly used prefixes complete the prefix sequence from $1000 \times$ to $\dfrac{1}{1000} \times$:

Less Common Metric Prefixes

$100 \times$	represented by hecto (abbreviated h)
$10 \times$	represented by deca (abbreviated da)
$\dfrac{1}{10} \times$	represented by deci (abbreviated d)

When one of these prefixes is added to a metric unit, the unit is mathematically multiplied by the prefix. For example, milliliter (abbreviated mL) literally means

$\dfrac{1}{1000} \times$ 1 liter or 0.001 liter. The prefix concept is illustrated in figure 2.4.

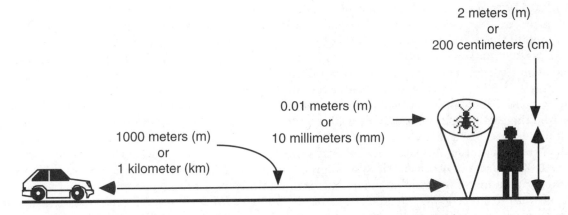

Figure 2.4   The Use of Metric Prefixes

Since the conversion of one metric prefix modifier to another metric prefix modifier always involves multiplication or division by factors of ten, metric/metric prefix conversions simply involve decimal point movement. The following schematic can be used to visualize this process:

Metric Prefix Conversions

kilo (k)	hecto (h)	deca (da)	unity (none)	deci (d)	centi (c)	milli (m)
1000	100	10	1	$\dfrac{1}{10}$	$\dfrac{1}{100}$	$\dfrac{1}{1000}$

1.23 km = 123,000 cm (*i.e.* 5 decimal point shifts to right)
5.34 mL = 0.00534 L (*i.e.* 3 decimal point shifts to left)

Notice from the two examples given above that the correct mathematical conversion can be visualized by realizing that movement from one column to the next corresponds mathematically to the movement of the decimal point one place in the direction of the movement. Thus, the conversion of mL to L involves movement from the milli column to the unity column—a movement of 3 decimal point shifts to the left.

Now that the prefixes have been illustrated, a particularly convenient feature of the metric system should be mentioned. In order to facilitate the conversion of cubic linear volume units and defined volume units, the metric system was originally set up so that the following relationship would hold:

$$1 \text{ cubic centimeter (cm}^3) = 1 \text{ milliliter (mL)}$$

Because of an error in one of the early definitions of the meter, this relationship was for years only approximately true. Since the nature of this error gives some insight into the intent of the founders of the metric system, it is instructive to consider the details of this story. The story also gives some insight into the basic problems associated with the construction of a fundamental system of measurement.

The founders of the metric system wanted the unit of length to be a fundamental unit from which the unit of volume and the unit of mass could be derived. They also wanted the unit of length to be related to some fundamental measurement that would be unrelated to national boundaries. Originally, two length standards were considered. The Acadèmie des Sciences de Paris in France debated using the length of a pendulum with a one second period of oscillation or some fraction of the earth's circumference. Thomas Jefferson was enthused about the adoption of an international metric system based on a seconds pendulum, but when the Acadèmie established the standard as one ten-millionth of the earth's quadrant, Jefferson lost his enthusiasm (Section 2.2). In 1798, after determining the distance between Dunkerque in France and Barcelona in Spain, the earth's quadrant was determined to be 5,130,740 toises. One ten-millionth of this distance was 3 pieds and 11.296 lignes—the meter. Toises, pieds, and lignes were French units used prior to the development of the metric system. In order to relate the units of volume and mass to the meter, the original metric system defined the kilogram as the mass of one cubic decimeter of water at 4 degrees Centigrade and the liter as the volume of one kilogram of water at 4 degrees Centigrade.

But why did the definition of the meter require French units used prior to the development of the metric system? The answer to this question is related to the fact that the original standard for the meter was the earth itself! Clearly, the earth had to be measured in older units of measurement, and these units had to be used to construct prototype meters. In 1875, twenty-eight countries agreed to establish the International Bureau of Standards in Paris, France, to deal with the inconvenience of this earth standard and other metric system problems. One of the first tasks of the Bureau was to construct and archive the official standard meter and kilogram. A prototype meter and a prototype kilogram were constructed as the official standards (*Mètre et Kilogramme des Archives*). The metric system was now free of the troublesome earth standard, but did the new *Mètre et Kilogramme des Archives* in Paris actually conform to the original definitions? Between 1895 and 1907, the Bureau determined that one kilogram of water at 4 degrees Centigrade occupied a volume of 1.000028 cubic decimeters. The *Mètre et Kilogramme des Archives* did not conform to the original definitions.

In 1960, the liter was redefined by international agreement so that one cubic centimeter (cm³) was exactly equal to one milliliter (mL). Thomas Jefferson would be pleased to know that the meter was also redefined using an oscillation standard. The meter was redefined in terms of the wavelength of a characteristic color of light emitted by the element krypton. The relationship between the milliliter and the cubic centimeter that was salvaged by the redefinition of the liter is a very useful relationship. Its utility is perhaps best illustrated by a question. How many gallons are there in one cubic foot? This English system conversion obviously requires a conversion table. In the metric system, cubic linear and volume measurements are easily inter-converted.

EXAMPLE I

Metric System Estimation

Estimate the length of a United States dollar bill.
a. 1.5 m
b. 0.15 m
c. 0.35 m

SOLUTION: b

As you can see from the choices given, you are only expected to make a ballpark estimate. In order to make this type of estimate, a common sense touchstone conversion needs to be memorized for each of the metric units (*e.g.* one meter is approximately equal to one yard).

2.8 Cause and Effect

We live in an orderly universe, and the events that occur in our universe seem to us to be connected. We see lightning, and we wait for thunder. We eat a piece of candy, and we fully expect the sweet taste. Such obvious connections are termed **cause and effect** relationships, and the belief that the entire universe ticks by understandable cause and effect relationships is inherent in the assumptions of natural science (Section 1.2). This belief is by no means universally accepted by human beings. An increasing interest in magic, the occult, and pop science attest to this fact. But the basic belief in a rational cause and effect universe is all too human a characteristic, and the fear and uncertainty that drives many to magic, the occult, or pop science cannot fully extinguish this flame of rationality. Consider a simple thought experiment.

The diagram below represents a box that is closed and cannot be opened. Protruding from each end of the box is a length of string (Fig. 2.5).

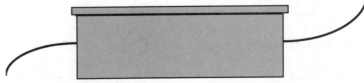

Figure 2.5   The Mystery Box

Now imagine that someone pulls the length of string on the right side of the box toward the right, and that you observe that the length of string on the left side of the box simultaneously begins to disappear into the box. In the next paragraph, you are going to be asked a question. At this point, however, stop reading for about ten seconds.

O.K., here is the question. What color is the box? Of course, if your mind was even partially occupied with thoughts of the box during the past ten seconds, color was probably not a major concern. In fact, most people confronted with the box experiment will do something that is really quite human. They will attempt to imagine a box contents consistent with the string observation. Even the most ardent believer of the daily horoscope (the ultimate in scientific irrationality) will find it difficult not to ponder the contents of the box and establish a rational explanation for the cause and effect observation. Color indeed! Here is the answer that most human minds formulate without being asked the question directly (Fig. 2.6).

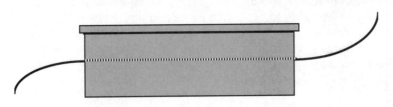

Figure 2.6   Inside the Mystery Box

A string passing continuously through the box is not the only rational explanation, but it is a good starting point.

The box experiment is given to illustrate that the belief in an orderly universe of cause and effect is really quite natural to most people. The experiment, however, is more relevant to understanding the basic nature of chemistry than you might think. Chemists see matter of one identity entering a process known as chemical change. They then observe

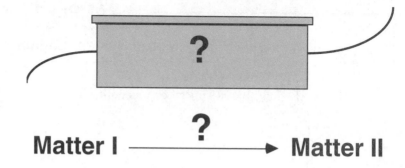

Figure 2.7 An Analogy to the Study of Chemistry

matter with a completely new identity emerging from the process. Just as you were driven to imagine the hidden workings of the box, so also is the chemist driven to imagine the hidden workings of chemical change (Fig. 2.7).

2.9 Techniques of Cause and Effect

The picture suggested for the workings of the box in the last section is based on one observation. Since the box cannot be opened, we can never be certain that we have imagined the correct point of view. We can, of course, test our imagined picture. This is accomplished by doing another very human thing: playing with the box. Stated in a more formal manner, we collect more cause and effect data. Consider a particularly revealing experiment.

The length of string on the right side of the box is pulled so that exactly one meter of new string emerges from the box. At the same time, it is observed that exactly two meters of string disappear into the box on the left side. Clearly we must abandon our original picture and attempt to imagine alternative points of view. For example, the strings might be attached to a lever system that could explain the quantitative data. Each alternative will suggest additional experimentation, and with each additional experiment, we will be closer to the "truth."[3] In all of this we are simply using the hypothetico-deductive method described in Chapter 1, but we have armed ourselves with a particularly powerful tool: quantitative cause and effect observation.

Since the box experiment is being used here as an analogy to the essence of chemistry, the inference is clear. Understanding chemistry demands some facility with quantitative cause and effect relationships, but please read on.

2.10 Arithmetic of Cause and Effect

Many students approach the quantitative aspects of chemistry with much fear and trepidation. For the most part, this response is an unnecessary one. This is particularly true in a nonscience major chemistry course where all of the quantitative aspects of the course can be dealt with by using very basic algebra. This is such an important point that it should be restated more forcefully. If you have had a basic course in algebra, then you already know how to do all of the computations that will be used in this book. Further, in many cases you will be able to deal with these computations intuitively. This point is best demonstrated by means of an example.

Consider that you are the proud owner of a basket that is capable of holding five apples, no more, no less. At a garage sale, you purchase four identical baskets, so that you now own a total of five baskets. If you fill all of your baskets with apples, how many apples will you have in the five baskets?

---

3.   The reference here is to scientific "truth" and not the ultimate truth of theology and philosophy.

The answer is twenty-five, and the problem is ridiculously simple. Yet the arithmetic involved represents essentially all the arithmetic required for an understanding of basic chemistry, including the quantitative reasoning involved in the discovery of atomic theory. So why do some students still have a problem with the arithmetic of chemistry? Well, consider another example.

Imagine that you live on the edge of a small pond. A scientist comes to visit you, and requests permission to study the frogs in your pond. After receiving your permission, the scientist informs you that the first step will be to count the number of frogs living in the pond. Since the frogs in question have caused you many a sleepless night, you watch with interest as the scientist sets out to count the little devils. Here is how the scientist does it.

On day one of the study, the scientist takes a walk around the edge of the pond. Every time a frog is sighted on the edge of the pond the frog is picked up and placed into a sack. At the end of the walk, the frogs in the sack are counted, and bands are placed around a leg of each frog. The scientist then releases all of the banded frogs, and they return to the pond. In all, twenty-five frogs are banded and released. Having worked very hard on day one, the scientist heads for the golf course.

On day two, the scientist, looking very sun-tanned, returns to work. The activities of the previous day are repeated. Upon completing the journey around the pond, the scientist informs you that there are a total of twenty frogs in the sack, and that five of these frogs bear the band of the previous day. Since the total number of frogs living in the pond is now known, the scientist decides to call it a day and heads for the tennis courts.

Well, how many frogs live in the pond? You probably are not answering this problem as quickly as you did the apple/basket problem, and yet the two problems are virtually identical. Why do many people have difficulty with the second problem? An analysis of the answer to this question may help you in your attempts to deal with chemistry at an arithmetic level.

The difficulty that some people have with the frog/pond problem is twofold. The first problem stems from the fact that most people solve simple problems of this type (*i.e.* the apple/basket problem) without a conscious awareness of the step by step procedure (algorithm) they are using. In trivial problems, this lack of awareness creates no serious problem. The minute that an unusual problem is encountered, however, the lack of a formal algorithm simply gets in the way of a direct problem solution. A second problem that is encountered is the fact that many problems involving real life situations require certain assumptions prior to solution. For many people, this is a tremendous barrier since assumptions infer the possibility of an incorrect solution. Both of these problems can be dealt with, the former by learning the algorithm, the latter by learning this simple maxim: In natural science, failure is often a necessary step toward a proper solution.

2.11 Linear Reasoning

The apple/basket problem and the frog/pond problem are both examples of the application of simple linear algebra. In this textbook the application of simple linear algebra is called *linear reasoning*. The purpose of this section is to examine this type of problem and to detail the step by step solution procedure (algorithm). To begin, consider a familiar example.

The figures shown below represent five geometric shapes that you will recognize as circles (Fig. 2.8).

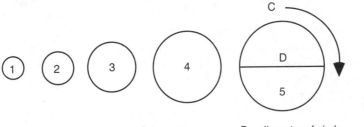

D = diameter of circle

C = circumference of circle

Figure 2.8   Five Circles

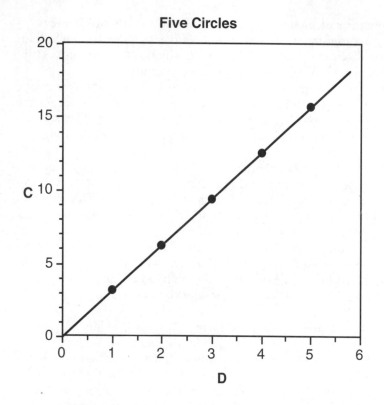

Five Circles

D Diameter	C Circumference
circle 1 - 1m	3.14m
circle 2 - 2m	6.28m
circle 3 - 3m	9.42m
circle 4 - 4m	12.57m
circle 5 - 4m	15.71m

The circumference and the diameter of a circle are linearly related or directly proportional.

Figure 2.9   The Relationship between Circumference and Diameter

Each circle has two characteristic measurements, the distance around the circle (circumference, C) and the distance across the center of the circle (diameter, D). A quantitative investigation of these five circles leads very quickly to the following observations about the relationship between the circumference of a circle and the diameter of a circle (Fig. 2.9).

It is clear from both the tabular observations and the graphical observations that as the diameter of a circle increases, the circumference also increases, and it increases in a regular (linear) fashion. The mathematical relationship between the circumference and diameter of a circle is a special case of the algebraic equation for a straight line (y=mx + b, y intercept [b] = 0 and slope [m] = 3.14). At this point, the example could become entrenched in the intricacies of linear algebra. This approach will not be taken, however, and the student need only appreciate the following aspects of the example:

Characteristics of Linearly Related Quantities

1. Measurable quantities that are related as the circumference of a circle and the diameter of a circle are said to be **linearly related** or **directly proportional**.
2. There is a simple test for the direct proportionality of two measurable quantities. If increasing one of the quantities by a certain multiple increases the other quantity by the same multiple (*e.g.* doubling the diameter of a circle doubles the circumference), then the two quantities are directly proportional. Sometimes, but not always, the direct proportionality of two quantities is intuitively apparent.
3. When two measurable quantities are directly proportional, the following symbolism is used to express this fact:

$$C \propto D$$

READ: C is directly proportional to D

4. There is a simple arithmetic consequence of direct proportionality. When two measurable quantities are directly proportional, the ratio of one to the other in a given situation is constant. For example:

$$C \propto D$$

$$\text{Infers} \frac{C}{D} = \text{Constant}$$

This is a consequence of the algebraic equation for a straight line (y=mx + b, y intercept [b] = 0 and slope [m] = constant). In some cases, the actual value of this constant is important. In other cases, the actual value is unimportant. In the example cited above, the constant is the familiar pi (3.14).[4]

5. Things equal to the same thing are equal to each other. One ratio of a directly proportional pair is, therefore, equal to any other ratio of the directly proportional pair. For example:

$$\frac{C}{D} = \text{Constant (Circle \#1)}$$

$$\frac{C'}{D'} = \text{Constant (Circle \#2)}$$

THEREFORE

$$\frac{C}{D} = \frac{C'}{D'}$$

C' and D' are read C prime and D prime. This designation simply indicates a pair of values from a second circle.

With these five points in mind, let us consider five examples of problems involving linear reasoning. In the first two examples (II and III), a six step formal procedure for solving this type of problem will be illustrated. We will begin with the problem that you have already solved (apple/basket), but we will take great pains to articulate each step in the solution.

EXAMPLE II

Linear Reasoning: The Apple/Basket Problem

Solve the apple/basket problem presented in Section 2.10.

SOLUTION: 25 apples

STEP I: Does the problem involve two measurable quantities that are linearly related?

---

4.   The expression C/D= π is a very familiar equation to most students, and its familiarity is the reason it is being used as an example. If you think that this expression has always been obvious to humans, see the description of Solomon's Temple in the Old Testament (2 Chronicles 4:1–2).

In this case, the answer is clearly yes. In the context of the problem, apples and baskets are directly proportional by item #2 above. An experiment is not required to verify this fact. You know intuitively that doubling the number of baskets will also double the number of apples, and you can, therefore, write:

$$\text{Apples} \propto \text{Baskets}$$

STEP II: What are the arithmetic consequences of the linear relationship?

By item #5 above, we can write:

$$\frac{\text{Apples}}{\text{Baskets}} = \frac{\text{Apples'}}{\text{Baskets'}}$$

At this point, a linear reasoning problem will always take the form:

$$\frac{X}{Y} = \frac{X'}{Y'}$$

Recall from algebra that it is equally valid to write:

$$\frac{Y}{X} = \frac{Y'}{X'}$$

$$\frac{\text{Baskets}}{\text{Apples}} = \frac{\text{Baskets'}}{\text{Apples'}}$$

STEP III: Does the wording of the problem supply one complete ratio? A complete ratio is a pair of numbers that are logically related.

All linear reasoning problems must supply one complete ratio. If they don't, they cannot be solved. It's as simple as that. In this case, the given complete ratio is:

$$\frac{5 \text{ Apples}}{1 \text{ Basket}}$$

These two numbers are logically related because 5 apples are contained in 1 basket.

STEP IV: Does the problem supply an incomplete ratio?

Once again, all linear reasoning problems must supply an incomplete ratio. The object of such problems is to find the missing number. In this case, the incomplete ratio is:

$$\frac{X \text{ Apples}}{5 \text{ Baskets}}$$

STEP V: Equate the complete ratio to the incomplete ratio as suggested by STEP II.

$$\frac{5 \text{ Apples}}{1 \text{ Basket}} = \frac{X \text{ Apples'}}{5 \text{ Baskets'}}$$

STEP VI: Solve the resulting equation for the missing quantity (X).

At this point, it is assumed that the student is familiar with the method by which such equations are solved. For those who are rusty, the procedure is illustrated below:

$$\frac{5 \text{ Apples}}{1 \text{ Basket}} = \frac{X \text{ Apples'}}{5 \text{ Baskets'}}$$

Cross multiply and equate the cross products.[5]

$$(1)(X) = (5)(5)$$

Divide both sides of the resulting equation by the coefficient of X.

$$\frac{(1)(X)}{(1)} = \frac{(5)(5)}{(1)}$$

$$X = 25 \text{ Apples}$$

This final step involves an arithmetic algorithm for the solution of a direct proportion problem. If you are a bit rusty on the details of this algorithm, see performance objective 2.3 at the end of this chapter for suggestions. A calculator algorithm that can be used to solve this type of arithmetic problem is also listed under performance objective 2.3.

All of this is belaboring the obvious. You arrived at this answer intuitively. But that's exactly the point. In cases where the problem involves unfamiliar situations, the solution is not intuitively obvious, and consequently a step by step approach is required. To see how this works, let us turn our attention to the frog/pond problem.

EXAMPLE III

Linear Reasoning: The Frog/Pond Problem

Solve the frog/pond problem presented in Section 2.10.

SOLUTION: 100 total frogs

STEP I: Does the problem involve two measurable quantities that are linearly related?

There is no question that this is the hardest part of the problem. The reason for this is that in order to identify the linear relationship, assumptions must be made. Even after the assumptions are made, the linear relationship is not necessarily obvious. In the chemical problems that you encounter, it is important to realize that an *ex post facto* (after the fact) understanding is a significant achievement. You may not recognize a linear relationship in chemistry until it is pointed out to you, but if you understand the relationship after it is pointed out, you are learning chemistry! In the case of the frog/pond problem, the linear relationship is:

Banded Frogs in A Collected Sample = (BF)
Total Frogs in A Collected Sample = (TF)

After Day One: BF $\propto$ TF

---

5.   In a direct proportion problem, the units of the answer can be deduced from the problem format. In this case, the format $\frac{\text{Apples}}{\text{Baskets}} = \frac{\text{Apples'}}{\text{Baskets'}}$ demands that X = Apples. This approach allows the units to be deleted from the computation.

Assuming:

>Homogeneous Mixing of Banded Frogs with Nonbanded Frogs
>Stable Frog Population

Can you identify other assumptions being made here?

Once the linear relationship is identified, the rest of the problem is very methodical.

STEP II: What are the arithmetic consequences of the linear relationship?

$$\frac{BF}{TF} = \frac{BF'}{TF'}$$

STEP III: Does the wording of the problem supply one complete ratio?

$$\frac{5\ BF}{20\ TF}$$

This is the sample collected on day two of the study.

STEP IV: Does the problem supply an incomplete ratio?

$$\frac{25\ BF}{X\ TF}$$

This represents the sample that the scientist never collected—all the frogs in the pond!

STEP V: Equate the complete ratio to the incomplete ratio as suggested by STEP II.

$$\frac{5\ BF}{20\ TF} = \frac{25\ BF'}{X\ TF'}$$

STEP VI: Solve the resulting equation for the missing quantity (X).

$$\frac{5\ BF}{20\ TF} = \frac{25\ BF'}{X\ TF'}$$

$$(5)\,(X) = (20)\,(25)$$

$$\frac{(5)\,(X)}{(5)} = \frac{(20)\,(25)}{(5)}$$

$$X = 100\ TF$$

Linear reasoning is to be found everywhere in human experience. The arithmetic techniques of linear reasoning are applicable wherever constant ratios are encountered. To further illustrate this important type of calculation, three additional examples are given for you to solve. In each example, the constant ratio is identified for you. Although you might be able to solve the problems intuitively, attempt to articulate the steps as in the first two examples.

EXAMPLE IV

Linear Reasoning: Percent Problem

A certain mixture is 20.0% carbon by mass. How much of the mixture is required to produce 85.0 grams of carbon. NOTE: In this problem, as in all simple percentage problems, the constant ratio is the percentage itself. In this case, 20.0% literally means:

$$\frac{20.0 \text{ mass units of carbon}}{100 \text{ mass units of mixture}} = \text{Constant}$$

SOLUTION: 425 grams of the mixture

Although most adults solve problems of this type without articulating a formal structure, the following linear reasoning structure is suggested for percent problems encountered in this chemistry text:

$$\frac{X}{Y} = \frac{X'}{Y'}$$

or for percent problems

$$\frac{\text{part}}{100 \text{ total}} = \frac{\text{part}'}{\text{total}'}$$

$$\frac{20.0 \text{ g carbon}}{100 \text{ g mixture}} = \frac{85.0 \text{ g carbon}}{X}$$

$$X = 425 \text{ g mixture}$$

EXAMPLE V

Linear Reasoning: Conversion Factor Problems

A certain book is 30.0 cm long. If the conversion factor relating cm to inches is 2.54 cm/in, how long is the book in inches? NOTE: In this problem, as in all conversion factor problems, the constant ratio is the conversion factor itself. In this case, 2.54 cm/in literally means:

$$\frac{2.54 \text{ cm}}{1.00 \text{ in}} = \text{Constant}$$

SOLUTION:[6] 11.8 in

---

6.   In any natural science, there are rigid rules dealing with the subject of rounding off calculated answers. In this book, only the spirit of these rules is important: Do not allow a calculated answer to be more accurate than an original measurement. Hence, although the calculator display for this problem indicates 11.811024, the answer is rounded off to reflect the accuracy of the original measurements. In nonscience major chemistry, this is not a major issue. To simplify matters on this issue, most problems presented in this text will involve measurements with three significant figures and answers will be rounded off appropriately.

Again, most adults solve problems of this type without articulating a formal structure, the following linear reasoning structure is suggested for conversion factor problems encountered in this chemistry text:

$$\frac{X}{Y} = \frac{X'}{Y'}$$

or for unit conversion problems

$$\frac{\text{unit system A}}{\text{unit system B}} = \frac{\text{unit system A'}}{\text{unit system B'}}$$

$$\frac{2.54 \text{ cm}}{1.00 \text{ in}} = \frac{30.0 \text{ cm}}{X}$$

$$X = 11.8 \text{ in}$$

Although the procedure outlined in the previous three examples is the procedure that will be emphasized in this text, it might be useful for some students to consider an alternative approach to solving linear reasoning problems. This alternative approach is based on the concept of starting the solution procedure at the final step of the six step procedure outlined in examples II and III. To see how this is done, consider the following representation of this final step (X unknown):

$$\frac{X}{Y} = \frac{X'}{Y'}$$

solving for X unknown

$$X = Y \frac{X'}{Y'}$$

This last expression represents the final step in any simple linear reasoning problem, and if there is a logical way to set it up directly, then this expression can become the starting point for solving this type of problem. Notice that the right hand side of this expression can be described as the given number from the incomplete ratio multiplied times the complete ratio. If the solution to a linear reasoning begins with this expression, however, there must be some logical way of deciding between the two reciprocal values of the complete ratio. There is a logical way of doing this. Let the units of the numbers in the ratio dictate which of the two reciprocal values to use. Although it is technically a misnomer, this approach is sometimes called "dimensional analysis," and it is illustrated below by reconsidering example V.

EXAMPLE V

Linear Reasoning: Conversion Factor Problems
"Dimensional Analysis"
An Alternative Approach

A certain book is 30.0 cm long. If the conversion factor relating cm to inches is 2.54 cm/in, how long is the book in inches?

SOLUTION: 11.8 in

Since a unit conversion problem is a simple linear reasoning problem, the final solution will take the final form:

$$X = Y \, \frac{X'}{Y'}$$

In the case of the stated problem, a decision must be made between the following two reciprocal variations of this final form:

$$X = 30.0 \text{ cm} \, \frac{2.54 \text{ cm}}{1.00 \text{ in}} \quad \text{or} \quad X = 30.0 \text{ cm} \, \frac{1.00 \text{ in}}{2.54 \text{ cm}}$$

If the units are included in both of these calculations, then only one of these two alternatives produces an answer with the correct units – inches in this case:

$$X = 76.2 \, \frac{\text{cm}^2}{\text{in}} \quad \text{or} \quad X = 11.8 \text{ in}$$

Thus, including the units in the calculation provides a logical method for deciding to use the ratio that leads to 11.8 inches. This shortcut method for solving linear reasoning problems is particularly useful in multi-step problems where the solution to one direct proportion calculation is used in a subsequent calculation. Since it is a shortcut method, it will not be the problem solution method that is emphasized in this text. Some students, however, may be familiar with some variation of this method from previous science courses. (Other terms used to describe this type of approach include the unit factor method and factor label method.) Students who feel comfortable with some variation of "dimensional analysis" are encouraged to use this method. To help students who elect to use this method, the textbook will occasionally show dual solutions to problem examples. In the presentation of these dual solutions, this alternative approach will be referred to as the **factor label method**.

EXAMPLE VI

Linear Reasoning: Equation Problems

The volume of a cylinder is given by the equation $V = \pi r^2 h$ where "V" represents the cylinder's volume, "r" represents the radius of the cylinder's circular base, "h" represents the cylinder's height, and π is the constant 3.14. Calculate the height of a cylinder with a volume of 925 cm³ and a base radius of 8.75 cm.

SOLUTION: 3.847654 cm rounds off to 3.85 (three significant figures)

The equation $V = \pi r^2 h$ indicates that V is linearly related to r^2 and h. In this text, all algebraic problems resulting from linear relationships will be of the one equation, one unknown type. When solving problems of this type, transpose equation terms if necessary, and then fit the physical situation to the equation by "plugging" the appropriate values into the equation. Although this approach is sometimes criticized, the act of "plugging" real world numbers into an equation properly demonstrates an understanding of the real world situation.

$$V = \pi r^2 h$$

after transposition

$$h = \frac{V}{\pi r^2}$$

$$h = \frac{925 \text{ cm}^3}{(3.14)(8.75 \text{ cm})^2}$$

$$h = 3.85 \text{ cm}$$

In this type of problem, it is useful to include all of the measured units in the computation. Notice that the inclusion of the units in the above calculation allows a confirmation that the units of the answer should be centimeters.

2.12 A Final Note

One of the purposes of this chapter has been to review the technique and language of basic linear algebra. The appropriateness of this language to the study of chemistry will become apparent in latter chapters. In the meantime, it is important to keep math anxiety in proper perspective. If you suffer from this strange affliction, don't panic. Quantitative reasoning is a fully human instinct, and any reasonably intelligent person can learn to apply it to new situations. The trick is to believe in your ability.

"The essence of belief is the establishment of a habit." – Charles S. Peirce, American Mathematician

"Do not worry about your difficulties in mathematics; I can assure you that mine are still greater." – Albert Einstein, German/American Physicist

Chapter Two
Performance Objectives

P.O. 2.0

Review all of the boldfaced terminology in this chapter, and make certain that you understand the use of each term.

cause and effect   directly proportional   gram
factor label method   linearly related   liter
mass   meter   metric prefixes
metric system   second   Système International
volume   weight

P.O. 2.1

You must be able to make a gross metric estimation of length, mass, or volume by learning a common sense touchstone for each of the metric units.

EXAMPLE:
Estimate the width of a United States dollar bill.

a) 6.4 cm   b) 25 cm   c) 75 cm

SOLUTION: a

Textbook Reference: Section 2.3

ADDITIONAL EXAMPLE:
Which of the following objects has the greatest mass?

a) a one pound brick   b) a roll of 50 pennies (USA)   c) a 1000 gram brick

ANSWER: c

P.O. 2.2

You must be able to make intra-metric conversions using the metric prefix system.

EXAMPLE:
How many liters are there in 121 milliliters?

SOLUTION: 0.121 L

Textbook Reference: Section 2.7

ADDITIONAL EXAMPLE:
How many grams in 3.56 kilograms?

ANSWER: 3,560 g

P.O. 2.3

You must be able to do the arithmetic of linear reasoning (*i.e.* direct proportion). This is a diagnostic objective that is included to help students identify arithmetic difficulties.

EXAMPLE:
Solve for Y in the following problem:

$$\frac{122}{(5.12)\,(Y)} = \frac{13.5}{57.8}$$

SOLUTION: 102.019676 rounds off to 102 (three significant figures)

If you cannot solve this type of problem, you need to discuss this with the course instructor. The application of direct proportion arithmetic to real world problems (defined here as linear reasoning) is essential to understanding basic chemistry. Although some aspects of linear reasoning come naturally to almost all students, some students do not possess an adequate command of the direct proportion arithmetic that services linear reasoning. The fact that some college students have forgotten (or never learned) how to do this formal arithmetic is a real, but manageable, problem.

If you have forgotten the algebraic method that is used to solve direct proportion arithmetic problems, then you need to review Example II in Section 2.11. The following two-step calculator algorithm might also prove useful.

Direct Proportion Arithmetic—A Calculator Algorithm

STEP I: In all direct proportion arithmetic problems, it is possible to classify the known numbers as "complete" or "incomplete" cross product numbers according to the following scheme:

complete cross #   incomplete cross #

$$\frac{122}{(5.12)\,(Y)} = \frac{13.5}{57.8}$$

incomplete cross #   complete cross #

In this scheme, the incomplete cross numbers are part of the cross product that contains the unknown, Y.

STEP II: Once each of the given numbers is classified, the series of algebraic calculator key strokes shown below will yield the correct answer.
In this calculator algorithm, the notations [×], [÷], and [=] indicate the calculator's arithmetic keys:

[complete #][×][other complete #][÷][incomplete #][÷]
[other incomplete #][=]

ADDITIONAL EXAMPLE:
Solve the following problem for X:

$$\frac{(2.00)\,(X)}{(3.00)\,(16.0)} = \frac{125}{214}$$

ANSWER: 14.018692 rounds off to 14.0 (three significant figures)

P.O. 2.4

You must be able to solve a simple linear reasoning problem.

EXAMPLE:
In a certain city 80.0% of the people have the name Smith. If 400 Smiths live in the city, what is the total population of the city?

SOLUTION: 500 total population of city

In this problem, as in all percentage simple problems, the constant ratio is the percentage itself. In this case, 80.0% literally means:

$$\frac{80 \text{ Smiths}}{100 \text{ Total Population}} = \text{constant for the city}$$

Textbook Reference: Section 2.11

ADDITIONAL EXAMPLE:
A certain chemical mixture is found to be 22.0 percent carbon. What mass of this mixture would be required to obtain 255 grams of carbon?

ANSWER: 1159.090909 rounds off to 1160 (three significant figures)

P.O. 2.5

You must be able to solve a problem involving the transposition of terms in a simple algebraic equation.

EXAMPLE:
The conversion between the Fahrenheit and centigrade temperature scale is represented by the following equation:

$$F = \frac{9}{5} C + 32$$

In this equation, F represents the Fahrenheit temperature, and C represents the corresponding centigrade temperature. Use this equation to convert a temperature of 75.0 degrees Fahrenheit into centigrade degrees.

SOLUTION: 23.888889 rounds off to 23.9 (three significant figures)

ADDITIONAL EXAMPLE:
The circumference of a circle may be expressed by the equation $C = 2\pi r$ where "C" represents the circle's circumference, "r" represents the circle's radius, and "π" is the constant 3.14. Calculate the radius of a circle that has a circumference of 300 meters.

ANSWER: 47.770701 rounds off to 47.8 (three significant figures)

43

Chapter Two
Problems

Metric Estimation

1. Estimate length of a standard piece of typing paper. Express this estimate in centimeters and meters.

    STUDENT SOLUTION:

     11 cm

2. Estimate the volume of a tennis ball. Express your answer in liters, milliliters, and cubic centimeters.

    STUDENT SOLUTION:

3. Estimate the mass of a 12 oz can of soda. Express your answer in grams and kilograms.

    STUDENT SOLUTION:

4. The distance run during one type of common foot race is 10,000 meters. Estimate this distance in miles. There are about 1,600 yards in a mile.

    STUDENT SOLUTION:

5. The gasoline tank in a Toyota Tercel holds 12 gallons. Estimate this volume in liters.

    STUDENT SOLUTION:

Metric Prefix Conversions

6. Use your knowledge of the metric prefixes to make the following metric/metric conversions:

   a) 1250 cm = ? km
   b) 235 g = ? kg
   c) 2570 mL = ? L

   d) 674 cg = ? g
   e) 5280 mm = ? cm
   f) 1.23 g = ? mg

   STUDENT SOLUTION:

7. A certain object weighs 24.5 grams. Express the mass of this object in milligrams and kilograms.

   STUDENT SOLUTION:

8. A bowling ball has a diameter of 0.210 meters. Express this length in mm and cm.

   STUDENT SOLUTION

Direct Proportion Arithmetic

9. Solve for X in the following problems:

   a) 125/X = 232/340
   b) 432/125 = 2.00X/234
   c) 5.00X/17.5 = 59.2/29.8

   d) 62.1/91.7 = 74.2/4.00X
   e) 1250/2.00X = 2350/9990
   f) 1.47X/2.57 = 1.70/6.71

   STUDENT SOLUTION:

Linear Reasoning

10. A chemistry student takes an exam consisting of 35 questions and gets 16 of these correct. What percentage did this student get *incorrect*?

    STUDENT SOLUTION:

54%

11. A student's score on a chemistry exam was 80.0 percent. The student had five questions wrong. How many questions were on the chemistry exam?

    STUDENT SOLUTION:

12. Chicken contains about 3.80% fat. If a person were to eat 9.67 ounces (oz) of chicken, how many ounces of fat were consumed?

    STUDENT SOLUTION:

    0.36702

13. How many grams of fat were consumed in problem the previous problem? The gram/ounce conversion factor is 28.4 g/oz.

    STUDENT SOLUTION:

14. The recommended daily allowance (RDA) of protein is 46.0 grams for young female adults. Mary (a young female adult) decides to go on a diet consisting of only whole wheat bread which contains 10.5% protein. How many pounds (lb) of whole wheat bread does she need to consume a day to attain the RDA? The gram/pound conversion factor is 454 g/lb.[7]

    STUDENT SOLUTION:

15. A cylindrical jar holds a large number of marbles. The jar is one meter high, and the marbles fill it completely. After removing 125 marbles from the jar, the cylindrical pile of marbles remaining in the jar has a height of 0.750 meters. Estimate the total number of marbles in the jar.

    STUDENT SOLUTION:

---

7. This is a conversion factor involving a mass unit (grams) and a weight unit (pounds). The relationship between mass and weight is discussed in Chapter 7. Although this type of conversion is not common in scientific laboratories, mass/weight conversions are often found in commerce.

16. If the 125 marbles from the previous question are found to have a total mass of 5.00 kilograms, what is the mass of one marble? Assume all of the marbles are identical. Express your answer in kilograms and grams.

    STUDENT SOLUTION:

17. Suppose that a cube 2.00 meters on an edge has a mass of 10.0 kilograms. Calculate the mass of a cube 3.00 meters on an edge, but made of the same material as the cube 2.00 meters on an edge.

    STUDENT SOLUTION:

18. In a linear reasoning problem, the identification of the two measurable quantities that are linearly related (directly proportional) requires insight. As pointed out in Chapter 1, human insight is far too complex to study in detail. It does not reduce to an algorithm. All people possess it, however. Consider the following problem as an exercise in human insight:

    > In the thirteenth century, the Italian mathematician Fibonacci discovered the following series of numbers: 0,1,1,2,3,5, ... The series is an infinite series in which each number is mathematically related to the previous numbers. Do you see a trend? Can you supply the next three numbers of the series?

    After solving this problem, it might be interesting to discuss how you solved the problem with another person who also solved the problem. During this discussion, each person might learn something about their own human insight.

    The series of Fibonacci, by the way, is encountered in the natural sciences. For example, the arrangement of leaves on the stems of certain plants is related to the Fibonacci series. In this regard, it might be interesting to consider the original question that led to the Fibonacci series:

    > How many pairs of rabbits can be produced from a single pair in a year, if (a) each pair begets a new pair every month, which from the second month on becomes productive, (b) deaths do not occur?

    STUDENT SOLUTION:

Solving Equations

19. The area of a circle is given by the equation $A = \pi r^2$ where "A" represents the circle's area, "r" represents the circle's radius, and "π" is the constant 3.14. Calculate the area of a circle with a diameter of 2.50 meters.

    STUDENT SOLUTION:

20. The circumference of a circle may be expressed by the equation C = 2πr, and the area of a circle is given by the equation A = πr². In these equations, "C" represents the circle's circumference, "A" represents the circle's area, "r" represents the circle's radius, and "π" is the constant 3.14. Calculate the area of a circle that has a circumference of 30.0 meters.

    STUDENT SOLUTION:

21. Neglecting air resistance, all objects fall to earth at a constant acceleration of 9.80 m/s². This value is known as the gravitational acceleration, g, and it is equal to 32.2 ft/s² in the English system of units. The distance, d, traveled by the object and the time, t, it takes for the object to fall is given by the following equation:

$$d = \frac{gt^2}{2}$$

    A penny falls from a hot-air balloon and reaches the ground in 15.2 seconds. Neglecting air resistance, how high is the balloon in meters and in feet?

    STUDENT SOLUTION:

22. The 110-story Sears Tower in Chicago, IL, is 443 meters high. Using the same equation given in the previous problem, calculate the time it would take for a marble to fall to the ground if dropped from the top of the Tower.

    STUDENT SOLUTION:

Problems Involving Household Chemistry and Science

A Demonstration of the Acidic Nature of Aspirin:

23. In problem twelve at the end of Chapter 1, the reaction between baking soda and vinegar was investigated. This reaction involved the neutralization reaction of an acid (vinegar) and a base (baking soda). Acids and bases will be discussed in more detail in Chapter 6, but at this point, a simple experiment involving acid/base chemistry might prove instructive. Add two oz of unsweetened grape juice to one cup of water and mix. Set up three small juice glasses and add four oz of the diluted grape juice to each. Add a level ⅛ teaspoon of baking soda to two glasses and note the color change compared to the third glass. Add two tablespoons of white vinegar to one glass containing baking soda and note the reaction and color change. Add five crushed, unbuffered aspirin tablets to the other glass containing baking soda. After a few minutes of stirring, the aspirin should produce an effect similar to the vinegar. The reactions observed in

this experiment involve acid/base neutralization, with the grape juice acting as an acid/base indicator. Note that vinegar and aspirin have similar chemical properties in that both are acids. They are, however, very different chemical compounds. Vinegar is not a headache remedy.

What would happen if aspirin powder from a crushed aspirin tablet was mixed with an equal volume of solid baking soda? Try this experiment, and after mixing the two solids thoroughly, add two or three drops of water to the solid mixture.

STUDENT SOLUTION:

A Calculation Involving Aspirin:

24. Why do most aspirin tablets contain 325 mg of aspirin? Why not 300 mg or 350 mg? Why 325 mg? The answer lies in an older unit of measurement called the "grain," which was used primarily in pharmaceuticals. One grain is equal to 64.8 mg. Calculate the number of grains of aspirin in a tablet containing 325 mg of aspirin. Now does the quantity 325 mg make sense?

STUDENT SOLUTION:

Demonstration of a Physical Difference between Diet and Regular Cola:

25. Place a 12 oz can of regular cola into a large bowl or bucket of water. Does the can sink or float? Repeat the experiment with a can of diet cola. Formulate a hypothesis to explain the observed density difference. In developing hypotheses, scientists often need to make certain assumptions about the systems that they are studying. What assumptions were made during the formulation of your cola density hypothesis?

STUDENT SOLUTION:

A Calculation Involving Diet and Regular Cola:

26. The artificial sweetener, aspartame (NutraSweet®), is 180 times more sweet than sucrose. If a 12 oz can of regular soda contains 37.0 grams of sucrose, how much aspartame should a 12 oz can of diet soda contain to give the same level of sweetness as the regular soda?

STUDENT SOLUTION:

Library Problems

27. The ability to recognize quantitative relationships in nature is essential to understanding chemistry. In the article "Student Conceptions and Competence Concerning Quantitative Relationships between Variables" (*J. Chem. Educ.* **1991**, 68, 370), Selvaratnam and Kumarasighe suggest a ten question test to help chemistry students identify misconceptions associated with the recognition of quantitative relationships. Although some of the questions in this test are beyond the scope of the text, read this article and attempt to answer the first three test questions. In question one, what would the values of Y have to be if Y were directly proportional to X? In question two, what would the values of Y have to be if Y were inversely proportional to X? Question three involves a possible misconception associated with the concept of density. The concept of density is discussed in Chapter 3 of this text.

STUDENT SOLUTION:

28. Speaking of artificial sweeteners (question 26), one of the sweetest chemical compounds ever synthesized by chemists is called sweetening agent P4000. This chemical compound has an interesting history related to the fact that its discovery resulted from a misunderstood translation of a German language research directive issued during World War II. The story of this misunderstanding is described in the first part of the article "Preparation of the Sweetening Agent P4000. A Student Project" (de Koning, A. J. *J. Chem. Educ.* **1976**, 53, 521). After reading the first part of this article, describe what the research directive really instructed the chemists to synthesize. What mass of sweetening agent P4000 is required to give the same sweetness level as 1 gram of sucrose?

STUDENT SOLUTION:

Formal Laboratory Exercise

Substance Identification by Physical Measurement
A Useful Linear Relationship

Introduction

For a pure substance, the relationship between mass and volume is linear at constant temperature and pressure. Using the symbolism of this chapter, this relationship can be expressed as follows:

$$\text{Mass} \propto \text{Volume}$$
$$\text{Infers: Mass/Volume} = \text{Constant}$$

This constant ratio for a given substance at a given temperature is a nearly unique number called density that can be used to identify the substance. The nearly unique constancy of the mass/volume ratio is illustrated graphically for the metals nickel, zinc, aluminum, and magnesium in Figure 2.10. The utility of an algebraic approach to density is discussed in Chapter 3. The graphical approach that is used in this formal laboratory exercise will serve as an introduction to that discussion.

In 1982, the U.S. Mint changed the composition of a penny from 95% copper to 2.4% copper. While the pre 1982 penny is almost pure copper, the post-1982 penny is a less expensive metal clad in a very thin covering of almost pure copper. This less expensive metal is one of the four metals represented in Figure 2.10. If experimental measurements of mass versus volume for post-1982 pennies are added to Figure 2.10, the resulting line should allow for the identification of the major component of these pennies. The purpose of this formal laboratory exercise is to use a graphical approach to identify the major component of post-1982 pennies. Although this graphical approach is rather laborious, the procedure will introduce several important quantitative measurement techniques that will be used in subsequent laboratory exercises.

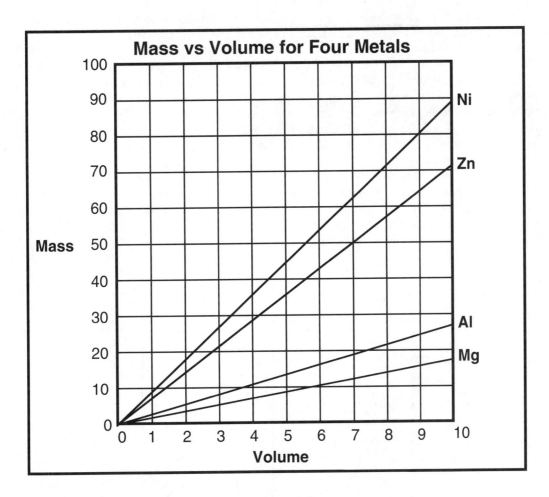

Figure 2.10   The Constancy of Mass/Volume

Procedural Overview

The graphical determination of the density of the metal that is used to make a penny requires the measurement of mass and volume for three different samples. The first sample is the average mass and volume of a penny supplied by the U.S. Mint. These are literature values, and they are already printed in the data table. The second sample is a ten penny stack. The mass of this sample will be determined by direct mass measurement on the laboratory balance. The volume of this sample will be calculated from the measured dimensions of the cylindrical stack of pennies. Recall that the volume of a cylinder is given by the following equation:

$$\text{Volume} = (\pi)(\text{Radius})^2(\text{Height})$$

The final sample is a 20 penny stack. The mass of this sample will be determined by direct mass measurement on the laboratory balance. The volume of this sample will be determined by measuring the volume of water that the 20 pennies displace when the entire stack is submerged in water. After the mass and volume data for the three samples is plotted on the graph in Figure 2.10, the identity of the major metallic component of post-1982 pennies can be determined.

This laboratory exercise requires the direct measurement of mass, liquid volume, and length. Although the techniques associated with these measurements will be demonstrated by the laboratory instructor, each of these measurement techniques will be addressed briefly in this overview. A careful reading of this section is an important prerequisite to the instructor's demonstration.

For this and subsequent experiments the electronic balance will be used to measure mass. The balance is capable of measuring masses to the nearest 1 mg (0.001 g). This delicate and expensive instrument must be treated with great care. Before using the balance set the balance to zero by depressing the "tare" button. After a few seconds, the digital readout will indicate 0.000 g. Taring is also used to subtract the mass of the container or weighing paper when weighing a sample. First, place the container on the balance, then depress the tare button. After the balance reads zero, place the object to be weighed in the container. The readout will indicate the mass of the object. When weighing powders, always use glassine weighing paper. Clean up any spills that may occur on or around the balance immediately. The use of the balance will be demonstrated by the laboratory instructor at the beginning of the laboratory period.

A graduate cylinder is used to measure the volume of a liquid sample. When placed in a cylinder, most liquids form a curved surface called the meniscus. The bottom of the meniscus is read to obtain the volume. It is also necessary to estimate the distance that the meniscus lies between the graduation marks. In the Figure 2.11 below, the volume is read as 8.85 mL by estimating to the nearest 5/10 of the smallest division on the scale.

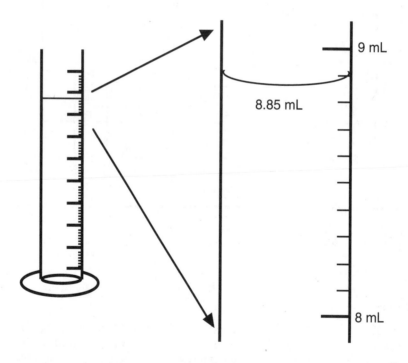

Figure 2.11   Reading Volume from a Graduated Cylinder

The use of a metric ruler to measure length is fairly straightforward. As in the case of reading the volume scale on a graduated cylinder, lengths on a metric ruler are estimated to the nearest 5/10 of the smallest division on the scale.

Materials

graduated cylinder, metric ruler

Chemicals

post-1982 pennies

Procedure

Note: Use only post-1982 pennies in the following procedure.

1. Using the balance, measure the mass of ten pennies. Record this mass in the data table.

2. Arrange the ten pennies in a cylindrical stack and use the metric ruler to measure the diameter and height of the stack. Record these measurements in the data table.

3. After calculating and recording the radius of the cylindrical stack of pennies, calculate and record the volume of the pennies using the equation for the volume of a cylinder.

4. Using the balance, measure the mass of twenty pennies. Record this mass in the data table.

5. Add about 20 mL of water to the 100 mL graduated cylinder and record the volume to the nearest 0.5 mL. Record this volume in the data table.

6. Place the previously weighed twenty pennies into the graduated cylinder and record the new volume level of the water meniscus in the data table.

7. Calculate the volume of the twenty pennies by calculating the water volume displaced by the pennies. Record this calculated volume in the data table.

8. Use the mass-volume data from the data table to plot a mass-volume point for each of the three penny samples on the graph in Figure 2.10. After the mass and volume data for the three samples is plotted, the identity of the major metallic component of post-1982 pennies can be determined.

9. Ask the laboratory instructor to check the graph and the data table entries. Based on this check, the instructor may request that some measurements be repeated.

10. Use a paper towel to dry the pennies, and return them to the reagent table.

11. Use the information in the data table to answer the questions in the conclusion and discussion section.

Data Sheet

mass of 1 penny measured by instructor	2.51 g
volume of 1 penny measured by instructor	0.40 cubic centimeters
mass of 10 penny stack grams measured	
diameter of 10 penny stack centimeters measured	
radius of 10 penny stack centimeters calculated	
height of 10 penny stack centimeters calculated	
volume of 10 penny stack cubic centimeters calculated	
mass of 20 penny stack grams measured	
initial volume of water milliliters measured	
final volume of water milliliters measured	
volume of 20 penny stack milliliters calculated	

Conclusion and Discussion

1. As mentioned in the introduction, the constant slope of the mass versus volume plot is called density. Determine slope of the mass versus volume line for post-1982 pennies.

2. What is the identity of the major component of post-1982 pennies?

3. Use the water displacement method (procedure steps 4–7) to determine the mass and volume of 15 five cent coins (nickels).

    Mass of 15 nickels: _____

    Volume of 15 nickels: _____

4. Plot a mass-volume point for 15 nickels on the graph in Figure 2.10. Based on the information in the graph, what does this point seem to suggest about the identity of the major metallic component of a nickel?

    Any answer to this question based only on the limited graphical data given in Figure 2.10 will be incorrect. The formal laboratory exercise associated with the next chapter will explore this point further.

Chapter Two Notes

Chapter Two Notes

Chapter Three

Matter and Chemical Change _____

3.1 Introduction

In Chapter 1, chemistry was defined as an attempt to understand by rational thought processes the ability of matter to change its identity. If chemists wish to understand the ability of matter to change its identity, they must have some way to recognize this type of change (chemical change). This obviously demands the ability to distinguish between various types of matter. It should come as no surprise, therefore, that attempts to classify matter have always played an important role in the study of chemistry. The purpose of this chapter is to study some of the methods that chemists use to classify and characterize matter.

3.2 The Chemical Classification of Matter

In any area of study, the act of classification is simply an attempt to organize and make manageable vast amounts of information. For example, the public library makes it easy for you to locate a Sherlock Holmes novel by placing the novel in the fiction section filed alphabetically by the author's last name. Classification of books can be accomplished by many different methods (Library of Congress, Dewey Decimal, etc.), but the public library has one major objective: easy access to books by the general reading public. Most small public libraries, therefore, use a variation of the Dewey Decimal system suited to this objective. As a result, even a small child can locate *Green Eggs and Ham* by Dr. Seuss, in a library containing thousands of volumes.

In a similar manner, the immense varieties of matter can be classified many different ways. In selecting a particular classification system, the chemist must keep an important chemical objective in mind. Since the chemist wishes to understand the process of chemical change, it is not surprising that the chemical classification system is related to chemical change.

DEFINITION: A **chemical change** is any change in matter that involves a change in identity.

EXAMPLE: When iron rusts, a chemical change takes place. (Fig. 3.1)

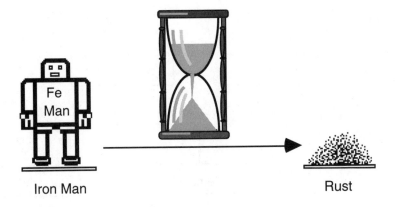

Iron Man   Rust

Figure 3.1 A Chemical Change

Although the classification scheme that chemists use is based on chemical change, it is important to recognize that matter undergoes another type of change. All matter is capable of experiencing physical change, and this type of change is as important to the chemist as chemical change.

DEFINITION: A physical change is any change in matter that does not involve a change in identity.

EXAMPLE: When water boils, a physical change takes place. In Figure 3.2, the water vapor and the water liquid have the same chemical composition. The vapor and the liquid represent different physical phases of water.

The schematic shown below represents a chemical classification of matter. In the following sections, each term used in the schematic (Fig. 3.3) will be carefully defined.

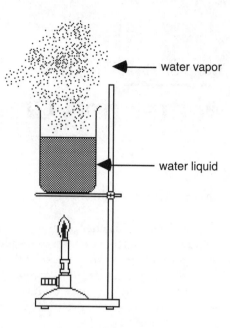

water vapor

water liquid

Figure 3.2 A Physical Change

3.3 Matter

Matter is usually defined as anything that occupies space and possesses mass. It is these two characteristics that are used to quantitatively measure matter (grams and liters). Even a casual inspection of the matter that you find around you will reveal that all matter can be placed in one of two qualitative categories. Water, for example, seems to appear uniform throughout its entire bulk. The water at the top of a glass appears identical to the water at the bottom. Matter that is uniform throughout its bulk is said to be **homogeneous**. Actually, when chemists use the word homogeneous,

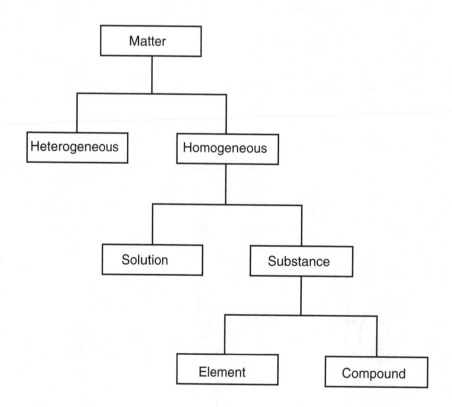

Figure 3.3 The Chemical Classification of Matter

they mean uniform throughout even when subjected to microscopic examination. Hence, although you may think that a piece of blackboard chalk is homogeneous, microscopic examination would reveal the presence of several different kinds of materials. What is important here is not your ability to evaluate a particular bit of matter for homogeneity, but rather your ability to understand what a chemist means by the term homogeneous. Matter that is not homogeneous (*e.g.* blackboard chalk) is said to be **heterogeneous**.

Although the terminology above may seem to be trivial information, it accomplishes for us a very important objective. Just as the librarian divides all books into fiction or nonfiction, so also the chemist divides all matter. The immense variety of matter is becoming more manageable.

3.4 Mixtures

The terms *homogeneous* and *heterogeneous* are precisely defined technical terms, and it is important that you learn their exact meanings. It is tempting to use the common word *mixture* to replace the technical word *heterogeneous*, but the word *heterogeneous* does not mean mixture. Let us take a closer look at the word *mixture* and see how it relates to our original organizational scheme.

Close examination of a piece of granite reveals that it is a mixture of two visibly different crystalline materials (Fig. 3.4). A microscope is not even required for this observation. In the simplest type of granite, the two crystalline materials are quartz and mica. Granite is, therefore, heterogeneous, not because it is a mixture, but because it fits the precise technical definition of heterogeneous. To better appreciate this subtle point, consider the material quartz, one of the components of granite. A microscopic examination of pure quartz reveals that it is uniform throughout a given crystal. By the technical definition of homogeneous, quartz is homogeneous. But here is an interesting fact about quartz. By appropriate techniques, quartz can be separated into the two materials silicon and oxygen. Quartz is a mixture. Nature evidently can hide certain mixtures, and it is through these hidden mixtures that chemistry truly reveals itself. Nature hides mixtures in two different ways. Understanding the difference between these two types of hidden mixtures is central to understanding the basic nature of chemistry. Before we can deal with the subject of hidden mixtures, however, we must first consider the characterization of matter.

Left: granite historical marker near Normal, IL

Right: close up view of granite marker

Figure 3.4 Granite: A Heterogeneous Mixture

3.5 The Characterization of Matter

All samples of matter have a unique set of characteristics called properties. The table below (Fig. 3.5) represents a partial list of properties for the homogeneous materials table salt, water, sulfur, and zinc.

Characteristics of Four Homogeneous Materials

Substance	Color	Electrical Conductivity	Taste	Combustion
table salt	white	yes (pure liquid) no (pure solid)	"salt"	no
water	colorless	no	none	no
sulfur	yellow	no	none	burns in air to form colorless gas
zinc	gray	yes	none	burns in air to form white solid

Figure 3.5

All of the properties on this chart fall into one of two categories. They are either chemical properties or physical properties.

DEFINITION: A **chemical property** is any characteristic that involves chemical change.

EXAMPLE: Solid sulfur burns in air to produce a new gaseous material. This fact is a chemical property of sulfur.

DEFINITION: A **physical property** is any characteristic that does not involve chemical change. A physical property may, however, involve physical change.

EXAMPLE: A physical property of sulfur is that it is a yellow solid.

Both chemical and physical properties are important to chemists. A given material's set of chemical and physical properties is used to characterize or identify that material. Hence table salt literally signals its presence by manifesting its chemical and physical properties. If a given sample of matter looks like table salt and reacts like table salt, it must be table salt. Of course, the more properties the chemist examines, the greater the certainty of this conclusion. And just as fingerprints are more powerful identifiers of human beings than hair colors, some properties are also more powerful identifiers of matter than others. The details of this may seem very complicated, but the basic idea is really very simple. If it looks like table salt and reacts like table salt, it must be table salt.

3.6 Hidden Mixtures

In the previous section, some of the chemical and physical properties of four different homogeneous materials were presented (Fig. 3.5). Let us consider some mixtures that could be made from these materials.

To begin, consider a mixture of table salt and water. This is probably a mixture you have made many times. What are the characteristics (properties) of this mixture? Listed below are some of the more obvious properties of a table salt/water mixture:

Characteristics of a Salt/Water Mixture

1. The mixture is homogeneous (hidden mixture).
2. The mixture looks and feels like water.
3. The mixture tastes like table salt.
4. The mixture conducts electricity (not obvious, but true).

An important aspect of this partial list is that each component of the mixture is signaling its presence through its properties. To be sure, the mixture has some unique properties all its own. For example, it boils at a higher temperature than pure water. These unique properties, however, cannot hide the fact that both table salt and water are present in the mixture. Another important thing to note about this mixture is that the components can be separated simply by evaporating the water (physical change). Homogeneous mixtures (hidden mixtures) that possess these characteristics are termed solutions.

DEFINITION: A **solution** is a homogeneous mixture. It retains some of the characteristics of each of its components, and the components can be separated by making use of physical properties.

EXAMPLE: A mixture of sugar and water forms a solution.

Now consider mixing the other two homogeneous materials on our original list (Fig. 3.5). Although you have probably never mixed sulfur and zinc yourself, the result of this mixing is easy to describe. If equal volumes of sulfur and zinc powders are mixed together very carefully, the result is quite obviously a heterogeneous mixture. Individual particles of both sulfur and zinc are visible to the naked eye. In this respect, the mixture is similar to granite. As in the case of solutions, the components of this mixture can be separated by making use of physical properties. For example, you could use a tweezers to pick the yellow sulfur out of the mixture.

If, however, the heterogeneous mixture of sulfur and zinc is struck by a spark, the situation changes drastically. An explosion takes place, and a white mushroom cloud rises into the air. A chemical reaction has taken place, and the only material product of this reaction is a white solid (mushroom cloud) that is completely homogeneous. Nature has indeed formed another hidden mixture, but this time it is a hidden mixture with a vengeance. No physical property of either component provides a means to undo the mixing. To be sure, the sulfur and zinc can become unmixed, but this can only be achieved by running another chemical reaction. In this particular case, it would take several chemical reactions to isolate the sulfur and zinc. Such an intimate homogeneous mixture is called a chemical compound.

DEFINITION: A **chemical compound** is a homogeneous mixture that has its own unique characteristics. Its components can be separated by chemical means (chemical reaction) only.

The components of chemical compounds are called elements. Nature provides ninety elements from which an immense number of compounds can be made.[1]

DEFINITION: An **element** is the simplest type of homogeneous material. One of its characteristics is that it cannot be broken down into simpler components by chemical means. Elements unite in the chemical reaction process to form chemical compounds.

EXAMPLE: Water is a chemical compound because it is formed by a chemical reaction that unites the elements hydrogen and oxygen (Fig. 3.6).

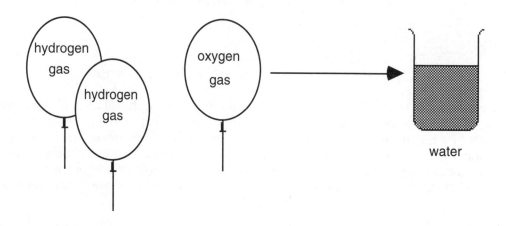

Figure 3.6 Water: A Chemical Compound

Although Western cultural belief in elemental substances originated in ancient Greece, the modern definition of a chemical element was not fully understood until the seventeenth century. In 1661, the English natural scientist Robert Boyle published *The Sceptical Chymist*. In this publication, Boyle rejected the philosophical four elements of the ancient Greeks, and clearly defined the elemental concept in modern terms. Boyle's definition required an understanding of the very subtle hidden mixtures called chemical compounds. The recognition of the elemental nature of some substances represented a milestone in the development of chemistry.

The name of each element can be represented by an internationally recognized **symbol**. In this text, two lists of chemical elements are provided for your convenience. Figure 3.7 represents an alphabetical listing by element name. This list contains the symbol, the atomic mass, and the atomic number for each element. The concept of atomic mass is discussed in Chapter 4, and the concept of atomic number is discussed in Chapter 6. Figure 3.8 represents an alphabetical listing of the elements by internationally recognized symbol. Note that the first letter of an element's symbol is always capitalized. If the element has a two letter symbol, the second letter of the symbol is never capitalized.

Chemical compounds and elements are known collectively as substances. Since this latter term has a common meaning almost synonymous with the word *matter*, it is important that you particularly note its technical definition. A substance is an element or a compound.

DEFINITION: A **substance** is an element or a compound.

EXAMPLE: Water, hydrogen, and oxygen are all substances.

---

1. Although nature provides only ninety elements, there are a number of synthetic elements. The synthetic elements listed in Fig. 3.7 are marked with an asterisk(*).

The Elements: Alphabetical List by Name

Name	Symbol	#	Mass	Name	Symbol	#	Mass
Actinium	Ac	89	[227]	Mendelevium*	Md	101	[258]
Aluminum	Al	13	27.0	**Mercury**	Hg	80	201
Americium*	Am	95	[243]	Molybdenum	Mo	42	95.9
Antimony	Sb	51	122	Neodymium	Nd	60	144
Argon	Ar	18	39.9	**Neon**	Ne	10	20.2
Arsenic	As	33	74.9	Neptunium*	Np	93	[237]
Astatine	At	85	[210]	**Nickel**	Ni	28	58.7
Barium	Ba	56	137	Niobium	Nb	41	92.9
Berkelium*	Bk	97	[247]	**Nitrogen**	N	7	14.0
Beryllium	Be	4	9.01	Nobelium*	No	102	[259]
Bismuth	Bi	83	209	Osmium	Os	76	190
Bohrium*	Bh	107	[262]	**Oxygen**	O	8	16.0
Boron	B	5	10.8	**Palladium**	Pd	46	106
Bromine	Br	35	79.9	**Phosphorus**	P	15	31.0
Cadmium	Cd	48	112	**Platinum**	Pt	78	195
Calcium	Ca	20	40.1	**Plutonium***	Pu	94	[244]
Californium*	Cf	98	[251]	**Polonium**	Po	84	[209]
Carbon	C	6	12.0	**Potassium**	K	19	39.1
Cerium	Ce	58	140	Praseodymium	Pr	59	141
Cesium	Cs	55	133	Promethium*	Pm	61	[145]
Chlorine	Cl	17	35.5	Protactinium	Pa	91	[231]
Chromium	Cr	24	52.0	**Radium**	Ra	88	[226]
Cobalt	Co	27	58.9	**Radon**	Rn	86	[222]
Copper	Cu	29	63.5	Rhenium	Re	75	186
Curium*	Cm	96	[247]	Rhodium	Rh	45	103
Dubnium*	Db	105	[262]	Rubidium	Rb	37	84.5
Dysprosium	Dy	66	162	Ruthenium	Ru	44	101
Einsteinium*	Es	99	[252]	Rutherfordium*	Rf	104	[261]
Erbium	Er	68	167	Samarium	Sm	62	150
Europium	Eu	63	152	Scandium	Sc	21	45.0
Fermium*	Fm	100	[257]	Seaborgium*	Sg	106	[263]
Fluorine	F	9	19.0	Selenium	Se	34	79.0
Francium	Fr	87	[223]	**Silicon**	Si	14	28.1
Gadolinium	Gd	64	157	**Silver**	Ag	47	108
Gallium	Ga	31	69.7	**Sodium**	Na	11	23.0
Germanium	Ge	32	72.6	**Strontium**	Sr	38	87.6
Gold	Au	79	197	**Sulfur**	S	16	32.1
Hafnium	Hf	72	178	Tantalum	Ta	73	181
Hassium*	Hs	108	[265]	Technetium*	Tc	43	[98.0]
Helium	He	2	4.00	Tellurium	Te	52	128
Holmium	Ho	67	165	Terbium	Tb	65	159
Hydrogen	H	1	1.01	Thallium	Tl	81	204
Indium	In	49	115	Thorium	Th	90	[232]
Iodine	I	53	127	Thulium	Tm	69	169
Iridium	Ir	77	192	**Tin**	Sn	50	119
Iron	Fe	26	55.8	**Titanium**	Ti	22	47.9
Krypton	Kr	36	83.8	**Tungsten**	W	74	184
Lanthanum	La	57	139	**Uranium**	U	92	[238]
Lawrencium*	Lr	103	[260]	**Vanadium**	V	23	50.9
Lead	Pb	82	207	**Xenon**	Xe	54	131
Lithium	Li	3	6.94	Ytterbium	Yb	70	173
Lutetium	Lu	71	175	Yttrium	Y	39	88.9
Magnesium	Mg	12	24.3	**Zinc**	Zn	30	65.4
Manganese	Mn	25	54.9	Zirconium	Zr	40	91.2
Meitnerium*	Mt	109	[266]	* = Synthetic Elements			

Symbols for **boldfaced** elements should be memorized.

Figure 3.7

The Elements: Alphabetical List by Symbol

Symbol	Name	Symbol	Name
Ac	Actinium	Mn	Manganese
Ag	Silver	Mo	Molybdenum
Al	Aluminum	Mt	Meitnerium*
Am	Americium*	N	Nitrogen
Ar	Argon	Na	Sodium
As	Arsenic	Nb	Niobium
At	Astatine	Nd	Neodymium
Au	Gold	Ne	Neon
B	Boron	Ni	Nickel
Ba	Barium	No	Nobelium*
Be	Beryllium	Np	Neptunium*
Bh	Bohrium*	O	Oxygen
Bi	Bismuth	Os	Osmium
Bk	Berkelium*	P	Phosphorus
Br	Bromine	Pa	Protactinium
C	Carbon	Pb	Lead
Ca	Calcium	Pd	Palladium
Cd	Cadmium	Pm	Promethium*
Ce	Cerium	Po	Polonium
Cf	Californium*	Pr	Praseodymium
Cl	Chlorine	Pt	Platinum
Cm	Curium*	Pu	Plutonium*
Co	Cobalt	Ra	Radium
Cr	Chromium	Rb	Rubidium
Cs	Cesium	Re	Rhenium
Cu	Copper	Rf	Rutherfordium*
Db	Dubnium*	Rh	Rhodium
Dy	Dysprosium	Rn	Radon
Er	Erbium	Ru	Ruthenium
Es	Einsteinium*	S	Sulfur
Eu	Europium	Sb	Antimony
F	Fluorine	Sc	Scandium
Fe	Iron	Se	Selenium
Fm	Fermium*	Sg	Seaborgium*
Fr	Francium	Si	Silicon
Ga	Gallium	Sm	Samarium
Gd	Gadolinium	Sn	Tin
Ge	Germanium	Sr	Strontium
H	Hydrogen	Ta	Tantalum
He	Helium	Tb	Terbium
Hf	Hafnium	Tc	Technetium*
Hg	Mercury	Te	Tellurium
Ho	Holmium	Th	Thorium
Hs	Hassium*	Ti	Titanium
I	Iodine	Tl	Thallium
In	Indium	Tm	Thulium
Ir	Iridium	U	Uranium
K	Potassium	V	Vanadium
Kr	Krypton	W	Tungsten
La	Lanthanum	Xe	Xenon
Li	Lithium	Y	Yttrium
Lr	Lawrencium*	Yb	Ytterbium
Lu	Lutetium	Zn	Zinc
Md	Mendelevium*	Zr	Zirconium
Mg	Magnesium	* Synthetic Elements	

Figure 3.8

3.7 A Historical Note

So far, this chapter has been devoted to developing the semantics or terminology of chemistry. Although the terms element and compound are easy to define and understand, it is instructive to consider a practical problem that early chemists had in distinguishing these two types of substances. Consider the following illustration.

If the homogeneous material copper is heated to a high temperature, the red metal is converted to a black solid. A chemical reaction has clearly taken place since there has been a change in identity. Now, using only the definitions of element and compound given in the last section, determine if the black solid is an element or a compound. It is tempting to use the smugness of hindsight and declare that of course copper is an element and the black solid is a compound of copper and the element oxygen which is contained in the air, but this was far from obvious to early chemists. In fact, there is not enough information given to make the decision. In order to determine whether the black substance is an element or a compound, it must be determined if the chemical reaction represents a decomposition into simpler substances or a building up into a more complex substance. Schematically, these two alternatives can be presented as follows:

Reaction 3.A
COPPER + SUBSTANCE X → BLACK SOLID

Reaction 3.B
COPPER → BLACK SOLID + SUBSTANCE Y

Ignoring for the moment the mystery substances, X and Y, the problem reduces to one of two possible situations. In the case of reaction 3.A (building up), the black solid will have a greater mass than the original copper sample, and it *will* be a compound. It has a greater mass because it is a mixture, albeit a very special intimate mixture, of the original copper and the mystery substance, X. On the other hand, in the case of reaction 3.B (decomposition), the black solid will have a lesser mass than the original copper sample, and it *might* be an element. It has a lesser mass because it is only one component of a mixture.

Hopefully, all of the above points are, at least in retrospect, painfully clear. But here is an interesting point you may have missed. In order to determine greater or lesser mass, someone had to think to measure the mass of the copper before the chemical reaction and the mass of black solid after the chemical reaction (Fig. 3.9).

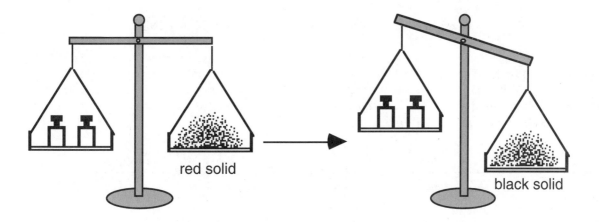

Figure 3.9 Mass Analysis of a Chemical Reaction

65

Now understand something about mass measurement and the device used for measuring mass, the balance. In the history of humanity, the balance is very old, but its utility to early humans was as an instrument of commerce. "I'll trade my corn for an equal quantity of your rice," is a statement that demanded mass measurement. A similar situation occurred in the history of astronomy. The telescope was around for a long time as an instrument of navigation and commerce before Galileo thought to point it toward the heavens and tell the world what he saw. In that act, by one man, modern astronomy was born. Modern chemistry had a similar birth. In the next chapter, we will look at chemistry's Galileo in greater detail. For the time being, you should appreciate that the completion of the chemical scheme of matter classification demanded that the balance be applied to the study of chemical reactions. This did not happen until the end of the eighteenth century, the century that ushered in modern chemistry.

3.8 A New Look at Chemical Change

In Chapter 1, chemical change was represented schematically as follows:

$$\text{MATTER I} \rightarrow \text{MATTER II}$$

At this point, let us see how the terminology of the previous sections allows a more sophisticated view of the chemical reaction process. All chemical reactions must fit one of the following reaction types:

Simple Types of Chemical Reactions

1. ELEMENTS → COMPOUND
2. COMPOUND → ELEMENTS (reverse of 1)
3. COMPOUND → SIMPLER COMPOUNDS
4. SIMPLER COMPOUNDS → COMPOUND (reverse of 3)
5. COMPOUND → COMPOUND OF EQUAL COMPLEXITY

By convention, substances entering a chemical reaction are called **reactants**, and the substances formed by the chemical reaction are called **products**.

It should also be pointed out that various combinations of the above five basic reactions are also possible. For example, the reason the black solid in reaction 3.B *might* be an element is because of the following possible combination of reaction types 2 and 3:

$$\text{COMPOUND} \rightarrow \text{SIMPLER COMPOUND} + \text{ELEMENTS}$$

Even in a decomposition reaction, the black solid could still be a compound. The important point here, however, is that the classification scheme not only makes the immense variety of matter more manageable, but it also makes the immense variety of chemical reactions more manageable. There is order in nature's chemistry. And where nature has order, nature also has reasons for order. Finding the order is the first step toward finding the reasons.

EXAMPLE I

Chemical Terminology

A black solid is heated in an oven for one hour. At the end of this time, a new white solid remains. The new white solid is found to be more massive than the original black solid. Examination of both solids indicates that they are pure substances. Did a chemical reaction take place? Can the original black solid be identified as an element or a compound? Can the new white solid be identified as an element or a compound?

SOLUTION:

An ability to correctly use the terminology of Chapter 3 allows the following answers: 1) A chemical reaction did take place. 2) The original black solid could be either an element or a compound. 3) The new white solid must be a compound.

3.9 A Closer Look at Chemical and Physical Properties

All chemical compounds have a large number of chemical and physical properties. Some of these are very subtle. For example, the unique way in which some compounds absorb light is a physical property that can actually "fingerprint" these compounds. At this point in the course, it is not necessary to detail all the various types of physical and chemical properties, but one point should be detailed. Both chemical and physical properties can be quantified. Remember that the birthplace of modern chemistry was on the merchant's balance. In anticipation of the importance of this, let us look at a typical example of a quantitative physical property and a quantitative chemical property.

3.10 A Quantitative Physical Property

Which weighs more, a pound of feathers or a pound of iron?[2] The person who answers this classic trick question correctly has an intuitive feel for one of the basic properties of matter. The correct answer is that a pound of feathers and a pound of iron weigh exactly the same. The trick is only that a pound of feathers would make a larger pile (larger volume) than a pound of iron. There is something deceiving about volume as a measure of how much matter is located in one place.

Another way to express this using the principles of Chapter 2 is as follows:

Expression 3.A

For feathers: $\text{Mass} \propto \text{Volume}$

Recall what this expression means is that if the volume of feathers in a given pile is doubled, the mass of the pile will also double. Most people feel this to be true intuitively, and experiment verifies that Expression 3.A is indeed a fact. Recall also that an arithmetic consequence of Expression 3.A can be stated as an equation:

Expression 3.B

For feathers: $\dfrac{\text{Mass}}{\text{Volume}} = \text{Constant}$

For a given substance, this constant is a quantitative physical property characteristic of that substance. The constant is called the substance's **density**. Perhaps the best way to illustrate density as a useful physical property is to consider a historical example:

---

2.   For reasons discussed in Chapter 2, chemists often use the verb "weight" to express the determination of mass and the determination of weight.

DEFINITION:

$$\text{Density (D)} = \frac{\text{Mass of a sample of matter (M)}}{\text{Volume of a sample of matter (V)}}$$

or

$$D = \frac{M}{V}$$

3.11 The Case of the Dishonest Goldsmith

In the third century BC, the Greek scholar Archimedes was presented with an intriguing problem. His friend and distant relative Hieron, king of Syracuse, suspected that a gold crown which had been fabricated from a gold bar did not contain the correct quantity of gold. Specifically, although the crown had the proper mass, Hieron suspected that the goldsmith had profited by removing gold from the crown and adding an equal mass of silver. Since a pound of gold is more valuable than a pound of silver, the goldsmith would experience a nice net profit. The problem that Hieron presented to Archimedes was to determine if the crown was made of pure gold.

This story comes to us from the Roman writer Vitruvius, and since Archimedes was somewhat of a folk hero, it is probable that some of the details of the story have been embellished. In detailing Archimedes' solution to the problem, let us add one more embellishment for the sake of clarity. Let us imagine that Archimedes measured and reasoned in the metric system.

To understand this solution, you must understand that Archimedes understood perfectly the concept of density presented in the last section. At least, he understood this concept in the Greek system of measurement used at the time. Expressed in modern terms, Archimedes knew the following:

Density of Gold = 18.9 gram/milliliter (abbreviated g/mL)[3]

Density of Silver = 10.5 g/mL

Hieron was no dummy, and he knew this also. Further, both Archimedes and Hieron knew that the authenticity of the crown could be tested by two measurements and one calculation. Specifically, by applying Expression 3.B to the crown, they could calculate the density of the crown and compare this value to the densities of gold and silver.

EXAMPLE II

Density Calculation

Using the following data, calculate the density of Hieron's crown:

Mass of Hieron's Crown = 7500 grams (g)
Volume of Hieron's Crown = 500 milliliter (mL)

SOLUTION: 15 g/mL

---

3. When calculating the actual value of a constant ratio, the units of the constant must be recorded with the value of the constant. The abbreviation g/mL is read grams per milliliter, and these units dictate the units that must be used in the density equation.

$$\text{Density of Crown} = \frac{\text{Mass of Crown}}{\text{Volume of Crown}}$$

$$\text{Density of Crown} = \frac{7500 \text{ g}}{500 \text{ mL}}$$

$$\text{Density of Crown} = 15.0 \text{ g/mL}$$

The fact that the density obtained in the above calculation was less than the density of gold but more than the density of silver indicated that the crown was not pure gold (much, we are sure, to the dismay of the goldsmith), but rather that it was a mixture of gold and silver. Although Archimedes is usually thought of as a mathematician, this is a beautiful example of the use of a quantitative physical property for the purpose of chemical identification. It also probably represents the first recorded example of the application of chemistry to a criminal investigation (forensic chemistry).

Although the preceding paragraph tells the story of Archimedes' solution of a chemical problem by the application of the concept of density, it does not really tell the story of Archimedes' great achievement. The Greeks, you see, only knew how to measure the volume of liquids and certain regular geometric solids. For example, if the crown had been in the shape of a cube, then Hieron himself could have solved the problem without Archimedes' help. But Hieron did not know how to find the volume of a crown without melting it to form a liquid or beating it into a cube. This was the real problem that he gave Archimedes. "Without destroying my crown, can you tell me if it is made of pure gold?"

At first, Archimedes was totally perplexed by the problem. Finally, we are told by Vitruvius, while relaxing in one of Syracuse's public baths, Archimedes discovered how to measure the volume of a crown. As he eased down into the bath, some of the bath water slopped over the side of the bath. Archimedes realized that when an irregular solid is immersed in a liquid (water) it displaces a volume of that liquid equal to its own volume. All one needs to do is measure the volume of the displaced liquid, which the Greeks could do, and that would be equal to the volume of the solid. We are told that Archimedes was so overjoyed with his discovery that he leaped out of the bath and ran naked through the streets of Syracuse to find Hieron, all the while screaming, "I found it! I found it!" He found it indeed, and he found it by applying mathematics to chemistry.

3.12 A Quantitative Chemical Property

The chemical compound water can be chemically decomposed into the elements hydrogen and oxygen. This fact can be represented as follows:

The Decomposition of Water into Its Elements
WATER → HYDROGEN + OXYGEN

When a compound is subjected to the chemical reaction process in order to determine its component elements, it is said to undergo **qualitative analysis**. The results of such a qualitative analysis constitute a chemical property of the compound. By using a balance, it is easy to convert this qualitative analysis into a **quantitative analysis**. Consider the case of water:

The Quantitative Analysis of Water
WATER → HYDROGEN + OXYGEN (QUALITATIVE)

50.0 g WATER → 5.60 g HYDROGEN + 44.4 g OXYGEN (QUANTITATIVE)

The quantitative data above is an example of a quantitative chemical property. Quantitative analysis data is very often expressed in terms of percent. This percent calculation is illustrated below using the method suggested in Chapter 2 (Example IV):

EXAMPLE III

Percent Calculation

Using the data given in the quantitative analysis of water shown above, calculate the percentage of oxygen contained in water.

SOLUTION: 88.8% oxygen

All percentage problems involve linear reasoning. Using the approach suggested in Chapter 2, it can be shown that the arithmetic relationship that applies to all simple percentage problems is:

$$\frac{Part}{Whole} = \frac{Part'}{100 \text{ Units of Whole}}$$

Recognizing that the compound (water) corresponds to the whole and that the element (oxygen) corresponds to the part, we have:

$$\frac{44.4 \text{ g oxygen}}{50.0 \text{ g water}} = \frac{X \text{ g oxygen}}{100 \text{ g water}}$$

$$x = 88.8 \text{ g of oxygen}[4]$$

Since percent (%) literally means per 100, we say that water is 88.8% oxygen. Although this approach may seem naive to a mathematically sophisticated person, it does have certain advantages. First, we are not pushing decimal points around, and hence we avoid the possibility of an inadvertent decimal point error. Second, this approach involves no computational ambiguity. If you can keep the *part* separate from the *whole*, then you can solve the problem.

Chemical substances have many different types of chemical properties. In the next chapter, you will see that the quantitative analysis of chemical compounds played a major role in the development of chemistry, and hence it is the only quantitative chemical property that we need to consider at this time. The elemental quantitative analysis is a unique property of a chemical compound. Although the simple chemical property which is illustrated in Example III may seem to be a trivial bit of information, it is through such mass bookkeeping that nature reveals to us a most profound secret. We shall see more of this in the next chapter.

"When I use a word,
It means what I choose it to mean—
Neither more nor less."

Humpty Dumpty, English Egg
Lewis Carroll
(Charles Dodson, English Mathematician)

---

4. As pointed out in Chapter 2 (Example II), the position of x in the format of the direct proportion dictates the units of x. In this case, x must equal grams of oxygen.

Chapter Three
Performance Objectives

P.O. 3.0

Review all of the boldfaced terminology in this chapter, and make certain that you understand the use of each term.

chemical change	chemical compound	chemical property
density	element	heterogeneous
homogeneous	matter	physical change
physical property	products	qualitative analysis
quantitative analysis	reactants	solution
substance	symbol	

P.O. 3.1

You must learn the symbols and names of the boldfaced elements listed in Figure 3.7.

P.O. 3.2

You must be able to use the chemical terminology defined in Chapter 3 and summarized in Figure 3.3 in the text book.

EXAMPLE:
A blue solid is heated in an oven for one hour. At the end of this time, a new black solid remains. The new black solid is found to be less massive than the original blue solid. Examination of both solids indicates that they are pure substances. Did a chemical reaction take place? Can the original blue solid be identified as an element or a compound? Can the new black solid be identified as an element or a compound?

SOLUTION:
An ability to correctly use the terminology presented throughout Chapter 3 allows the following answers: 1) A chemical reaction did take place. 2) The original blue solid is a compound. 3) The new black solid could be an element or a compound.

ADDITIONAL EXAMPLE:
Consider the following schematic of a chemical reaction:

$$\text{SUBSTANCE A} + \text{SUBSTANCE B} \rightarrow \text{SUBSTANCE C}$$

Is substance A an element or a compound?

ANSWER: insufficient information

P.O. 3.3

You must be able to solve a problem involving density. This problem requires use of the density equation: $D = \dfrac{M}{V}$

EXAMPLE:
    The density of elemental aluminum is known to be 2.70 g/mL. In an attempt to discover the identity of an unknown metal, a chemist determines the mass of a 150 mL sample of the metal. If the unknown metal is aluminum, what should the mass be?

SOLUTION: 405 grams

    Textbook Reference: Section 3.11

ADDITIONAL EXAMPLE:
    The density of aluminum is known to be 2.70 g/mL, and the density of magnesium is known to be 1.74 g/mL. If 1.50 mL of an unknown metal has a mass of 3.20 grams, which of the following statements is most correct?

    a. The unknown metal is most likely aluminum.
    b. The unknown metal is most likely magnesium.
    c. The unknown metal is not aluminum or magnesium.

ANSWER: c

P.O. 3.4

    You must be able to solve a percent linear reasoning problem involving the unique constant composition of a chemical compound.

EXAMPLE:
    A chemical compound contains the elements carbon and hydrogen only. A quantitative analysis of this compound indicates that it contains 25.0% hydrogen by mass. What mass of the compound would need to be decomposed in order to produce 900 grams of carbon?

SOLUTION: 1200 grams of compound

    This P.O. is almost identical to P.O. 2.3. In the following solution, however, notice that the first step requires an understanding of the term *chemical compound*. By the wording of the problem, 25.0% hydrogen implies 75.0% carbon.

$$\frac{X}{Y} = \frac{X'}{Y'}$$

$$\frac{Part}{100} = \frac{Part'}{Whole'}$$

$$\frac{75.0 \text{ g carbon}}{100 \text{ g compound}} = \frac{900 \text{ g carbon}}{X}$$

$$X = 1200 \text{ g compound}$$

ADDITIONAL EXAMPLE:
    A chemical compound contains the elements sulfur and oxygen only. A quantitative analysis of this compound indicates that it contains 40.0% sulfur by mass. What mass of the compound would need to be decomposed in order to produce 150 grams of oxygen?

ANSWER: 250 grams of compound

Chapter Three
Problems

Chemical Terminology

1. A white solid substance is heated in an oven for one hour. At the end of this time, a new white substance remains. The chemical and physical properties of the new white substance are significantly different than the original white substance. The mass of this new white substance is also more than the mass of the original substance. Based on this information only, answer the following questions:

   a. Has a chemical reaction taken place?
   b. Can the original white substance be identified as an element or a compound?
   c. Can the new white substance be identified as an element or a compound?

   STUDENT SOLUTION:

2. Consider the following chemical reaction schematic:

   Compound
   SUBSTANCE A → SUBSTANCE B + SUBSTANCE C

   Is substance A an element or a compound? What about substance B?

   STUDENT SOLUTION:

3. How could you determine whether a colorless, odorless, tasteless liquid is a pure substance or a solution?

   STUDENT SOLUTION:

4. In 1771, the French chemists Pierre Joseph Macquer and Antoine Laurent Lavoisier demonstrated that diamond is a form of the element carbon (charcoal). Explain how they might have demonstrated this fact.

   STUDENT SOLUTION:

Density Calculations

5. The mass of 2.00 mL of an ethanol/water solution is 1.87 grams. Calculate the density of this solution.

   STUDENT SOLUTION:

6. The density of the element mercury is 13.6 grams per milliliter. What is the mass of a liter (about a quart) of this liquid metallic element? Recalling that a United States nickel has a mass of about 5 grams, how many nickels would be required to balance a liter of mercury? NOTE: Since the density in this problem is expressed in g/mL, the mass must be expressed in grams and volume must be expressed in milliliters.

   STUDENT SOLUTION:

7. The density of a sugar solution is known to be 1.50 g/mL. How many mL of this solution should be measured out in order to obtain 140 grams of solution?

   STUDENT SOLUTION:

8. A pure liquid is thought to be either cyclohexanol (density = 0.962 g/mL), tetrachloroethylene (density = 1.63 g/mL), or chlorobenzene (density = 1.11 g/mL). A chemist determines that 8.46 mL of this unknown liquid has a mass of 8.14 grams. Which of the above compounds is this liquid likely to be?

   STUDENT SOLUTION:

9. A cube of lead is 5.34 cm on the edge. Calculate the mass of this cube in grams if the density of lead is 11.3 g/mL.

   STUDENT SOLUTION:

10. The density of aluminum is known to be 2.70 g/mL. The density of magnesium is known to be 1.74 g/mL. In an attempt to discover the identity of an unknown metal, a chemist determines the volume of a 15.3 g sample of the metal. If the unknown metal is aluminum, what should this volume be in milliliters?

STUDENT SOLUTION:

11. The densities of aluminum, magnesium and beryllium are 2.70, 1.74, and 1.85 g/mL, respectively. In an attempt to discover the identity of an unknown metal, a chemist determines that a 2.25 mL sample of the metal has a mass of 6.08 g. What is one possible identity of the unknown metal based on the given densities?

STUDENT SOLUTION:

12. Actually, the density of a substance is constant only at a given temperature. Density varies with temperature change. In general, as temperature decreases, the density of a substance will increase. What evidence exists that water might violate this rule in some temperature range?

STUDENT SOLUTION:

Chemical Compound Composition Problems

13. Qualitative chemical analysis indicates that a certain compound is composed of hydrogen and oxygen only. A quantitative analysis indicates that 150 grams of this compound contains 8.82 grams of hydrogen. What is the percentage of hydrogen and oxygen in this compound?

STUDENT SOLUTION:

14. Quantitative chemical analysis of a compound composed of only sulfur and oxygen indicates that 175 grams of the compound contain 70.1 grams of sulfur. Calculate the percentage of sulfur and oxygen in this compound.

STUDENT SOLUTION:

15. A certain chemical compound is found to be 80.0% copper by mass. What mass of the compound would need to be decomposed in order to obtain 235 grams of copper?

    STUDENT SOLUTION:

16. A chemical compound is known to contain only the elements carbon and hydrogen. A quantitative analysis of this compound indicates that it contains 92.3% carbon. What mass of this compound in grams would need to be decomposed in order to produce 67.3 grams of carbon?

    STUDENT SOLUTION:

17. A chemical compound contains only the elements carbon and hydrogen. A quantitative analysis of this compound indicates that it contains 25.0% hydrogen. What mass of carbon in grams would be required to synthesize 175 grams of this compound?

    STUDENT SOLUTION:

18. Sulfuric acid, an important industrial chemical, is synthesized from elemental sulfur which is found in underground mineral deposits. Sulfuric acid contains 32.7% sulfur. What mass of sulfur in kilograms needs to be purchased in order to manufacture 15,000 kg of sulfuric acid?

    STUDENT SOLUTION:

19. Tetraethyllead, an antiknock agent added to gasoline, contains 64.1% lead. Calculate the number of grams of lead in gasoline per gallon containing 3.00 grams of tetraethyllead per gallon of gasoline (a typical value before the Environmental Protection Agency (EPA) required reductions in the lead content of gasoline).

    STUDENT SOLUTION:

20. Qualitative analysis of a certain compound indicates that it is composed of iron and oxygen only. A quantitative analysis indicates that the compound is 30.0% oxygen by mass. What mass of the compound would need to be decomposed in order to obtain 190 grams of iron?

    STUDENT SOLUTION:

21. A certain chemical compound contains only the elements lead and sulfur. Quantitative analysis indicates that this compound contains 13.4% sulfur by mass. What mass of this compound in grams needs to be decomposed in order to produce 147 grams of lead?

    STUDENT SOLUTION:

22. Pentaborane is a compound used in rocket propulsion systems. It consists of only the elements boron and hydrogen and contains 14.4% hydrogen. Calculate the number of grams of boron which are needed to synthesize 125 grams of pentaborane.

    STUDENT SOLUTION:

Problems Involving Household Chemistry and Science

23. It is important to recognize the difference between chemical change and physical change. Antiseptic tincture of iodine is an alcohol/water solution that contains, among other substances, the element iodine. The iodine in tincture of iodine can be used in a simple experiment to illustrate a chemical and a physical change. Place one drop of tincture of iodine on a penny and one drop of tincture of iodine in the bottom of a white Styrofoam coffee cup. Note any change in the appearance of these drops during the first 60 seconds of the experiment. Periodically, observe any other changes in the appearance of these drops over a period of about 30 minutes. Although there is a change in the appearance of both drops, only one of these changes represents a chemical change. Which drop of iodine solution caused a chemical change to take place?

    There is a very simple chemical test for elemental iodine that will be used in a number of other experiments. Iodine forms a dark blue complex with starch. Spray both drop spots from this iodine experiment with household spray starch. Which spot turns blue indicating that the iodine was not chemically changed?

    STUDENT SOLUTION:

24. Due to the rise in the price of copper, the U.S. Mint changed the composition of the penny in 1983. The old penny was made up of 95% copper and 5% zinc, while the new penny is composed of 97.6% zinc and 2.4% copper. Copper and zinc have different densities, and, since pennies have remained the same size, these different compositions should result in different penny masses. Obtain 90 pennies with mint dates 1982 or earlier and 90 with dates 1983 or later. Weigh each set on a kitchen scale. Convert to grams (1 gram = 0.035 oz), and calculate the mass in grams per penny. Is there a difference in the mass of a penny from each set? Next, confirm that the size (volume) of the penny has not changed by adding each set of pennies to a graduated measuring cup containing exactly four oz of water. Is the volume of water displaced by both sets the same? A more exact determination of the volume of a penny can be made by measuring the height and diameter with a metric ruler graduated in mm and calculating the volume (refer to the example in Chapter 2). Which metal is more dense, copper or zinc?

    STUDENT SOLUTION:

Library Problems

25. The December 1985 issue of the *Journal of Chemical Education* includes a collection of articles relating to forensic chemistry. Although forensic chemistry has come a long way since Archimedes, forensic chemists still use the measurement of density as an analytical tool. Read the article "Focus on Forensic Experiments" (Berry, Keith O. *J. Chem. Educ.* **1985**, 62, 1060). Among other techniques, the article describes a procedure that can be used to determine the density of a small irregular solid object. Without going into mathematical detail, briefly describe the procedure that is used to measure density. How is this procedure related to the procedure used by Archimedes?

    STUDENT SOLUTION:

26. Chemists are not always serious. Occasionally, we like to have fun. In the December 1991 issue of the *Journal of Chemical Education*, an entire section is devoted to "A Winter's Entertainment." Although several of the offerings in this section might be of interest to beginning chemistry students, the article "A Divertimento on the Symbols of the Elements" (Earl, B. L. *J. Chem. Educ.* **1991**, 68, 1011) is particularly appropriate to Chapter 3 of this text. The third movement of B. L. Earl's divertimento will force you to learn chemical symbols. Isn't chemistry fun?

    STUDENT SOLUTION:
    Yes Teacher!

Chapter Three Notes

Chapter Three Notes

Nineteenth Century Atomic Theory ———

4.1 Introduction

Perhaps the greatest contribution of the eighteenth century chemists was the introduction of quantitative chemical analysis. In this regard, the work of the Scottish chemist Joseph Black and the French chemist Antoine Lavoisier was particularly noteworthy. By the beginning of the nineteenth century, enough quantitative data had been collected to allow for the formulation of the first fully quantitative theory of chemical change. This theory was the atomic theory of chemical change presented by the English chemist John Dalton.

The atomic theory of chemical change, as presented by John Dalton in 1803, has gone through a long and interesting evolution. The purpose of this chapter is to present the basic logic of this theory in essentially modern terms. Although this presentation involves some arithmetic, the mathematical details of the atomic theory are considered in the next chapter. Since the atomic theory of chemical change is the foundation of modern chemistry, the present chapter also attempts to detail some of the history associated with the development of the theory.

4.2 A Pre-Daltonian Theory of Chemical Action

Although there are many types of chemical change, the chemical transformation known as **combustion** (burning) played a major roll in the development of chemistry. Fire has always been of vital interest to humans, consequently it is not surprising that one of the earliest scientific theories of chemical change dealt with combustion. This theory was called the phlogiston theory, and the theory's basic ideas were introduced by the German chemist Johann Becher at the end of the seventeenth century. During the early part of the eighteenth century, the German chemist Georg Ernst Stahl developed phlogiston theory into a comprehensive chemical world view.

Simply stated, the theory recognized that the combustion process appeared to be a decomposition. One need only observe the burning of a piece of wood to reach this conclusion. The flames leaping from the surface of the wood even seem to be carrying something away. Stahl summarized these observations by stating that flammable bodies contained a principle of fire called **phlogiston**, which they release during the combustion process. Easily combustible materials (charcoal) must therefore contain a lot of phlogiston. According to this theory, charcoal was a "compound" of ash and phlogiston, and its combustion was represented as follows:

Reaction 4.A
[ASH/PHLOGISTON] → ASH + PHLOGISTON
charcoal

Eighteenth century chemists used the Greek letter phi (Ø) to represent phlogiston. Using this notation, the combustion of charcoal could be expressed by the following notation:

Reaction 4.A
[ASH/Ø] → ASH + Ø
charcoal

Metals also burn to produce metal ores. The rusting of iron is an example of the slow combustion of a metal. Early chemists called metal ores calx, and they represented the formation of iron ore (rust) as follows:

Reaction 4.B
[IRON ORE/PHLOGISTON] → IRON ORE + PHLOGISTON
         iron   rust

Notice that a metal was viewed as a "compound" of its ore (calx) and phlogiston. Since metals were not as combustible as charcoal, they were thought to contain less phlogiston.

Although it is tempting to laugh at this theory, the temptation comes only with the smugness of hindsight. The phlogiston theory was good science in every sense of the word. First, it was a rationalization consistent with most of the facts that the eighteenth century chemists had at their disposal. Second, it had practical applicability. For example, it supported and helped to develop the technology of metallurgy. Notice that the ancient process of winning iron from its ore could be viewed as a simple phlogiston exchange:

Reaction 4.C
[ASH/PHLOGISTON] + IRON ORE → [IRON ORE/PHLOGISTON] + ASH
    charcoal   rust   iron

The phlogiston exchange even took place in a logical direction, with a phlogiston-rich material (charcoal) being used as a phlogiston source. Such "understanding" is essential to the growth of technology.

Even in its death throes, the phlogiston theory was typical of good science. It was dogmatically (but not pigheadedly) retained. Its ultimate demise was caused by the quantitative measurement of the chemical reaction process that became popular toward the end of the eighteenth century. When quantitative analysis techniques were applied to reactions such as reaction 4.B (above), it was discovered that the metal ore (rust) weighed more than the original metal (iron) sample. The supposed decomposition appeared to be a building up process (synthesis).[1] But never underestimate the ingenuity of a dogmatic establishment. If phlogiston is given the quality of negative mass, then its levity (really!) would still make reaction 4.B consistent with the facts:

Reaction 4.B
A Quantitative Phlogistonist View
[IRON ORE/PHLOGISTON] → IRON ORE + PHLOGISTON
        iron   rust
       112 g   160 g   −48 g

Actually this concept, which was first introduced by the French chemist Gabriel Francois Venel in 1750, is not as ridiculous as it sounds. Negative mass does not violate the principles of classical (Newtonian) physics, although it would, if it existed, possess unusual characteristics. Ultimately the phlogiston theory became too complicated to explain existing facts, and a simpler theory of combustion prevailed. It is to this simpler theory that we now direct our attention.

4.3 The Beginning of Modern Chemistry

In 1774, the English chemist Joseph Priestley discovered that by heating mercury calx (mercury ore) he produced a gas which supported combustion much better than air. He called this gas

---

1.   This statement implies the understanding of a basic law of nature, the law of conservation of mass. Stated in chemical terms, this law demands that the mass of the products of a chemical reaction be equal to the mass of the reactants of the reaction. This law was generally accepted without proof by the chemists of the latter eighteenth century.

dephlogisticated air, and its formation and reactivity were perfectly consistent with the phlogiston theory. All one had to assume, logically enough, was that normal air contained a certain amount of phlogiston. After all, things were always burning and releasing phlogiston into the air. According to phlogiston theory, the formation of dephlogisticated air could be represented as shown in reaction 4.D below (compare reaction 4.C). In this reaction, the Greek letter phi (Ø) is used to represent phlogiston and the letters DPA are used to represent Priestley's dephlogisticated air:

Reaction 4.D
[DPA/Ø] + MERCURY ORE → [MERCURY ORE/Ø] + DPA
normal air   mercury

During the fall of the same year, Priestley visited the French chemist Antoine Laurent Lavoisier, who two years earlier had begun his own experiments on combustion. By applying mass measurement to the study of the combustion process, Lavoisier had become convinced that the combustion process involved synthesis (building up) not decomposition. Specifically, he felt that the phlogistonist reaction 4.B could be more logically viewed as follows:

Reaction 4.B
Phlogistonist View
[IRON ORE/PHLOGISTON] → IRON ORE + PHLOGISTON
iron   rust
112 g   160 g   –48 g

Reaction 4.E
Lavoisier View
IRON + MYSTERY SUBSTANCE → [IRON/MYSTERY SUBSTANCE]
112 g   48 g   160 g

According to this view, iron ore was a compound of iron and the mystery substance. Lavoisier's problem was that the mystery substance was every bit as elusive as phlogiston. During his visit, Priestley told Lavoisier about dephlogisticated air and its combustion-supporting ability. Lavoisier realized that Priestley's dephlogisticated air was, indeed, the mystery substance of his theory. He renamed the substance, calling it by the name *oxygine*, and he suggested that all combustion reactions involved a simple chemical combination between the combustible material and *oxygine* (*i.e.* oxygen). The word *oxygine* was constructed from Greek words meaning *acid former* under the mistaken impression that oxygen was an essential component of the class of chemical compounds called acids. In 1777, Lavoisier reported an elegant experiment involving the combustion of mercury. The quantitative details of this experiment proved that the combustion of mercury involved chemical combination with oxygen, a component of normal air, and that this combustion should be represented as follows:[2]

Reaction 4.F
The Combustion of an Element
MERCURY + OXYGEN → MERCURY/OXYGEN
element   element   compound
                              (called an oxide)

---

2.   In order for this reaction to take place, mercury must be heated to the temperature of glowing charcoal. If the resulting oxide is heated to a much higher temperature, then it decomposes and reforms the elements:

MERCURY/OXYGEN → MERCURY + OXYGEN
compound (oxide)   element   element

This reaction is the reverse of reaction 4.F, and it is the reaction that Priestley represented by reaction 4.D. Priestley obtained the high temperature required for this decomposition by focusing the sun's rays with a lens.

This reaction (reaction 4.F) was recognized as the prototype for the combustion of any element.

The presentation of this theory of combustion marked a turning point in the history of chemistry.[3] Indeed, many historians of chemistry identify Lavoisier as the father of modern chemistry. Because of the importance of this theory to the subsequent development of the science of chemistry, a closer look at the human side of the story is appropriate.

4.4 Giving Credit Where Credit Is Due

In the last section, we looked at a discovery of major importance to the development of the discipline of chemistry. By placing the focus of our attention on scientific achievement, there is a very real danger that an important point will be overlooked—scientists, like all men and women, are a fully human part of the society in which they live. For example, the first president of the State of Israel, Chaim Weizmann, was also an internationally renowned chemist, and Great Britain's Margaret Thatcher was originally trained as a chemist. If we allow the existence of academic disciplines to limit our thinking, then we will find ourselves saying such things as, "Why must I take a chemistry course—I'm a political science major?" This statement is as ridiculous as Chaim Weizmann saying, "How can I be the President of the State of Israel—I'm a chemist?" With these thoughts in mind, let us take a closer look at the human side of Lavoisier and Priestley.

Antoine Laurent Lavoisier (1743–1794) was the son of a wealthy Paris lawyer. Although his early schooling was directed toward the legal profession, his interest in natural science ultimately led him to the field of chemistry. In 1769, his position as a scientist solidified with his appointment to the French Academy of Science. In 1771, Lavoisier married Marie-Anne Paulze (age 14), the daughter of a French "tax farmer" or, more accurately, a tax collector who purchased from nobility the right to collect taxes. In 1776, Lavoisier was commissioned to inspect the manufacture of French gunpowder. Eventually, he established his laboratory in the national Arsenal, and under his direction, French gunpowder became far superior to British gunpowder, a fact that was to become important in the conflict known as the American Revolution.

In 1780, Lavoisier made two decisions that were destined to have a fatal impact on his life. First, using money that he had inherited from his father, he purchased the right to collect taxes. Second he acquired a dangerous enemy when his negative vote barred Jean-Paul Marat from membership in the French Academy of Science. Marat, originally trained in the medical sciences, became an influential politician during the French Revolution, and although Lavoisier's political and economic views were enlightened for the age, Marat was still able to get his revenge. In 1793, Lavoisier was arrested on the charge of having mixed water and other ingredients harmful to the health of citizens in snuff! During Lavoisier's internment, Marat was brutally assassinated, an event that helped trigger the Reign of Terror. On May 8, 1794, Lavoisier was guillotined, a victim of this bloody campaign.

Although the French "tax farmers" were notorious for extorting money from the poor, it must be reiterated that Lavoisier was an exception. His enlightened political and economic views were undoubtedly a part of his friendship with the French economist Pierre du Pont. As a result of this friendship, du Pont's son, Eléuthere Irénée, was apprenticed to Lavoisier at the national Arsenal. After Lavoisier's execution, the senior du Pont moved his family to the United States. Although Pierre du Pont later returned to France, his son remained in the United States to ply his trade as a gunpowder manufacturer. He established, on the banks of the Brandywine Creek in the State of Delaware, a gunpowder manufacturing firm that was to become the chemical company E.I. du Pont de Nemours, Inc.

Like all humans, Lavoisier had shortcomings. One particular character flaw was that he tended to underplay the importance of the work of other scientists to his own discoveries. Lavoisier was reluctant to give Priestley proper credit for his part in the discovery of oxygen. Over the protests of Priestley, Lavoisier attempted to ignore their fall meeting of 1774 (previous section). Since it was only Lavoisier who understood the importance of the discovery of oxygen to a unified theory of combustion,

---

3. Lavoisier's theory of combustion has, of course, become a modern fact. This aspect of natural science was considered in Chapter 1 (Problem 3).

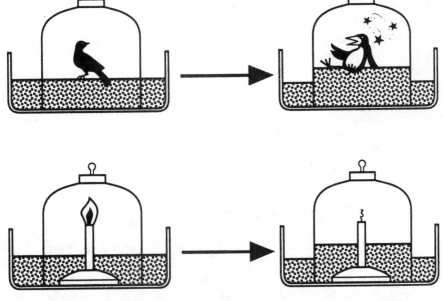

Lavoisier understood the unity of all combustion processes. The burning of a candle, the burning of an element, the burning of life itself, all involved chemical combination with the element oxygen.

Figure 4.1 Lavoisier's View of Combustion

modern chemists have tended to overlook this transgression (Fig. 4.1). In a memoir dated April 20, 1776, Lavoisier seemed to anticipate history's judgment.

"Perhaps strictly speaking, there is nothing in it of which Mr. Priestley would not be able to claim the original idea; but as the same facts have conducted us to diametrically opposite results, I trust that, if I am reproached for having borrowed my proofs from the works of this celebrated philosopher, my right at least to the conclusions will not be contested." – Antoine Lavoisier, French Chemist, *Mémoire sur l'Existence de l'Air dans l'Acide Nitreux*

The assignment of proper credit for scientific discovery is no small issue among scientists, and this fact is not totally a matter of professional vanity. Financial support for scientific work is very often tied to previous performance. Since scientific discoveries are very often the product of many minds, it is important for pragmatic reasons that credit be shared. Let us, therefore, take a closer look at the human side of Lavoisier's partner in discovery.

Joseph Priestley (1733–1804) was the son of an English cloth finisher. Orphaned at an early age, he was adopted by an aunt. He developed an early interest in theology, and rejected the Church of England to become a Unitarian, a fact that prevented him from attending Oxford or Cambridge University. He attended, instead, one of the "Dissenting Academies" for religious Nonconformists where he prepared for the ministry. Unorthodox behavior marked all facets of his life, political as well as religious. In 1751, Priestley married, and after a period in the ministry, he turned his attention to teaching. In 1761, he was appointed tutor in languages at the Dissenting Academy at Warrington.

About 1765, Priestley met Benjamin Franklin. Increasingly, Priestley directed his studies toward the natural sciences. The friendship with Franklin grew, and Priestley became a supporter of the American Colonies. In 1773, Priestley found a patron, Lord Shelburne, and scientific work became his major occupation.

Priestley's freely-expressed political and religious ideas were not well received in his native England. In 1794, after a particularly bad incident involving the destruction of his house and laboratory, Priestley immigrated to the United States where he settled on the banks of the

Susquehanna River. Never feeling fully a part of his new country, Priestley worked in his laboratory at Northumberland, Pennsylvania, until his death in 1804:

"I cannot refrain from repeating again, that I leave my native country with real regret, never expecting to find anywhere else society so suited to my disposition and habits." – Joseph Priestley, English/American Chemist

For the infant United States, it was a case of unrequited love. In 1794, Priestley was elected professor of chemistry at the University of Pennsylvania. He declined, saying that he never really gave much attention to the routine of chemistry. But we must give credit where credit is due, and to this end, one of the highest awards that the modern American Chemical Society can bestow is the Priestley Medal.[4]

4.5 The Laws of Chemical Action

By the beginning of the nineteenth century, the quantitative research efforts of Lavoisier and others had led to some clearly defined laws of chemical action. Scientific laws are concise summaries that describe nature's factual behavior. Since it is factual behavior that science seeks to rationalize by means of scientific theories, an understanding of the facts is essential to an understanding of the theory. Before we proceed to a discussion of John Dalton's atomic theory of chemical action, let us consider the laws of chemical action that John Dalton had at his disposal at the beginning of the nineteenth century.

First, the concept of element and compound as outlined in Chapter 3 was well understood. The early nineteenth century chemist had a fairly good comprehension of the matter classification system represented in Chapter 3 (Fig. 3.3). There was much confusion over specific details. Lavoisier, for example, had only identified thirty-three elements, and some of these were identified in error. Although there were specific problems, it is important to recognize that the early nineteenth century chemist understood the essence of the element/compound concept.

Second, the law of **conservation of mass**—matter cannot be created or destroyed—was accepted as being applicable to the process of chemical change. This was stated in chemical terminology as follows:

The mass of the products of a chemical reaction must equal the mass of the reactants of the chemical reaction.

As stated above, this law was generally accepted without proof by the chemists of the latter eighteenth century.[5]

Third, the law of **constant composition** of chemical compounds was recognized. This law, which was also accepted by many early chemists without proof, was stated with experimental verification by the French chemist Louis Joseph Proust in 1797. Simply stated, the law of constant composition says that a chemical compound will always have a definite elemental composition by mass. For example, if the quantitative analysis of one sample of water indicates that it is 88.8% oxygen by mass (Example III, Chapter 3), then all samples of water will yield this analysis. In other words, a chemical compound's quantitative elemental percent mass composition is a fixed chemical property of the compound.

---

4. The American Chemical Society is an international association of chemists, and it is the professional organization for American chemists. The Society was formed in 1876. In recognition of its service to the United States of America, the American Chemical Society was granted by Congress a National Charter in 1937. Now in its second century, the Society is the largest single discipline scientific organization in the world.
5. Since Lavoisier provided a well-defined experimental verification of the law of conservation of mass, he is usually given the credit for its discovery.

As a final note to this section, the importance of the work of Lavoisier should be reiterated. Lavoisier's explanation of the combustion process (Fig. 4.1) served as a foundation to support all of the above laws of chemical action. To the extent that these laws served as the basis for our modern theory of chemical action, Lavoisier must indeed be recognized as the father of modern chemistry.

4.6 The Mystery of the Law of Constant Composition

It is interesting that the essence of nineteenth century knowledge of the chemical reaction process is so simply stated—three short paragraphs in the previous section. Most natural scientists, however, hold a basic belief in the ultimate simplicity of nature. It may seem a contradiction to hold that nature is both infinitely complex and ultimately simple, yet this is the essence of the basic assumptions of the natural sciences (Chapter 1). With infinite complexity, nature subtly hides essence in the same way that an abstract painter hides subject. But the subtlety is not a permanent mask, rather it is a challenge to find hidden meaning. The following quote by Albert Einstein eloquently expresses this philosophical belief:

"Raffiniert ist der Herr Gott, aber boshaft ist Er nicht."
"God is subtle, but He is not mischievous." – Albert Einstein, German/American Physicist

The simple statement of the law of constant composition and its apparent self-evidence to many early chemists should not be allowed to hide the fact that it is really a very curious property of chemical compounds. As indicated in Chapter 3, chemical compounds are intimate homogeneous mixtures formed by the chemical union (chemical reaction) of elements. The law of constant composition, however, indicates that nature places a severe restriction on the mixing of elements to form compounds, and this restriction is a curious thing indeed. Let us consider this subtlety in slightly greater detail.

Many familiar mixing operations involve little or no restriction. For example, it is possible to dissolve sugar in water in almost any combination. One gram of sugar dissolved in ninety-nine grams of water forms a one percent solution, or two grams of sugar dissolved in ninety-eight grams of water form a two percent solution. In a similar manner, a large variety of sugar solutions, each differing in percent composition, can be made.

Now consider mixing the element beryllium (Be) with the element oxygen (O) to form the chemical compound beryllium oxide. If one gram (1.00 g) of oxygen is mixed with ninety-nine grams (99.00 g) of beryllium, a chemical reaction will take place, and the compound beryllium oxide will form. However, nature does not allow a one percent (1.00%) compound mixing of these two elements. By an experimental quantitative analysis, it can be determined that the constant composition of beryllium oxide is thirty-six percent (36.0%) beryllium and sixty-four percent (64.0%) oxygen. Therefore, when one gram (1.00 g) of oxygen is mixed with ninety-nine grams (99.00 g) of beryllium, nature will only allow a small portion (0.56 g) of the beryllium to react. After the reaction, the resulting sample would weigh 100.00 grams, but analysis of this sample would show it to be a mixture of unreacted beryllium and a smaller quantity of beryllium oxide.

Reaction 4.G
BERYLLIUM + OXYGEN → BERYLLIUM OXIDE
99.00 g   1.00 g   1.56 g

This reaction leaves 98.44 grams of beryllium unreacted, forming only enough beryllium oxide (1.56 g) to meet the 36.0% beryllium requirement:

$$\frac{\text{Part}}{\text{Whole}} = \frac{\text{Part'}}{100 \text{ Units of Whole}}$$

$$\frac{0.56 \text{ g beryllium}}{1.56 \text{ g beryllium oxide}} = \frac{X \text{ g beryllium}}{100 \text{ g beryllium oxide}}$$

X = 36.0 grams of beryllium or 36.0% beryllium

(see also Example III, Chapter 3)

Why does nature restrict the mixing of elements to form compounds as illustrated above? What hidden meaning lies behind this subtlety? This was the true mystery of the law of constant composition. A mystery that was ultimately solved by the English chemist John Dalton.

4.7 John Dalton and the Philosopher's Atom

John Dalton (1766–1844) was the son of a middle class English wool weaver. Both of his parents were Quakers, a fact which was to influence Dalton all of his life. His early schooling at the hand of Quaker teachers included arithmetic, Greek, Latin, and philosophy. At an early age, Dalton also developed an interest in natural philosophy (natural science), one of his earliest interests being meteorology (study of weather). Dalton's teaching career began at an early age (12 years old). Although he was an active researcher and scientific thinker, Dalton remained a teacher throughout his entire life.

One of John Dalton's early scientific investigations involved an attempt to explain the physical properties of gases. In order to explain the physical behavior of gases, Dalton revived an old hypothesis that had originally been suggested by Sir Isaac Newton. The essence of this hypothesis was that the physical behavior of gases could be explained (rationalized) by imagining that gases were composed of invisible ultimate particles. Although Newton suggested this point of view in the seventeenth century, the concept of invisible ultimate (indivisible) particles was first suggested by the ancient Greeks (Democritus, 400 B.C.). This Greek point of view was based not so much on empiricism (experimental observation) as on pure philosophical thought. To the ancient Greek philosophers, matter—a piece of paper for example—logically had to fall into one of two categories. Either the paper could be torn in half over and over again, endlessly, or the paper could be torn in half over and over again only until an ultimate bit of paper was obtained. Democritus and his followers believed in this latter point of view. The ultimate bits of matter were said to be *atomos*, the Greek word for indivisible.

Although Dalton's attempts to explain the physical properties of gases by extending the original scientific hypothesis of Newton are of interest to historians of chemistry, the details of these attempts are not important to the present discussion. Many of Dalton's arguments are faulty, and they would only confuse the present issue. One point, however, must be discussed, albeit not in detail. On October 21, 1803, Dalton read a paper to the Manchester Society. The paper read before this scientific organization described Dalton's efforts to explain the solubilities of different gases in water using the atomic hypothesis. The text of this paper ends with the following paragraph:

The greatest difficulty attending the mechanical hypothesis, arises from different gases observing different laws. Why does water not admit its bulk of every kind of gas alike? This question I have duly considered, and though I am not yet able to satisfy myself completely, I am nearly persuaded that the circumstance depends on the weight and number of the ultimate particles of the several gases: Those whose particles are lightest and single being least absorbable and the others more according as they increase in weight and complexity. *An inquiry into the relative weights of the ultimate particles of bodies is a subject, as far as I know, entirely new; I have lately been prosecuting this enquiry with remarkable success(!!).* The principle cannot be entered upon in this paper; but I shall subjoin the results, as far as they appear to be ascertained by my experiments.

Although the nonchemist may be somewhat confused by all of this, the essence of the above paragraph is that Dalton had reached an impasse in his physical atomic theory. For reasons that are not important to the present discussion, Dalton felt that he needed to know the weights of atoms in order to explain why different gases had different solubilities in water. But how to determine the weight of an invisible atom? *"I have lately been prosecuting this enquiry with remarkable success(!!)."* The exclamation points are not part of the original quote; rather, they are a mark of respect by these authors. For, indeed, John Dalton had found a way to determine the weight of an atom, and at the same time, he had solved the mystery of the law of constant composition.

4.8 The Daltonian Theory of Chemical Action

Although John Dalton's original interest in developing an atomic theory was related to explaining the physical behavior of gases, by 1803, he had also developed a chemical atomic theory. The chemical atomic theory was a response to information required by the physical atomic theory. Simply stated, Dalton felt that he needed information about the weights of atoms. Dalton's physical atomic theory is naive by modern standards, but his chemical atomic theory has stood the test of almost two centuries. It is the way the modern chemist "explains" the chemical reaction process. It is the answer to the question suggested in Chapter 1:

$$?$$
$$\text{MATTER I} \rightarrow \text{MATTER II}$$

The chemical atomic theory of John Dalton stated or implied the following points. For the sake of clarity, these points are stated in modern terms:[6]

1. All elements are composed of ultimate particles called **atoms**.
2. The atoms of a given element have a unique mass (**atomic mass**).
3. Atoms of different elements have different masses.
4. The ultimate particle of a compound, called a **molecule**, is formed from a fixed number of atoms of its component elements. The **molecular mass** of a molecule is the sum of the masses of its component atoms.
5. The atoms that compose a molecule are held together by attachments called **chemical bonds**.
6. The **chemical reaction** process involves the formation or alteration of chemical bonds.

Let us see how this theory "explains" the facts associated with reaction 4.G, the formation of beryllium oxide:

Reaction 4.G
BERYLLIUM + OXYGEN → BERYLLIUM OXIDE
99.00 g   1.00 g   1.56 g

One of the most important facts associated with this reaction is beryllium oxide's constant composition—*i.e.* beryllium and oxygen must react in a mass ratio of 36.0:64.0 (Be:O). At this point, it is important to note the following arithmetic consequences of the law of constant composition. For any compound, the law of constant composition demands that:

Expression 4.A
$$\frac{\text{The Combining Mass of Any Element}}{\text{The Combining Mass of Any Other Element}} = \text{Constant}$$

---

6.   Dalton, for example, did not use the term "molecule" in its modern context.

Expression 4.B

$$\frac{\text{The Combining Mass of Any Element}}{\text{The Mass of Compound Formed}} = \text{Constant}$$

For the sake of brevity, expressions 4.A and 4.B will be written as follows:

Expression 4.A

$$\frac{E_1}{E_2} = \text{Constant}$$

Expression 4.B

$$\frac{E}{C} = \text{Constant}$$

Expressions 4.A and 4.B are really alternative statements of the law of constant composition. For the compound beryllium oxide, therefore, the factual law of constant composition is expressed as follows:

$$\frac{36.0 \text{ g beryllium}}{64.0 \text{ g oxygen}} = \text{constant ratio for beryllium oxide}$$

$$\frac{36.0 \text{ g beryllium}}{100 \text{ g beryllium oxide}} = \text{constant ratio for beryllium oxide}$$

$$\frac{64.0 \text{ g oxygen}}{100 \text{ g beryllium oxide}} = \text{constant ratio for beryllium oxide}$$

These constant ratios can be expressed in any mass units, and they are, of course, subject to the full arithmetic of linear reasoning (Chapter 2):

$$\frac{X}{Y} = \frac{X'}{Y'}$$

Now consider the following atomic/molecular view of reaction 4.G (Fig. 4.2):

FACT:

$$\text{BERYLLIUM} + \text{OXYGEN} \rightarrow \text{BERYLLIUM OXIDE}$$

$$36.0 \text{ g} \qquad 64.0 \text{ g} \qquad 100 \text{ g}$$

THEORY:

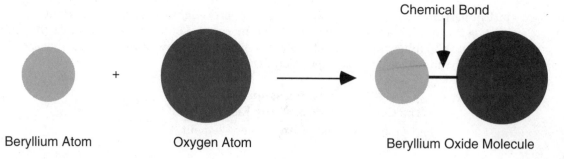

Figure 4.2 An Atomic View of Reaction 4.G

If the oxygen atom is arbitrarily assigned a mass of 16.0 theoretical mass units (called atomic mass units or amu), then the use of linear reasoning allows us to calculate a more complete "view" of the atomic world. The assignment of 16.0 amu as the mass of an oxygen atom establishes the oxygen atom as an atom mass standard (Fig. 4.3).[7]

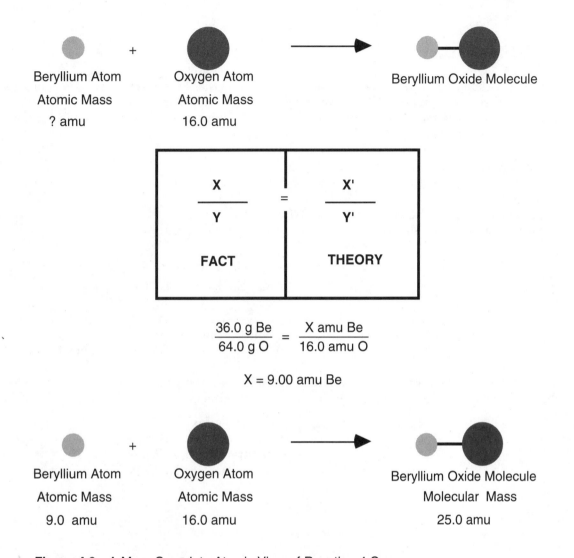

Figure 4.3 A More Complete Atomic View of Reaction 4.G

The use of the factual mass ratio to predict the theoretical (atomic) mass ratio forces the theoretical "view" to be consistent with the factual view. At the same time, the theoretical "view" becomes an explanation of the facts. Why do compounds have a constant composition? Because atoms have characteristic masses, and when forming molecules, they bond in fixed ratios. Why do chemical reactions obey the law of mass conservation? Because the total number of atoms does not change during a chemical reaction—only the chemical bonding arrangement changes.

The logic of this example is applicable to other compounds. John Dalton used this logical approach to determine the relative masses of the atoms of the various elements. This is what he meant by, *"I have lately been prosecuting this enquiry with remarkable success."*

---

7. The value 16 amu is not entirely arbitrary. This point will be discussed in greater detail in Chapter 5.

$$?$$
$$\text{MATTER I} \rightarrow \text{MATTER II}$$

$$? = \text{ATOMIC/MOLECULAR "VIEW"}$$

4.9 A Final Note

In the next chapter, we will look at the atomic/molecular theory of chemical reactions in greater detail. The purpose of the present chapter was to consider the basic logic of this theory. Using elegant arithmetic logic, John Dalton "saw" imaginary atoms and molecules. This was a milestone in human thinking, a thing of beauty.

"I can't believe that!"
"Can't you! Try again: draw a long breath, and shut your eyes."
"There's no use trying; one can't believe impossible things."
"I daresay you haven't had much practice. When I was your age, I
always did it for half-an-hour a day. Why, sometimes I've believed as
many as six impossible things before breakfast."

> Dialogue between:
> Alice, English Little Girl
> and
> The Queen, A Queen
> Lewis Carroll
> (Charles Dodgson, English Mathematician)

"And now here is my secret, a simple secret: It is only with the heart
that one can see rightly; what is essential is invisible to the eye."

> The Fox, A Fox
> to
> The little Prince, A Prince
> Antoine DeSaint-Exupery, French Aviator

Performance Objectives

P.O. 4.0

Review all of the boldfaced terminology in this chapter, and make certain that you understand the use of each term.

atom	atomic mass	chemical bond
chemical reaction	combustion	conservation of mass
constant composition	molecular mass	molecule
phlogiston		

P.O. 4.1

You must demonstrate an understanding of Georg Stahl's phlogiston theory of combustion.

EXAMPLE:
A candle burning in air trapped in a glass jar above water will extinguish. How would the phlogiston theory of Georg Ernst Stahl have been used to explain this result?

SOLUTION:
Phlogiston theory stated that the burning candle liberated phlogiston. When the air trapped in the jar became saturated with phlogiston, the candle extinguished because it could no longer liberate phlogiston.

ADDITIONAL EXAMPLE:
A mouse trapped in the same air space will die. How would the phlogiston theory of Georg Ernst Stahl have been used to explain this result?

ANSWER:
Phlogiston theory recognized that combustion and animal respiration were similar chemical processes.

P.O. 4.2

You must demonstrate an understanding of Antoine Lavoisier's oxygen explanation of combustion.

EXAMPLE:
A candle burning in air trapped in a glass jar above water will extinguish. How is Antoine Lavoisier's oxygen explanation used to explain this result?

SOLUTION:
Oxygen theory stated that the burning candle consumed oxygen that was a component of the air. When the oxygen was consumed, the candle could no longer burn.

ADDITIONAL EXAMPLE:
A mouse trapped in the same air space will die. How is Antoine Lavoisier's oxygen explanation used to explain this result?

ANSWER:
Oxygen theory also recognized that combustion and animal respiration were similar processes.

P.O. 4.3

You must demonstrate an understanding of the relationship between the law of constant composition and Daltonian atomic/molecular theory.

EXAMPLE:
In his original system, John Dalton represented the water molecule with one atom of hydrogen and one atom of oxygen. Dalton's atomic mass value for the hydrogen atom was 1.0 atomic mass units, and his atomic mass value for the oxygen atom was 8.0 atomic mass units. Using this point of view, calculate the percentage of oxygen in the water molecule (Fig 4.4).

SOLUTION: 89% (2 significant figures in this case)

Textbook Reference: Section 3.11

ADDITIONAL EXAMPLE:
Today we know that a molecule of water should be represented as containing two atoms of hydrogen and one atom of oxygen. Water is known to contain 88.8% oxygen. The modern atomic mass of hydrogen is 1.01 atomic mass units. Use this information to calculate the modern atomic mass of oxygen.

ANSWER: 16.0 atomic mass units

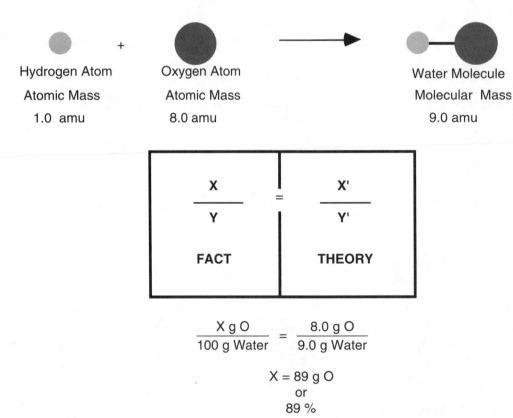

Figure 4.4 John Dalton's View of Water

Chapter Four
Problems

General Problems

1. There was no love lost between Lavoisier and Priestley, and Priestley remained an ardent phlogistonist throughout his entire life, referring at one point to Lavoisier's explanation of combustion as "that despicable French doctrine." In the year 1800, Priestley reported a series of experiments which he felt proved the phlogiston theory. One of these experiments was the isolation of a combustible gas obtained by burning charcoal (carbon). Under the system of Lavoisier, the burning of charcoal produced an oxide of carbon which was noncombustible. According to the system of Lavoisier, an oxide had to be noncombustible. Modern chemists recognize that this latter statement is not correct and that the combustion of charcoal takes place in stages:

Reaction 4.H
Carbon + Oxygen → Carbon Monoxide (Combustible)

Reaction 4.I
Carbon + Oxygen → Carbon Dioxide (Noncombustible)

Which gas is isolated depends on the conditions of combustion.[8] Priestley felt that his isolation of a combustible gas during the burning of charcoal not only disproved the theory of Lavoisier, but also proved the phlogiston theory. What was his reasoning?

STUDENT SOLUTION:

2. How do you think John Dalton would have written the formation of carbon monoxide and carbon dioxide (reactions 4.H and 4.I) using his atomic/molecular theory?

STUDENT SOLUTION:

3. When the element lithium burns, a single oxide is formed. How do you think John Dalton would have written the formation of this compound using his atomic/molecular theory? Actually, Dalton encountered a problem in drawing the atomic/molecular "picture" of the formation of compounds. Can you identify the problem?

STUDENT SOLUTION:

---

8. Glowing charcoal results in the formation of a considerable quantity of carbon monoxide, a highly poisonous gas. For this reason, charcoal "fires" must always be used with adequate ventilation. There may be a historical irony here. Prior to his death, Priestley manifested symptoms of carbon monoxide poisoning. Some historians of chemistry believe that Priestley, in his quest for phlogiston, became a victim of what he actually believed to be phlogiston (carbon monoxide).

4. When a sample of carbon weighing 1.50 grams is burned completely in an excess of air, 5.50 grams of carbon dioxide is formed. What is the percentage of oxygen in carbon dioxide? According to the law of constant composition, what mass of carbon would have to be burned in order to form 11.00 grams of carbon dioxide?

    STUDENT SOLUTION:

5. When a sample of pure carbon burns in an excess of air, the only product is gaseous carbon dioxide. When charcoal burns in an excess of air, a solid ash remains. The presence of this ash tended to confuse early attempts to explain combustion. How is charcoal manufactured? Why does the combustion of charcoal result in the formation of a solid ash?

    STUDENT SOLUTION:

6. When ordinary iron rust is mixed with charcoal and heated to a high temperature, elemental iron is formed. This is a simple statement of fact.

    A. Briefly explain how Georg Ernst Stahl explained this fact.
    B. Briefly explain how Antoine Lavoisier explained this fact.
    C. Using modern symbols and formulas, show how contemporary chemists use the atomic molecular theory of John Dalton to explain this fact.

    STUDENT SOLUTION:

7. What is an atom?

    STUDENT SOLUTION:

8. As noted in problem one, carbon can form two compounds with oxygen (carbon monoxide and carbon dioxide). Both of these compounds have different compositions yet each obeys the law of constant composition. Atomic theory explains the existence of these two compounds by assuming that the carbon and oxygen atom ratios in the molecules of the two atoms differ. This aspect of atomic theory should have been reflected in your answer to problem two. John Dalton realized that there was a factual consequence of this aspect of his atomic theory, and he stated this consequence as his law of multiple proportions. This law states that when two or more compounds are formed from the same two elements, the masses of one element that combine with a fixed mass of a second element are in a ratio of small whole numbers.

    If carbon monoxide contains 42.9% by weight of carbon and carbon dioxide contains 27.3% by weight of carbon, show that these facts are consistent with the Law of Multiple Proportions.

Hint: Calculate the mass of oxygen necessary to react with 100 g of carbon for each compound. Is the ratio of these masses a whole number?

STUDENT SOLUTION:

9. Carbon and hydrogen can also form different compounds called hydrocarbons. Three of these hydrocarbons are methane, ethylene, and acetylene which contain 75.0%, 85.7%, and 92.3% carbon, respectively. Use these data to demonstrate the Law of Multiple Proportions.

    STUDENT SOLUTION:

10. In the nineteenth century, J. J. Berzelius, a Swedish chemist, used an oxygen standard atomic weight system. For reasons that are not entirely clear, he assigned oxygen an atomic weight of 100. (Dalton had already expressed a preference for a system in which hydrogen would equal 1.) Using the data for carbon monoxide in problem number 8, calculate the atomic weight of carbon under the system of Berzelius.

    STUDENT SOLUTION:

Problems Involving Household Chemistry and Science

11. It is quite easy to generate pure oxygen and to demonstrate its ability to support combustion. Add about 2 oz of household hydrogen peroxide (3%) to a juice glass. Be sure to use 3% hydrogen peroxide and not a product with a higher concentration. Add about ¼ teaspoon of dried yeast to the 2 oz of hydrogen peroxide in the glass. The yeast contains an enzyme that causes hydrogen peroxide to decompose into oxygen gas and water. Cover the glass with a piece of cardboard. While the hydrogen peroxide is reacting with the yeast, use a burning match to ignite the end of a toothpick. After the hydrogen peroxide reaction has subsided, remove the cover, blow out the flame on the burning toothpick, and insert a glowing toothpick into the glass without immersing it into the liquid. Repeat the experiment with 2 oz of white vinegar and one teaspoon of baking soda. Does the gas carbon dioxide which is formed in this reaction support combustion? How would Georg Ernst Stahl have interpreted these results in terms of phlogiston theory?

    STUDENT SOLUTION:

12. Although they used different theories, both Priestley and Lavoisier understood that the respiration and combustion were similar chemical processes. Figure 4.1 shows an idealized experiment that demonstrates this similarity. Demonstrate that the corrosion of iron is chemically similar to respiration and combustion by conducting the following experiment.

Thoroughly moisten an SOS steel wool pad. Remove as much of the soap as possible. After most of the soap has been removed, soak the steel wool in vinegar for fifteen minutes. Use the steel wool, a baking pan, and a drinking glass to assemble the apparatus shown in Figure 4.5.

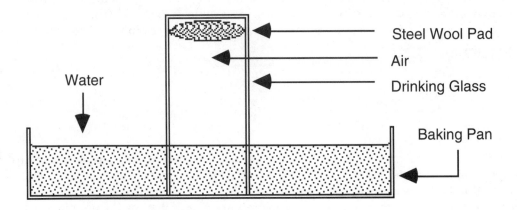

Figure 4.5 The Corrosion of Iron

Allow the apparatus to stand for one hour. As the iron combines with oxygen to form iron oxide, the water level in the glass should rise. After one hour, calculate the approximate percent volume reduction of gas in the drinking glass. Compare this figure to the known percentage of oxygen in air. Although the experiments shown in Figure 4.1 are idealized, these experiments could actually be performed. The experiments represented in Figures 4.1 and 4.5 would all result in different percent volume reductions of gas in the drinking glass. Which experiment do you think would more accurately measure the percent of oxygen in air? Explain your answer.

If the steel wool pad is not soaked in vinegar, then the formation of iron oxide requires a much longer period of time to reach completion. Chemically, vinegar is a dilute solution of acetic acid. The reaction of the acetic acid with the steel wool produces a catalyst which increases the speed of this chemical reaction. Chemical catalysts are very important in all aspects of applied chemistry, and they are discussed in Chapter 5 of this text. Repeat the above experiment using untreated steel wool. How long does the reaction take to reach completion without the catalytic effect?

STUDENT SOLUTION:

Library Problems

13. The idealization of actual experiments is a common textbook technique. By presenting the essence of an experiment through an idealized view, textbook authors can avoid discussing confusing technical problems associated with the process of experimentation. Although this is a valuable textbook technique, a basic course in chemistry should expose students to the problems of experimental complexity. Use the following two articles from the *Journal of Chemical Education* to investigate the details of Lavoisier's experiments with oxygen in the air:

    "An Excerpt from Lavoisier's Laboratory Journal," Oesper, R. E. *J. Chem. Educ.* **1941**, 18, 85

    "The Chemical Revolution—the Second Phase," French, Sidney J. *J. Chem. Educ.* **1950**, 27, 83

    Give a brief description of Lavoisier's actual experiments which were idealized in Figure 4.1. Compare Lavoisier's approach to the procedure used in question 12 above.

    STUDENT SOLUTION:

14. In the article by R. E. Oesper (question 13), you learned that Marie Lavoisier made her own contributions to the development of modern chemistry. She also played a major role in a tragic love story that was ultimately to affect the development of the chemical industry in the United States. The details of this love story can be found in "The du Ponts and the Lavoisiers: A Bit of Untold History, with an Accent on America," French, Sidney J. *J. Chem. Educ.* **1979**, 56, 791. Briefly describe how the woman whom Pierre du Pont describes as ". . . her who most interests my intelligence, my chivalry and my heart . . ." may have influenced the development of E.I. du Pont de Nemours, Inc. in the United States.

    STUDENT SOLUTION:

Formal Laboratory Exercise

Determination of an Atomic Mass
A Quantitative Chemical Analysis

Introduction

The calculation of the atomic mass of a metal according to the method described in section 4.8 requires the experimental determination of the constant composition of the metal oxide. For many metal elements, the oxide can be formed by simply heating the metal in air. By measuring the mass of the original metal sample and the mass of the metal oxide formed, a constant composition of the oxide can be determined according to the following synthetic analysis:

Metal	+	Oxygen	→	Metal Oxide
mass 1		(mass 2 – mass 1)		mass 2

The purpose of this formal laboratory exercise is to use this synthetic analysis approach to determine the constant composition of magnesium oxide. This constant composition will then be used to calculate the atomic mass of magnesium according to the method of section 4.8.

The metal element magnesium can be converted to magnesium oxide at reasonably low laboratory temperatures. One problem associated with the quantitative study of magnesium oxide formation is that magnesium also reacts chemically with nitrogen in the atmosphere. Fortunately, the nitrogen compound formed by this reaction, magnesium nitride, is readily converted to magnesium oxide by reaction with water followed by further heating. The complete series of chemical reactions is outlined below:

1.   Magnesium + Oxygen → Magnesium Oxide

2.   Magnesium + Nitrogen → Magnesium Nitride

3.   Magnesium Nitride + Water → Magnesium Hydroxide + Ammonia

4.   Magnesium Hydroxide → Magnesium Oxide + Water

Reactions (1) and (2) are carried out simultaneously by heating magnesium metal in air. After the initial heating, a small quantity of water is added to the reaction mixture to form magnesium hydroxide and ammonia by reaction (3). Finally, the reaction mixture is heated again, and the magnesium hydroxide is converted to magnesium oxide and water by reaction (4). This all may seem very confusing to a newcomer to the study of chemistry, and in Chapter 5 atomic molecular theory will be used to reduce this confusion. For the time being, it is only important to realize that the entire series of reactions quantitatively converts magnesium into magnesium oxide.

In order to calculate the atomic mass of magnesium according to the method described in section 4.8, the combining ratio of magnesium atoms to oxygen atoms is required. As suggested by problem 3 in the chapter end problem set, the method for determining this ratio is not obvious. This problem is explored in more detail in Chapters 5 (section 5.6) and 7 (section 7.15) of the textbook. For the time being, we will anticipate the results of this future study and use the known combining ratio of 1:1 in the atomic mass calculation.

Procedural Overview

The formation of magnesium oxide is carried out by heating a sample of magnesium metal in a porcelain crucible supported by a ring stand over a Bunsen burner. This apparatus is shown in Figure 4.6. Proper flame adjustment and positioning are essential for the success of this experiment. The use of the Bunsen burner will be discussed by the laboratory instructor at the beginning of the laboratory period, but two safety issues associated with this apparatus should be emphasized in this procedural overview. First, a properly adjusted burner flame can be hard to see. Exercise due care when working with an active burner. Second, heated porcelain cools very slowly. Crucible tongs must be used for the manipulation of the heated crucible and crucible lid. These safety issues will also be discussed by the laboratory instructor.

This experiment involves several crucible heatings followed by cooling and mass determination. The cooling process can be facilitated by the following procedure. Turn off the burner and allow the crucible to partially cool on the ring stand apparatus. After about 5 minutes, use crucible tongs to move the crucible and lid from the ring stand apparatus to a wire gauze. Allow the crucible and lid to cool for an additional 5 minutes on the wire gauze.

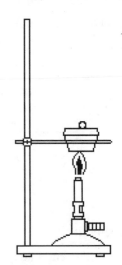

Figure 4.6 Formation of Magnesium Oxide

The use of the electronic balance was discussed in the formal laboratory exercise associated with Chapter 2. In the present experiment, all mass determinations are performed by weighing the crucible and its contents. The original mass of magnesium is measured by a technique that chemists find quite useful in quantitative experiments. According to this procedure, the mass of the empty crucible is first determined, and then an estimated quantity of magnesium is added to the crucible. The crucible and magnesium are weighed after this estimated addition, and the mass of the magnesium is determined by difference. All of the mass determinations for this experiment are described in detail in the experimental procedure below.

Materials

porcelain crucible and lid, crucible tongs, clay triangle, wire gauze, ring stand and ring, Bunsen burner, laboratory balance

Chemicals

magnesium metal, water

Procedure

1. In preparation for an initial heating to dry the crucible, place the empty crucible and crucible lid on the clay triangle support as shown in Figure 4.6.

2. Use the Bunsen burner to heat the empty crucible and crucible lid at full heat for 5 minutes. After 5 minutes, turn off the burner and allow the crucible to cool for 10 minutes as described in the procedural notes.

3. While the crucible is cooling, measure 25 cm of magnesium ribbon and cut the ribbon into small pieces (approximately 2 cm) into a small dry beaker. After the crucible and lid have cooled, determine the mass of the empty crucible and lid. Use the crucible tongs to manipulate the crucible and lid. Record this mass in the data table.

4. Add the magnesium ribbon pieces to the crucible and determine the mass of the crucible, lid, and magnesium. Record this mass in the data table.

5. Place the crucible containing the magnesium on the clay triangle support and carefully position the lid on the crucible so that it is slightly ajar. Use the Bunsen burner to heat the crucible, lid, and magnesium at full heat for 10 minutes. After 10 minutes, turn off the burner and allow the crucible to cool for 5 minutes on the ring stand apparatus. After this 5 minute cooling period, use the crucible tongs to remove the lid and add 20 drops of water to the crucible contents. Carefully, note the odor of ammonia after this addition.

6. Place the crucible lid on the crucible so that it is slightly ajar and heat the crucible, gently at first, for an additional 10 minutes. The last 5 minutes should be at the full heat of the Bunsen burner.

7. Turn off the burner and allow the crucible to cool for 10 minutes as described in the procedural notes.

8. After the crucible has cooled, determine the mass of the crucible, lid, and magnesium oxide. Use the crucible tongs to manipulate the crucible and lid. Record this mass in the data table.

9. Ask the laboratory instructor to check the crucible contents and the data table entries. Based on this check, the instructor may request that the experiment be repeated

10. After the instructor checks the crucible contents, the magnesium oxide should be flushed down the drain. The crucible should be washed well with water.

11. Use the information in the data table to answer the questions in the conclusion and discussion section.

Data Table

mass of empty crucible & lid grams measured	
mass of crucible, lid, & magnesium grams measured	
mass of crucible, lid, & magnesium oxide grams measured	
mass of magnesium reacted grams calculated	
mass of magnesium oxide formed grams calculated	
mass of oxygen reacted grams calculated	

Conclusion and Discussion

1. Calculate the percentage of magnesium in magnesium oxide.

2. Calculate the atomic mass of magnesium according to the method of section 4.8. Assume that the combining atom ratio of magnesium and oxygen is 1:1. Use oxygen atoms equal to 16.0 amu as a standard for this calculation.

3. Repeat the above atomic mass calculation, but assume the combining atom ratio of magnesium and oxygen is 1:2 (*i.e.* two atoms of oxygen for every atom of magnesium).

Chapter Four Notes

Chapter Four Notes

Chapter Five

Chemical Calculations _____

5.1 Introduction

Since the Daltonian atomic theory is based on the quantitative behavior of mass during the chemical reaction process, the theory is capable of quantitative prediction. Prediction is the desired characteristic of any scientific theory, and it is prediction of mass behavior during chemical reactions that makes Daltonian atomic theory a valuable modern tool.

The purpose of this chapter is to introduce certain calculation types (chemical calculations) that capture the essence of the predictive nature of Daltonian atomic theory. The student is not expected to develop these computation skills to the same degree as a chemistry major, but even a casual understanding of modern chemistry demands a familiarity with these skills. Since the emphasis in this chapter will be placed on the modern usage of Daltonian atomic theory, the first task will be to develop the modern formalisms of this theory.

5.2 Chemical Symbols

The use of symbols to represent chemical substances was introduced into the field of chemistry in ancient times. Originally, the symbols were part of the mystery and mystic of alchemy (early chemistry), with each alchemist (early chemist) using their own secret symbols. Eventually, some symbols became accepted in a common usage. For example, most alchemists equated the seven metal elements known to the ancients with the seven known planets, assigning each metal the astrological symbol for its planet (Fig. 5.1).[1]

With the advent of his atomic theory, John Dalton realized that a unique symbol would be convenient representation for the atom of an element. He introduced a system that retained some of the flavor of the alchemical symbols. According to the Daltonian scheme, each element was represented by a circular symbol, with interior markings designating the particular element (Fig. 5.2).

It was soon realized that this system was quite cumbersome, since there was no easy way to remember the symbols. Finally, the Swedish

The Seven Ancient Metals

Metal	Planet	Symbol
Copper	Venus	♀
Gold	Sun	☉
Iron	Mars	♂
Lead	Saturn	♄
Mercury	Mercury	☿
Silver	Moon	☾
Tin	Jupiter	♃

Figure 5.1 Alchemist's Symbols for Seven Ancient Metals

---

1. The Ptolemaic model of the solar system regarded the sun and the moon as planets. The word "planet" comes from the Greek word meaning "to wander," and all seven of these heavenly objects seem to wander against the background of fixed stars.

Element	Symbol
Hydrogen	⊙
Nitrogen	⊖
Carbon	●
Oxygen	○
Sulfur	⊕
Phosphorus	⊗

Figure 5.2   Daltonian Symbols

S = Sulfur

S = ○ A Sulfur Atom

S = 32.1 amu (Fig 3.7)

Figure 5.3   The Meaning of a Symbol

chemist Jons Jacob Berzelius introduced a system of alphabetical symbols. Although there were minor modifications, the Berzelius system was eventually adopted by all chemists.[2] This system was introduced in Chapter 3 (Fig. 3.7).

For the purpose of understanding the present chapter, it is important to recognize that the modern **symbol** for an atom can mean any one of three things to a chemist. First, it can represent the name of the element. Second, it can represent one atom of the element. Finally, it can represent the **atomic mass** of the element (Fig. 5.3).

This may seem a bit confusing, but the context in which the symbol is used always makes the meaning clear. In this text, all three meanings will find common usage.

5.3 Chemical Formulas

Since Daltonian atomic theory viewed the ultimate particles of compounds as aggregates (molecules) composed of the atoms of the compound's component elements, it was possible to use the symbols of these elements to write a symbolic representation for the molecule of a compound. For example, John Dalton viewed the molecule of the chemical compound nitrous oxide (laughing gas) as an aggregate of two nitrogen atoms and one oxygen atom, thus he would have represented this molecule as follows (Fig. 5.4):

Figure 5.4   Nitrous Oxide Molecule

---

2.   Dalton, who probably selected his circular symbols because he thought of atoms as spheres, never accepted the Berzelius system. He regarded the new symbols as "horrifying."

Such a symbolic representation is called a **formula**. Using the system of Berzelius as it was finally accepted by world chemists, the formula of nitrous oxide is written N_2O (Fig. 5.5):

$$N_2O$$

Figure 5.5   Nitrous Oxide Formula

The formula for nitrous oxide is read N-two-O (en-too-o). The subscript following the symbol for a given element indicates how many atoms of that element are contained in the molecule, with the subscript one always understood in the absence of another integer. Hence the formula N_2O indicates a molecule that contains two atoms of nitrogen and one atom of oxygen.

The early nineteenth century chemist had no knowledge of atomic arrangements within molecules; the actual atomic arrangements were investigated toward the end of the nineteenth century. These investigations will be discussed in Chapter 8. For the time being, it is important to recognize that the modern chemist uses parentheses to incorporate atomic arrangement information into chemical formulas. For example, the formula of the compound aluminum sulfate is written $Al_2(SO_4)_3$. This formula indicates that there are three SO_4 atom groupings contained within the molecule.[3] In decoding the atom count indicated by this formula, each subscript within the parentheses is multiplied by three. Hence the formula $Al_2(SO_4)_3$ represents a molecule that contains two aluminum atoms, three sulfur atoms, and twelve oxygen atoms. If there is no concern about atomic arrangements, then it is perfectly legitimate to write this formula $Al_2S_3O_{12}$. The modern chemist, however, always uses the former representation.

Just as the symbol can mean one of three things to a chemist, so also a formula has three meanings. First, it can represent the name of the compound. Second, it can represent one molecule of the compound. Finally, it can represent the **molecular mass** of the compound (Fig. 5.6). The systematic naming of chemical compounds is discussed in Chapter 6. Calculation of molecular mass is illustrated in Example I below.

$$Al_2(SO_4)_3 = \text{Aluminum Sulfate}$$
$$Al_2(SO_4)_3 = \text{An Aluminum Sulfate Molecule}$$
$$Al_2(SO_4)_3 = 342 \text{ Atomic Mass Units}$$

Figure 5.6   The Meaning of a Formula

EXAMPLE I

Calculation of Molecular Mass

Calculate the molecular mass of $Al_2(SO_4)_3$. Atomic Masses: Al = 27.0   S = 32.1   O = 16.0 amu.

SOLUTION: 342.3

---

3.   For reasons that are not important to the present discussion, the modern chemist would refer to $Al_2(SO_4)_3$ as a formula unit of aluminum sulfate, not a molecule. In this text, the word molecule will be used to describe the ultimate particles of all compounds.

Since each atom in this molecule has a characteristic mass, the molecule as a whole has a mass that is calculated as follows:

$$
\begin{array}{rclcl}
2 \times Al & = & 2 \times 27.0 & = & 54.0 \text{ amu} \\
3 \times S & = & 3 \times 32.1 & = & 96.3 \text{ amu} \\
12 \times 0 & = & 12 \times 16.0 & = & \underline{192.0 \text{ amu}} \\
& & & & 342.3 \text{ amu}
\end{array}
$$

In this explanation, the symbols of the elements are being used to represent the atomic masses of their respective atoms. These masses are obtained from Figure 3.7.

———————————————————□———————————————————

As is the case with symbols, the context in which a formula is used always makes its meaning clear.

5.4 Chemical Equations

Chemical reactions involve the interaction and transformation of elements and compounds. Since elements and compounds can be represented by symbols and formulas, respectively, chemical reactions can be represented by appropriate combinations of symbols and formulas. For example, if the chemical compound calcium carbonate ($CaCO_3$) is heated to a high temperature, it decomposes to form the chemical compounds calcium oxide (CaO) and carbon dioxide (CO_2). This entire chemical reaction can be summarized by the following statement, called a **chemical equation** (Fig. 5.7):

$$CaCO_3 \rightarrow CaO + CO_2$$

Figure 5.7 A Chemical Equation

The arrow ($\rightarrow$) is read *reacts to form*. Notice that the equation is an atomic/molecular description of the chemical reaction. It shows the chemist exactly how the atoms and molecules are being altered during the chemical reaction. This latter point is not a trivial one. The nonchemist might be tempted to view a chemical equation as merely a secretarial shorthand. It is a shorthand, but it is much more. It is a theoretical description of what is occurring during a chemical reaction. It is the answer to the question proposed in Chapter 1:

Chemical Equation 5.A
$$CaCO_3 \rightarrow CaO + CO_2$$

?
$$\text{MATTER I} \rightarrow \text{MATTER II}$$

The chemical reaction used in the above illustration results in a particularly simple chemical equation (Chemical Equation 5.A). Consider now a second illustration that demonstrates two important aspects of chemical equations in general.

If the element sodium is added to water (H_2O), the resulting chemical reaction produces the element hydrogen and the compound sodium hydroxide (NaOH). This reaction is represented in Chemical Equation 5.B.

Chemical Equation 5.B
$$Na + H_2O \rightarrow H_2 + NaOH$$

Two features of this chemical equation merit further discussion.

First, although the element sodium is represented by its symbol (Na), the element hydrogen is represented by a subscripted symbol (H_2). The reason for this is that certain elements occur in nature as **diatomic molecules** of their atoms. Since experimental evidence indicates that hydrogen is a diatomic molecular element, this must be represented in the chemical equation. Although the discovery of the existence of diatomic molecules is an interesting story, at this point it is only important to appreciate the meaning of the subscript two (Fig. 5.8):

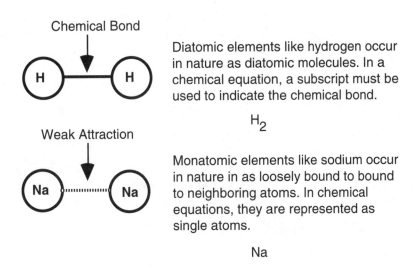

Diatomic elements like hydrogen occur in nature as diatomic molecules. In a chemical equation, a subscript must be used to indicate the chemical bond.

$$H_2$$

Monatomic elements like sodium occur in nature in as loosely bound to bound to neighboring atoms. In chemical equations, they are represented as single atoms.

Na

Figure 5.8   Diatomic and Monatomic Elements

Of the more common elements, only seven occur as diatomic molecules. These elements are H, N, O, F, Cl, Br, and I. Whenever these elements are represented in a chemical equation, they should be written as diatomic molecules: H_2, N_2, O_2, F_2, Cl_2, Br_2, I_2.

The second aspect of Chemical Equation 5.B that merits further discussion is the fact that, as it stands, it violates the law of conservation of mass. Notice that on the left hand side of the equation (reactants) a total of two hydrogen atoms are represented—they both appear in the molecule of water. On the other side of the equation (products) a total of three hydrogen atoms are represented—two appear in the diatomic molecule of hydrogen and one appears in the molecule of sodium hydroxide. As it stands, the equation suggests that an atom of hydrogen was created during the chemical reaction, and this is not possible. The way out of this dilemma is to recognize that chemical reactions do not necessarily involve interactions between single ultimate particles (atoms and molecules). The appropriate number of ultimate particles is determined by a trial and error process called balancing the equation. This process is illustrated below in Example IIA.

EXAMPLE IIA

Balancing an Equation A

Balance the following equation: $Na + H_2O \rightarrow H_2 + NaOH$

SOLUTION: $2\,Na + 2\,H_2O \rightarrow H_2 + 2\,NaOH$

1. Count the total number of atoms of each element on each side of the equation, and identify the elements that are out of balance.

    In this case, only H is out of balance with a total of two atoms appearing on the left and a total of three atoms appearing on the right.

2. Select one of the unbalanced elements, and *attempt* to bring it into balance by increasing the number of one of the ultimate particles on the deficient side. Increasing the number of an ultimate particle must adhere to the following rules:

    a. DO NOT CHANGE EXISTING SUBSCRIPTS!
    b. Increase ultimate particles by placing coefficients in front of symbols (for elements) or formulas (for compounds). For example, by placing the coefficient 2 in front of the formula for water ($2\,H_2O$), we are really writing $H_2O + H_2O$.
    c. DO NOT INSERT COEFFICIENTS INTO A FORMULA! Therefore, changing H_2O to $H_2 2O$ is *not* allowed

    In this case, the coefficient 2 must be placed in front of the formula for water:

    $$Na + 2\,H_2O \rightarrow H_2 + NaOH$$

    Since $2\,H_2O$ means $H_2O + H_2O$, the left hand side of the equation now represents a total of four hydrogen atoms. Also, the number of oxygen atoms has been unbalanced. This is no problem, since the original instruction was to *attempt* to bring the selected element into balance.

3. Check to see if all elements are in balance.

    In this case they are not.

4. Repeat steps 1, 2, and 3 until all elements are in balance.

    In this case, this results in the following series:

    $$Na + 2\,H_2O \rightarrow H_2 + 2\,NaOH$$

    $$2\,Na + 2\,H_2O \rightarrow H_2 + 2\,NaOH \text{ (balanced)}$$

An equation is not a correct representation of a chemical reaction until it is balanced. There is some trial and error associated with the balancing procedure illustrated above. But for simple equations, the procedure is easily mastered. Since balancing an equation is an important fundamental skill, the consideration of a second example is appropriate. In Example IIB, an atom tally for the reactants and products of each partially balanced equation is represented below the equation. Note that each atom tally suggests the next course of action.

EXAMPLE IIB

Balancing an Equation B

Balance the following equation: $C_2H_6 + O_2 \rightarrow CO_2 + H_2O$

SOLUTION: $2\,C_2H_6 + 7\,O_2 \rightarrow 4\,CO_2 + 6\,H_2O$

Unbalanced Equation

$$C_2H_6 + O_2 \rightarrow CO_2 + H_2O$$

C = 2	C = 1
H = 6	H = 2
O = 2	O = 3

First Coefficient Addition

$$C_2H_6 + O_2 \rightarrow 2\ CO_2 + H_2O$$

C = 2	C = 2
H = 6	H = 2
O = 2	O = 5

Second Coefficient Addition

$$C_2H_6 + O_2 \rightarrow 2\ CO_2 + 3\ H_2O$$

C = 2	C = 2
H = 6	H = 6
O = 2	O = 7

Third Coefficient Addition

$$C_2H_6 + 7/2\ O_2 \rightarrow 2\ CO_2 + 3\ H_2O$$

C = 2	C = 2
H = 6	H = 6
O = 7	O = 7

With the addition of the third coefficient, the equation is balanced. The use of fractional coefficients is allowed, but it is sometimes considered poor form to leave fractional coefficients in an equation. Fractions can be cleared from equations by realizing that the balance of an equation is not destroyed by multiplying all of the coefficients by the same integer. In this case, therefore, if all of the coefficients are multiplied by 2, the final balanced equation is $2\ C_2H_6 + 7\ O_2 \rightarrow 4\ CO_2 + 6\ H_2O$.

5.5 Sympathy for the Alchemist

Since scientific theories are constructed from facts, they are valuable tools in the understanding of facts. Before we turn our attention to the quantitative predictive power of the atomic/molecular theory of chemical reactions, let us look at a qualitative aspect of this theory that helps explain a very old problem associated with the discipline of chemistry.

The early practice of chemistry is called alchemy, and the practitioners of this trade are called alchemists. The term "alchemist" carries a negative connotation. It suggests incompetence, and this is not entirely fair. After all, we do not call physicians who practiced medicine prior to the germ theory of disease "witch doctors." Even the term "witch doctor," which also carries somewhat of a negative connotation, is being replaced by "practitioner of traditional medicine." Still the term "alchemist" is historically useful, provided that its meaning is understood.

Stated simply, the alchemists held two beliefs. First, they believed that the study of matter undergoing changes in identity was a worthy study. In this respect, they were truly the predecessors of the modern chemist. Second, they believed that one of the changes that matter was capable of undergoing was the transformation of one metal into another. Expressed in modern terms, the alchemists believed in the following transformation:

Of course, being practical people, they were particularly intrigued by the possibility of converting inexpensive metals into gold—chemists have always been very practical. The conversion of one element into another is called **transmutation**, and the alchemists believed that transmutation by chemical reaction was a possibility. How did the alchemists come to believe in transmutation by chemical reaction? The answer to this question is simple. They believed in transmutation because they observed transmutation. This statement will take some explaining.

Consider the following chemical reactions carried out by the alchemists. If certain blue rocks that are found in nature are digested (dissolved by chemical reaction) in a solution of the chemical compound sulfuric acid, a blue solution forms. If powdered elemental iron is added to this solution, the iron quickly disappears and is replaced by powdered elemental copper. This chemical reaction is easily demonstrated in a classroom, and if you see it, you will swear that the iron has been transformed into copper. Through the process of random mixing of substances, early alchemists observed many chemical reactions similar to the one described above, a reaction where transmutation appears to be taking place. Of course, transmutation is not taking place, but this is only easy to explain to you because you have become conversant in the terminology of modern chemistry. The easiest way to explain what the early alchemist missed is to write the chemical equation for this chemical reaction (Chemical Equation 5.C).

Chemical Equation 5.C
$$3\ CuSO_4 + 2\ Fe \rightarrow Fe_2(SO_4)_3 + 3\ Cu$$
Blue
Solution

The early alchemist missed the point that the element copper was chemically bound in the blue rock in the form of a chemical compound.[4] The formation of the blue solution represents the formation of a new copper compound, $CuSO_4$, from which the copper is released by the chemical action of elemental iron. Because what is going on during this chemical reaction is so easy for us to express, it is also easy for us to understand. But the ease of expression is due solely to the modern language of chemistry and its supporting theory. The early alchemist cannot be faulted for a failure to understand a concept that could not be expressed.

5.6 Calculation of Atomic Mass

When John Dalton first proposed his chemical atomic theory, he suggested only one characteristic of atoms that was related to fact. This characteristic was the unique atomic mass of the atoms of a given element. Using the law of constant composition, Dalton demonstrated that a table of relative atomic masses could be computed for the chemical elements. This computation was illustrated in Chapter 4. The purpose of the present discussion is to review this computation and to consider some of the practical problems that the nineteenth century chemist faced in preparing a table of atomic masses.

First, let us review the logic of Dalton's approach to the determination of atomic mass. Recall that one consequence of the law of **constant composition** is that for a given chemical compound, the following ratio is always a constant:

$$\frac{\text{Combining Mass of Any Element}}{\text{Combining Mass of Any Other Element}} = \text{Constant}$$

---

4. Later alchemists did not miss this point, but once an idea (transmutation) becomes entrenched in a discipline, it resists change.

This constant ratio, which can be expressed in any mass units, is subject to the full arithmetic of linear reasoning. Hence for a given compound, the factual ratio obtained by quantitative analysis is used to calculate a theoretical atomic combining mass ratio (Fig. 5.9).

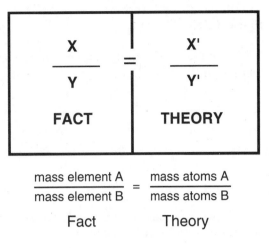

$$\frac{\text{mass element A}}{\text{mass element B}} = \frac{\text{mass atoms A}}{\text{mass atoms B}}$$

$$\qquad \text{Fact} \qquad\qquad \text{Theory}$$

Figure 5.9 Calculation of Atomic Mass

Although this basic logic seems simple enough, two problems encountered by the nineteenth century chemist merit further discussion.

The first and most obvious problem is that the theoretical ratio contains two unknowns, X' and Y'. At first glance, the solution to this problem seems simple enough—simply select an element as an atomic mass standard, assign its atom an arbitrary atomic mass, and use the factual constant composition of compounds containing the standard element to calculate a table of relative atomic masses. In the beginning of the nineteenth century, this is exactly what chemists set out to do. The problem was that chemists could not agree on which element to use as a standard. To make matters worse, if two chemists did agree on a given standard element, they did not always agree on the arbitrary mass value that should be assigned to that standard. This was a period of tremendous frustration in the history of chemistry. Chemists knew that they had a theory of chemical change that had an immense potential, but the use of the theory demanded a table of relative atomic masses. The atomic masses were the whole ball game, and chemists could not agree on the basic issue of an atomic mass standard.

Ultimately, the standard problem was solved, and it was solved most logically. Chemists finally agreed that the atomic mass standard should have the following two characteristics:

1. The standard should be a common element that forms binary (two element) compounds with most of the other elements.
2. The standard should be assigned an arbitrary mass that results in an atomic mass of approximately one amu for the element that has the least massive atom (hydrogen).

Once these two points were agreed on, the choice of standard was obvious. Of all the common elements, only oxygen fits characteristic 1 with a vengeance. And if its atom is assigned a mass of exactly 16.0 amu, then the atomic mass of hydrogen works out to approximately 1 amu.[5]

Now, by way of review, consider the calculation of the atomic mass of the element magnesium (Example IIIA).

EXAMPLE IIIA

Atomic Mass Calculation A

The formula of the compound magnesium oxide is MgO. A quantitative analysis of this compound reveals the following fact: Mg = 60.3% O = 39.7%. If the standard of the atomic mass system is O = 16.0 amu, calculate the atomic mass of magnesium.

---

5. In the twentieth century, the atomic mass system was subjected to a new crisis of standard. The reason for this crisis, which involved a dispute between chemists and physicists, will be discussed in Chapter 9. This dispute ended by compromise in 1961 when chemists and physicists agreed to a new standard (carbon) that changed the older chemical atomic masses only very slightly. For example, under the old system (oxygen standard), the chemical atomic mass of oxygen was 16.000 amu. Under the new system (carbon standard), the chemical atomic mass of oxygen is 15.999 amu. For chemical calculations, this difference is trivial.

SOLUTION: 24.3 amu

The law of constant composition demands that the combining mass ratio of magnesium to oxygen be constant:

$$\frac{60.3 \text{ g Mg}}{39.7 \text{ g O}} = \frac{X \text{ amu Mg}}{16.0 \text{ amu O}}$$

FACT   THEORY

$$X = 24.3 \text{ amu Mg}$$

The above calculation simply demands that the factual constant mass ratio apply to the theoretical (atomic/molecular) world.

———————————————————————□———————————————————————

Once a standard was agreed on, it would seem that all chemists had to do was to subject all of the elements to the procedure illustrated in Example IIIA and tabulate the resultant atomic masses. There was, however, a second problem, and this problem was much more difficult to solve than the atomic mass standard problem.

The nature of this second problem can be understood by considering the first sentence in Example IIIA. Think about it for a minute. This is not a statement of fact; it is a statement of theory. The sentence says that a molecule of the compound magnesium oxide contains one atom of magnesium and one atom of oxygen. In order to make this statement, you would have to be able to count the atoms in a molecule. In other words, in order to do the atomic mass calculation, the nineteenth century chemist needed a glimpse of the molecule. The problem was that no one in 1803 had any idea how to count the atoms in a molecule. No one was really sure that it was a possible task. Well, what was the alternative? The answer is simple—guess! And that is exactly what the early nineteenth century chemist did. This state of affairs obviously led to a great deal of confusion, and the chemists of this period cried for a resolution of the problem.

Finally, in 1860, the German chemist Friedrich August Kekule suggested that prominent world chemists meet to discuss the problem of calculating atomic masses. The meeting was scheduled in Karlsruhe, Germany, and this Karlsruhe Congress was the first international meeting of scientists. In modern times, international scientific meetings are a common occurrence, but in 1860, it was an extreme measure. It must be remembered, however, that the utility of the chemical atomic theory hinged on tabulating atomic masses. The atomic masses were the whole ball game. The problem was of the utmost importance.

At the Karlsruhe Congress, methods for making educated guesses about molecular compositions were discussed. Although the meeting ended with the chemists in disagreement, these methods were eventually accepted by world chemists. More important was the fact that these methods for catching a "glimpse" of a molecule were eventually proven to be valid. In fact, one of the methods involved the discovery of the diatomic elements. For the time being, we will anticipate the result of this method, and consider a second atomic mass calculation in which the atom ratio in the molecule is not one to one (Example IIIB):

EXAMPLE IIIB

Atomic Mass Calculation B

In 1818, Jons Jacob Berzelius represented the formula of red iron oxide (rust) as FeO_3. One of the atom counting techniques introduced at the Karlsruhe Congress predicted the correct formula to be

118

Fe_2O_3. The factual constant composition of red iron oxide is Fe = 69.9% and O = 30.1%. If the standard of the atomic mass system is O = 16.0 amu, calculate the atomic mass of iron.

SOLUTION: 55.7 amu

The law of constant composition demands that the combining mass ratio of hydrogen to oxygen be constant:

$$\frac{69.9 \text{ g Fe}}{30.1 \text{ g O}} = \frac{(2)(X) \text{ amu Fe}}{(3)(16.0) \text{ amu O}}$$

$$\text{FACT} \qquad \text{THEORY}$$

$$X = 55.7 \text{ amu Fe}$$

In the above calculation, X is equal to the atomic mass of an iron atom. The (2) (X) and (3) (16.0) reflect a red iron oxide molecule that contains two atoms of iron and three atoms of oxygen.

Once a valid table of atomic masses was assembled, the chemical atomic/molecular theory of John Dalton could be put to its full quantitative use. In the final sections of this chapter, other examples of the theory's quantitative utility are considered.

5.7 Calculation of Empirical Formula

One of the jobs of the chemist is to be an imitator of nature. Nature provides an abundance of chemical substances that have practical uses. For example, the bark of a willow tree produces a white substance that has been used for centuries for the treatment of pain and fever. Is it possible, the chemist asks, to synthesize this substance in the chemical laboratory? Is it possible to by-pass the willow tree? In order to use atomic/molecular theory to attack this problem, the chemist needs a "picture" of a molecule of the white substance. Although the modern chemist can actually obtain a three dimensional "picture" of a molecule, the starting point for obtaining these "pictures" is to calculate the empirical formula of a compound. The **empirical formula** of a compound is the formula derived from the compound's constant composition. The empirical formula is a code for the constant composition of a chemical compound.

Since the calculation of an atomic mass (previous section) required the combining mass ratio of the elements of a compound and the combining atom ratio in the molecule of the compound, a table of valid atomic masses allows this calculation to be reversed. Specifically, for any new compound that a chemist encounters, a quantitative analysis of the compound (combining mass ratio) allows calculation of the combining atom ratio in the molecule of the compound. This combining atom ratio is called the empirical formula.

Before the empirical formula calculation is illustrated, it is useful to consider a concept that allows the chemist to deal with a difficult problem. The goal of an empirical formula calculation is to determine atom ratios, but if atoms are invisible, how can these ratios be determined? How can atoms be counted? Well, in doing an empirical formula calculation, chemists do not actually count atoms, but they do the next best thing. They count equivalent piles of atoms. To see how this is accomplished, consider two kinds of marbles, one gram marbles and two gram marbles (Fig. 5.10).

1 gram marble   2 gram marble

Figure 5.10 Two Kinds of Marbles

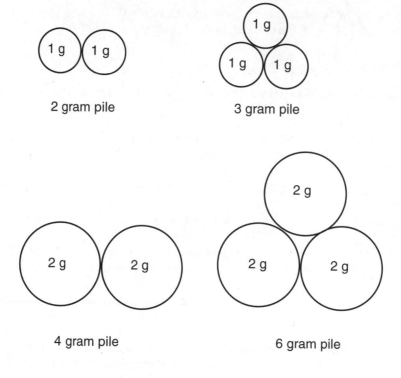

2 gram pile

3 gram pile

4 gram pile

6 gram pile

Figure 5.11 Equivalent Piles of Marbles

Now consider the construction of equivalent piles of the marbles in Figure 5.11.

Notice that as long as the piles of the two kinds of marbles are equivalent in number, the total mass ratio of the piles is equal to the mass ratio of the individual marbles. For example, the piles of two marbles have a mass ratio of 2:4, but this is equal to the individual marble ratio of 1:2. The piles of three marbles have a mass ratio of 3:6, but this is equal to the individual marble ratio of 1:2. Although the above illustration only shows two sets of equivalent piles, the general principle should be obvious. As long as the total pile masses for these two types of marbles are in a ratio of 1:2 (individual marble ratio), the piles of marbles will be equivalent in number. Further, the units of mass used to measure the pile masses do not affect the logic of this argument. Hence, a pile of one gram marbles with a total mass of two metric tons (a lot of marbles) will contain exactly the same number of marbles as a pile of two gram marbles with a total mass of four metric tons. If this principle is valid for marbles, it must also be valid for atoms. It is by counting equivalent piles of atoms that the chemist dodges the issue of counting individual atoms.

In order to ensure that piles of atoms are indeed equivalent, the chemist defines a standard pile of atoms for all elements. This is accomplished by using the principle outlined in the previous paragraph. By way of example, consider an oxygen atom and a sulfur atom (Fig. 5.12).

By the principle of the previous paragraph, a pile of oxygen atoms with a mass of 16.0 grams will contain the same number of atoms as a pile of sulfur atoms with a mass of 32.1 grams. By this simple operation, the chemist defines the mass of a standard pile of atoms. A standard pile of atoms is the atomic mass of the element expressed as grams. Since it is possible to form an infinite number of piles of atoms, the chemist gives this standard pile of atoms a special name. A standard pile

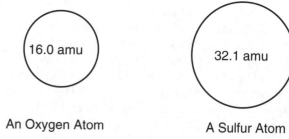

An Oxygen Atom

A Sulfur Atom

Figure 5.12 An Oxygen Atom and a Sulfur Atom

of atoms (the atomic mass of the element expressed as grams) is called a **mole**, from the Greek word meaning "pile." Since the atomic mass of sulfur is 32.1 amu, 32.1 grams of sulfur constitute one mole of sulfur (32.1 grams per mole or 32.1 g/mole). One mole of each of the elements contains the same number of atoms. In this text, the definition of a mole (standard pile) of atoms of a particular element will always be expressed by the ratio grams per mole (g/mole):

Expression 5.A
The Mole

$$\frac{\text{Atomic Mass of Element Expressed as Grams}}{1.00 \text{ Mole of Element}}$$

Since simple logic dictates that doubling the number of moles (standard piles) of an element would double the mass of the element, the above ratio is always constant and subject to the arithmetic of linear reasoning (Expression 5.B):

Expression 5.B
Counting Moles with Linear Reasoning

$$\frac{\text{Atomic Mass of Element as Grams}}{1.00 \text{ Mole}} = \frac{\text{Grams of Element}}{\text{Moles of Element}}$$

EXAMPLE IV

The Mole Definition and Counting Moles

Write the defining ratio for one mole of the elements sulfur and nitrogen. Calculate the number of moles of sulfur in a 45.5 gram sample of sulfur. Calculate the number of grams of nitrogen required to make up a 2.25 mole sample of nitrogen. Atomic Masses: S = 32.1   N = 14.0 amu.

SOLUTION: 45.5 g S = 1.42 moles S; 2.25 moles N = 31.5 g N

Definitions: $\dfrac{32.1 \text{ g S}}{1.00 \text{ mole S}}$ $\dfrac{14.0 \text{ g N}}{1.00 \text{ mole N}}$

For S: $\dfrac{32.1 \text{ g S}}{1.00 \text{ mole S}} = \dfrac{45.5 \text{ g S}}{X \text{ mole S}}$ X = 1.42 moles S

For N: $\dfrac{14.0 \text{ g N}}{1.00 \text{ mole N}} = \dfrac{X \text{ g N}}{2.25 \text{ mole N}}$ X = 31.5 g N

It is important to remember that one mole of all elements contains the same number of atoms. Ultimately, it is this concept that will prove most useful in solving chemical problems. But how many atoms are contained in a one mole sample? Using atomic theory, experiments can actually be designed to measure this number. One very approximate method involving the measurement of the dimensions of a monomolecular oil slick on water will be illustrated in one of the problems at the end of this chapter. Using more precise techniques, the number of atoms in one mole can be determined more accurately. Expressed to three significant figure accuracy, this number is 602000000000000000000000 (6.02×10^{23}). The number of atoms in one mole is called Avogadro's number, in honor of the nineteenth century Italian chemist Amedeo Avogadro. Avogadro was the first chemist to offer a solution to the atomic mass crisis discussed in Section 5.6. Avogadro's approach to this problem will be discussed in Chapter 7.[6]

EXAMPLE V

The Mole Definition and Counting Atoms and Molecules

The number of atoms or molecules in given masses of two or more elements or compounds may be compared. First, calculate the moles of each given mass using the atomic weight of the element or the formula mass of the compound. Since a mole is a standard pile, the greater number of moles contains the greater number of atoms. Which contains a greater number of atoms?

    (i) 150 grams of iron (Fe)   (ii) 150 grams of lead (Pb)

    Atomic Masses: Fe = 55.9   Pb = 207

SOLUTION: 150 g Fe = 2.68 moles Fe; 150 g Pb = 0.725 moles Pb. Therefore, 150 grams of Fe contains more atoms.

$$\text{For Fe:} \quad \frac{55.9 \text{ g Fe}}{1.00 \text{ mole Fe}} = \frac{150 \text{ g Fe}}{X \text{ mole Fe}} \quad X = 2.68 \text{ moles Fe}$$

$$\text{For Pb:} \quad \frac{207 \text{ g Pb}}{1.00 \text{ mole Pb}} = \frac{150 \text{ g Pb}}{X \text{ mole Pb}} \quad X = 0.725 \text{ moles Pb}$$

The question of how the chemist converts combining elemental mass ratios into combining atom ratios (empirical formulas) can now be addressed. The chemist solves this problem by recognizing that an elemental combining mole ratio is equal to the combining atom ratio. For example, if a molecule of water contains one atom of oxygen and two atoms of hydrogen, then one mole of oxygen will require two moles of hydrogen in the chemical reaction to form water. Since combining molar ratios can be determined by experiment, combining atom ratios can be determined. This calculation is illustrated in Example VI.

EXAMPLE VI

Empirical Formula Calculation

Quantitative analysis of a certain compound indicates that the compound has the following constant composition: C = 24.0%, F = 76.0%. Use this information to calculate the empirical formula of this compound (*i.e.* C_xF_y). Atomic Masses: C = 12.0 amu   F = 19.0 amu.

---

6.   The number of atoms in a mole was not determined until the end of the nineteenth century, well after Avogadro's death. Many chemists feel that this number should be named for the German chemist Josef Loschmidt, who first measured the number. Although American and Italian chemists generally refer to the number as Avogadro's number, in many parts of the world, the number is called Loschmidt's number.

SOLUTION: CF_2

The information given in this problem is the combining mass ratio of the elements. What is required is the combining atom ratio. The problem is solved by converting the mass ratio into a mole ratio:

$$\text{For C:} \quad \frac{12.0 \text{ g C}}{1.00 \text{ mole C}} = \frac{24.0 \text{ g C}}{X \text{ mole C}} \qquad X = 2.00 \text{ moles C}$$

$$\text{For F:} \quad \frac{19.0 \text{ g F}}{1.00 \text{ mole F}} = \frac{76.0 \text{ g F}}{X \text{ mole F}} \qquad X = 4.00 \text{ moles F}$$

If we now convert the mole ratio of these two elements into the simplest whole number ratio, this simplest whole number ratio of moles (standard piles) must equal the ratio of individual atoms in the molecule. The arithmetic trick for converting a series of numbers to the simplest whole number ratio is to divide each number in the series by the smallest number in the series. In this case, the operation is trivial:

$$2{:}4 \qquad \text{divide both numbers by 2} \qquad 1{:}2$$

Since the combining mole ratio is 1C:2F, the combining atom ratio must also be 1C:2F. The empirical formula of the compound is, therefore, CF_2.

Although the empirical formula of a compound is a far cry from a three dimensional "picture" of a molecule, it is an important step in the process of determining what a compound's molecule "looks like." When modern chemists discover a new chemical compound, one of the first things they do is determine the empirical formula of this compound. The process by which the modern chemist gets a more complete "view" of molecules will be considered in Chapter 8.

5.8 Calculations Using Gravimetric Ratios

Two examples of the predictive power of Daltonian atomic/molecular theory will be illustrated in this section. The calculations in these examples use something called a gravimetric ratio. **Gravimetric ratios** are constant mass ratios demanded by the logic of the atomic/molecular theory. These ratios are constant because of the law of constant composition. Two of the gravimetric ratios were introduced in Chapter 4 (Expressions 5.C and 5.D):

Expression 5.C
Formula Gravimetric Ratio

$$\frac{\text{Combining Mass of Any Element}}{\text{Combining Mass of Any Other Element}} = \text{Constant}$$

Expression 5.D
Formula Gravimetric Ratio

$$\frac{\text{Combining Mass of Any Element}}{\text{Mass of Compound Formed}} = \text{Constant}$$

The values of these ratios for a particular compound are coded in the formula of the compound as illustrated on the next page in Example VII.

EXAMPLE VII

Formula Gravimetric Ratio Definition

The formula of aluminum sulfate is $Al_2(SO_4)_3$. Express all of the gravimetric ratios implied by the formula $Al_2(SO_4)_3$. Atomic Masses: Al = 27.0   S = 32.1   O = 16.0 amu.

SOLUTION:

The masses of any two elements in a molecule form a constant ratio, also the mass of any element and the total mass of the molecule form a constant ratio. This information is coded in the formula $Al_2(SO_4)_3$.

$(2 \times Al)/(3 \times S)$ = 54.0 amu Al/96.3 amu S
$(2 \times Al)/(12 \times O)$ = 54.0 amu Al/192.0 amu O
$(3 \times S)/12 \times O)$ = 96.3 amu S/192.0 amu O
$(2 \times Al)/Al_2(SO_4)_3$ = 54.0 amu Al/342.3 amu $Al_2(SO_4)_3$
$(3 \times S)/Al_2(SO_4)_3$ = 96.3 amu S/342.3 amu $Al_2(SO_4)_3$
$(12 \times O)/Al_2(SO_4)_3$ = 192.0 amu O/342.3 amu $Al_2(SO_4)_3$

The ratios in Example VII express all of the combining mass relationships for a molecule of aluminum sulfate, $Al_2(SO_4)_3$. Since the concept of the mole can be extended to include chemical compounds, these ratios also express all of the combining mass relationships for a mole of $Al_2(SO_4)_3$. Using the same logic that was used to define a mole of an element, one mole of a chemical compound is defined as the molecular mass of the compound expressed as grams. The extension of the mole concept to include chemical compounds allows a formula and its gravimetric ratios to be interpreted several ways. The following interpretations of the gravimetric ratio expressing the quantity of oxygen in aluminum sulfate, $Al_2(SO_4)_3$, is presented to illustrate this point:

Four Interpretations of the Gravimetric Ratio $(12 \times O)/Al_2(SO_4)_3$

1. One molecule of aluminum sulfate contains twelve oxygen atoms.
2. A 342.3 amu sample of aluminum sulfate (one molecule) contains 192.0 amu of oxygen (twelve atoms).
3. One mole of aluminum sulfate contains twelve moles of oxygen atoms.
4. A 342.3 g sample of aluminum sulfate (one mole) contains 192.0 g of oxygen (twelve moles of oxygen atoms).

When chemists use the constant composition of a compound to calculate its empirical formula, they are really encoding the constant composition into the formula. Since the formulas incorporate this constant composition information into chemical equations, a third type of gravimetric ratio involving chemical equations can be expressed (Expression 5.E).

Expression 5.E
Equation Gravimetric Ratio

$$\frac{\text{Combining Mass of Any Reaction Component}}{\text{Combining Mass of Any Other Reaction Component}} = \text{Constant}$$

The values of these ratios for a particular chemical reaction are coded in the equation of the compound as illustrated on the next page in Example VIII.

EXAMPLE VIII

Equation Gravimetric Ratio Definition

The equation for the reaction of sodium metal with oxygen to form sodium oxide is shown below:

$$4\,Na + O_2 \rightarrow 2\,Na_2O$$

Express all of the gravimetric ratios implied by this equation. Atomic Masses: Na = 23.0
O = 16.0 amu

SOLUTION:
The masses of any two components in a chemical equation form a constant ratio. This information is coded in the chemical equation.

$$(4 \times Na)/O_2 = 92.0 \text{ amu Na}/32.0 \text{ amu } O_2$$
$$(4 \times Na)/(2 \times Na_2O) = 92.0 \text{ amu Na}/124.0 \text{ amu } Na_2O$$
$$(2 \times Na_2O)/O_2 = 124.0 \text{ amu } Na_2O/32.0 \text{ amu } O_2$$

Note the importance of including a component's coefficient in formulating a gravimetric ratio from a chemical equation.

As in the case of chemical compounds, the extension of the mole concept to include chemical reactions allows an equation and its gravimetric ratios to be interpreted several ways. The following interpretations of the equation representing the reaction of sodium with oxygen to form sodium oxide $(4\,Na + O_2 \rightarrow 2\,Na_2O)$ is presented to illustrate this point:

Four Interpretations of the Equation $4\,Na + O_2 \rightarrow 2\,Na_2O$

1. Four atoms of sodium react with one molecule of diatomic oxygen to form 2 molecules of sodium oxide.
2. A 92.0 amu sample of sodium (four atoms) reacts with a 32.0 amu sample of oxygen (one molecule) to form 124 amu of sodium oxide (two molecules).
3. Four moles of sodium atoms react with one mole of diatomic oxygen to form 2 moles of sodium oxide.
4. A 92.0 g sample of sodium (four moles) reacts with a 32.0 g sample of oxygen (one mole) to form 124 g of sodium oxide (two moles).

Three types of constant ratios (gravimetric ratios) can be formulated from atomic/molecular theory. Two of these ratios (5.C and 5.D) are formulated from chemical formulas, and the other (5.E) is formulated from chemical equations. These constant ratios are subject to the arithmetic of linear reasoning. They can be used to calculate the percent composition of a chemical compound (Example IX). They can also be used to make useful predictions about chemical compounds (Example X) and chemical reactions (Example XI). The application of gravimetric ratios to chemical calculations is called **stoichiometry**.[7]

---

7. Stoichiometry actually predates atomic theory. During the latter part of the eighteenth century, the German chemist Jeremias Benjanin Richter introduced the basic concept of the gravimetric ratio based on the factual tabulation of quantitative changes during the chemical reaction process. He named this quantitative study of chemistry stoichiometry from two Greek words meaning to measure the magnitude of something that cannot be divided.

EXAMPLE IX

Stoichiometric Calculation: Formula Gravimetric Ratio

The *formula* of sodium nitrate is $NaNO_3$. Calculate the percentage of *nitrogen* in *sodium nitrate*. Atomic Masses: Na = 23.0   N = 14.0   O = 16.0 amu.

SOLUTION: 16.5%

The first step in a problem of this type is to identify the gravimetric ratio that applies to the wording of the problem. In this problem, the *italicized* words indicate that the appropriate ratio will be of type 5.D. Specifically, the terms *nitrogen* and *sodium nitrate* indicate that the required ratio is $N/NaNO_3$ = 14.0 amu N/85.0 amu $NaNO_3$. The next step is to remember that percentage is units of a component (element) per 100 units of the whole (compound). The final step is to use linear reasoning to make theory predict fact:

$$\frac{14.0 \text{ amu N}}{85.0 \text{ amu NaNO}_3} = \frac{X \text{ g N}}{100 \text{ g NaNO}_3}$$

THEORY   FACT

X = 16.9 g of N/100 g of $NaNO_3$
or
X = 16.5%

EXAMPLE X

Stoichiometric Calculation: Formula Gravimetric Ratio

The *formula* of sulfuric acid is H_2SO_4. In theory, what mass of *sulfur* is required to manufacture 150 metric tons *sulfuric acid*?

SOLUTION: 49.0 metric tons of S

The first solution step of this problem is to recognize the gravimetric ratio that applies to the wording of the problem. As in the previous example, the *italicized* words indicate that the appropriate ratio will be of type 5.D. Thus, the terms *sulfur* and *sulfuric acid* indicate that the required ratio is S/H_2SO_4 = 32.1 amu S/98.1 amu H_2SO_4. The second step in the problem is to use linear reasoning to make theory predict fact:

$$\frac{32.1 \text{ amu S}}{98.1 \text{ amu H}_2SO_4} = \frac{X \text{ tons S}}{150 \text{ tons H}_2SO_4}$$

THEORY   FACT

X = 49.1 metric tons of S

EXAMPLE XI

Stoichiometric Calculation: Equation Gravimetric Ratio

The *chemical equation* for the combustion of one of the components of gasoline, C_8H_{18}, is represented below:

$$2\ C_8H_{18} + 25\ O_2 \rightarrow 16\ CO_2 + 18\ H_2O$$

In theory, what mass of *oxygen*, O_2, is required for the complete combustion of 1.60 grams of the *gasoline component*, C_8H_{18}?

SOLUTION: 5.61 g of O_2

As in the previous examples, the first step in this problem is to identify the gravimetric ratio that applies to the wording of the problem. The *italicized* words indicate that the appropriate ratio will be of type 5.E. Specifically, the ratio indicated is $25(O_2)/2(C_8H_{18}) = 800$ amu $O_2/228$ amu C_8H_{18}. Again, the second step in the problem is to use linear reasoning to make theory predict fact:

$$\frac{800 \text{ amu } O_2}{228 \text{ amu } C_8H_{18}} = \frac{X \text{ g } O_2}{1.60 \text{ g } C_8H_{18}}$$

THEORY   FACT

$$X = 5.61 \text{ g } O_2$$

Don't underestimate the importance of the kind of prediction being made in Examples X and XI. For example, sulfuric acid is an industrial chemical of major importance. Note, for example, the relationship that exists between the United States production of sulfuric acid and the United States Gross National Product (Fig. 5.13).

This linear relationship is typical of any industrial nation. Now consider yourself to be a manufacturer of sulfuric acid. As a businessperson, you find that you are sitting with orders for 150 metric tons of sulfuric acid. You know that sulfuric acid is manufactured from sulfur, which your company must purchase, but how much sulfur should be purchased? The purchase of either too much or too little can spell financial disaster. Using Daltonian atomic/molecular theory, the modern chemist can answer this financial question.

The illustration given above may be a bit simplistic, both economically and chemically, but it does illustrate an important point. Daltonian atomic/molecular theory provided chemists with a means of "understanding" the chemical reaction process, and the application of Daltonian atomic/molecular theory provided chemists with a service of value to society. Practical application through "understanding" is the desired end result of any good scientific theory.

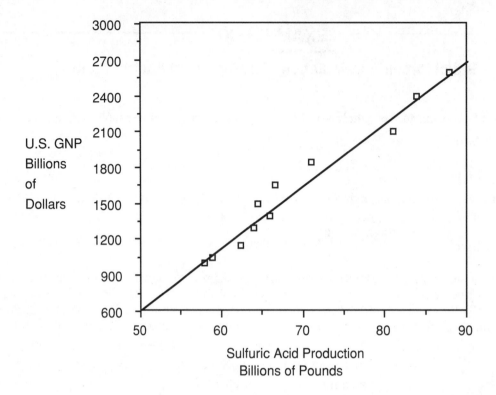

Figure 5.13 U.S. GNP as a Function of Sulfuric Acid Production 1970–1980

5.9 Stoichiometric Calculations Using the Factor Label Method

In the previous sections, several examples of stoichiometric calculations were presented using the linear reasoning approach presented in Chapter 2. In this section, the alternative factor label method will be used to solve several stoichiometric problems. (See Example V in Section 2.11.) The examples in this section are presented for students who elect to use the factor label method. Students who do not elect to use the factor label approach to stoichiometry may skip this section without loss of continuity.

EXAMPLE XII

The Mole Definition and Counting Moles
Factor Label Method

Write the defining ratio for one mole of the elements carbon and oxygen. Calculate the number of moles of carbon in a 225 gram sample of carbon. Calculate the number of grams of oxygen required to make up a 5.12 mole sample of oxygen. Atomic Masses: C = 12.0   O = 16.0 amu.

SOLUTION: 225 g C = 18.8 moles C; 5.12 moles O = 81.9 g O

$$\text{Definitions:} \qquad \frac{12.0 \text{ g C}}{1.00 \text{ mole C}} \qquad \qquad \frac{16.0 \text{ g O}}{1.00 \text{ mole O}}$$

Since a mole conversion problem is a simple linear reasoning problem, the final solution will take the following form where $\frac{X'}{Y'}$ is one of two reciprocal forms of the mole definition:

$$X = Y \frac{X'}{Y'}$$

In the conversion of 225 g of carbon to moles of carbon, a decision must be made between the following two reciprocal variations:

$$X = 225 \text{ g C} \frac{12.0 \text{ g C}}{1.00 \text{ mole C}} \quad \text{or} \quad X = 225 \text{ g C} \frac{1.00 \text{ mole C}}{12.0 \text{ g C}}$$

If the units are included in both of these calculations, then only one of these two alternatives produces an answer with the correct units—moles of C in this case:

$$X = 225 \text{ g C} \frac{1.00 \text{ mole C}}{12.0 \text{ g C}} \qquad X = 18.8 \text{ mole C}$$

In the conversion of 5.12 moles of oxygen to grams of oxygen, a decision must be made between the following two reciprocal variations:

$$X = 5.12 \text{ mole O} \frac{16.0 \text{ g O}}{1.00 \text{ mole O}} \quad \text{or} \quad X = 5.12 \text{ mole O} \frac{1.00 \text{ mole O}}{16.0 \text{ g O}}$$

If the units are included in both of these calculations, then only one of these two alternatives produces an answer with the correct units—grams of O in this case:

$$X = 5.12 \text{ mole O} \frac{16.0 \text{ g O}}{1.00 \text{ mole O}} \qquad X = 81.9 \text{ g O}$$

EXAMPLE XIII

Stoichiometric Calculation: Formula Gravimetric Ratio
Factor Label Method

The *formula* of aluminum oxide is Al_2O_3. In theory, what mass of *aluminum* could be produced by the complete chemical decomposition of 375 grams of *aluminum oxide*?

SOLUTION: 199 g Al

The first step in this problem is to recognize the mole ratio that applies to the wording of the problem. The *italicized* words indicate the appropriate ratio. Thus, the terms *aluminum* and *aluminum oxide* indicate that the required ratio is as follows:

$$\frac{2 \text{ mole Al}}{1 \text{ mole Al}_2\text{O}_3}$$

Using the mole definitions for Al and Al_2O_3 leads to the following gravimetric conversion factor for aluminum oxide:

$$\frac{(2\ \text{mole Al})\ \dfrac{27.0\ \text{g Al}}{1.00\ \text{mole Al}}}{(1\ \text{mole Al}_2O_3)\ \dfrac{102\ \text{g Al}_2O_3}{1.00\ \text{mole Al}_2O_3}}$$

$$\frac{2(27.0)\ \text{g Al}}{102\ \text{g Al}_2O_3}$$

This is a formal factor label verification of one of the formula gravimetric ratio interpretations presented in Section 5.8. For the calculation at hand, a decision must be made between the following two reciprocal variations:

$$X = 375\ \text{g Al}_2O_3\ \frac{2(27.0)\ \text{g Al}}{102\ \text{g Al}_2O_3} \qquad \text{or} \qquad X = 375\ \text{g Al}_2O_3\ \frac{102\ \text{g Al}_2O_3}{2(27.0)\ \text{g Al}}$$

If the units are included in both of these calculations, then only one of these two alternatives produces an answer with the correct units—grams of Al in this case:

$$X = 375\ \text{g Al}_2O_3\ \frac{2(27.0)\ \text{g Al}}{102\ \text{g Al}_2O_3} \qquad\qquad X = 199\ \text{g Al}$$

EXAMPLE XIV

Stoichiometric Calculation: Equation Gravimetric Ratio Factor Label Method

The *chemical equation* for the combustion of the butane in a butane lighter, C_4H_{10}, is represented below:

$$2\ C_4H_{10} + 13\ O_2 \rightarrow 8\ CO_2 + 10\ H_2O$$

In theory, what mass of *carbon dioxide*, CO_2, is formed by the complete combustion of 175 grams of the *butane*, C_4H_{10}?

SOLUTION: 530 g of CO_2

The first solution step of this problem is to recognize the mole ratio that applies to the wording of the problem. The *italicized* words indicate the appropriate ratio. Thus, the terms *carbon dioxide* and *butane* indicate that the required ratio is as follows:

$$\frac{2\ \text{mole } C_4H_{10}}{8\ \text{mole } CO_2}$$

Using the mole definitions for C_4H_{10} and CO_2, leads to the following gravimetric conversion factor for aluminum oxide:

$$\frac{(2 \text{ mole } C_4H_{10}) \dfrac{58.1 \text{ g } C_4H_{10}}{1.00 \text{ mole } C_4H_{10}}}{(8 \text{ mole } CO_2) \dfrac{44.0 \text{ g } CO_2}{1.00 \text{ mole } CO_2}}$$

$$\frac{2\,(58.1) \text{ g } C_4H_{10}}{8\,(44.0) \text{ g } CO_2}$$

This is a formal factor label verification of one of the equation gravimetric ratio interpretations presented in Section 5.8. For the calculation at hand, a decision must be made between the following two reciprocal variations:

$$X = 175 \text{ g } C_4H_{10} \; \frac{2\,(58.1) \text{ g } C_4H_{10}}{8\,(44.0) \text{ g } CO_2} \qquad \text{or} \qquad X = 175 \text{ g } C_4H_{10} \; \frac{8\,(44.0) \text{ g } CO_2}{2\,(58.1) \text{ g } C_4H_{10}}$$

If the units are included in both of these calculations, then only one of these two alternatives produces an answer with the correct units—grams of CO_2 in this case:

$$X = 175 \text{ g } C_4H_{10} \; \frac{8\,(44.0) \text{ g } CO_2}{2\,(58.1) \text{ g } C_4H_{10}} \qquad X = 530 \text{ g } CO_2$$

5.10 Skill Review

The arithmetic problems that you encountered in this chapter fell into two categories. There were problems that were designed to show you exactly how a modern chemist uses Daltonian atomic theory in an industrial setting, and there were problems that simply illustrated basic skills. For the sake of narrative continuity, no attempt was made to separate the two types of problems. The purpose of this review section is to isolate the basic skills that you will need to master before you attempt to solve the problems in Chapter 5.

SKILL I   READING A FORMULA

EXAMPLE: $Al_2(SO_4)_3$
This formula represents a "molecule" that contains two aluminum atoms, three sulfur atoms, and twelve oxygen atoms.

SKILL II   CALCULATING A MOLECULAR MASS

EXAMPLE: $Al_2(SO_4)_3$
Since each atom in the above "molecule" has a characteristic mass, the "molecule" as a whole has a mass that is calculated as follows:

$2 \times Al$	=	2×27.0	=	54.0 amu
$3 \times S$	=	3×32.1	=	96.3 amu
12×0	=	12×16.0	=	192.0 amu
				342.3 amu

SKILL III   EXPRESSING A GRAVIMETRIC RATIO FOR A FORMULA

EXAMPLE: $Al_2(SO_4)_3$

The masses of any two elements in a molecule form a constant ratio, also the mass of any element and the total mass of the molecule form a constant ratio. All of these constant ratios are subject to the full arithmetic linear reasoning. Listed below are all but two of the constant ratios implied by the formula $Al_2(SO_4)_3$:

$$(2 \times Al)/(3 \times S) = 54.0 \text{ amu Al}/96.3 \text{ amu S}$$
$$(2 \times Al)/(12 \times O) = 54.0 \text{ amu Al}/192.0 \text{ amu O}$$
$$(3 \times S)/12 \times O) = 96.3 \text{ amu S}/192.0 \text{ amu O}$$
$$(2 \times Al)/Al_2(SO_4)_3 = 54.0 \text{ amu Al}/342.3 \text{ amu } Al_2(SO_4)_3$$

Identify the two ratios that are missing above.

SKILL IV   EXPRESSING THE MOLE/MASS RATIO FOR EACH ELEMENT

EXAMPLES: S & O

For each element, the chemist defines a standard pile of atoms (mole) by the following operation. This operation leads to a constant ratio which is subject to the full arithmetic of linear reasoning:

$$S = 32.1 \text{ amu. Therefore by definition } 32.1 \text{ g S}/1 \text{ mole S}$$
$$O = 16.0 \text{ amu. Therefore by definition } 16.0 \text{ g O}/1 \text{ mole O}$$

SKILL V   READING A CHEMICAL EQUATION

EXAMPLE: $4 \text{ Na} + O_2 \rightarrow 2 \text{ Na}_2O$

This equation represents a chemical reaction in which 4 atoms (or moles) of the element sodium react with 1 molecule (or mole) of the element oxygen to form 2 molecules (or moles) of the compound Na_2O.

SKILL VI   EXPRESSING A GRAVIMETRIC RATIO FOR A CHEMICAL EQUATION

EXAMPLE: $4 \text{ Na} + O_2 \rightarrow 2 \text{ Na}_2O$

The masses of any two components in a chemical equation form a constant ratio. All of these ratios are subject to the full arithmetic of linear reasoning. Listed below are all of the constant ratios implied by the sample equation:

$$(4 \times Na)/O_2 = 92.0 \text{ amu Na}/32.0 \text{ amu } O_2$$
$$(4 \times Na)/2Na_2 O = 92.0 \text{ amu Na}/124.0 \text{ amu } Na_2O$$
$$O_2/2Na_2O = 32.0 \text{ amu } O_2/124.0 \text{ amu } Na_2O$$

5.11 Chemical Kinetics—Atoms and Molecules in Transition

The quantitative laws of chemical action discussed in Section 4.5 describe the mass changes that are observed after a chemical reaction has taken place. These changes can be observed by a quantitative analysis of the reactants and products of a chemical reaction. Atomic theory explains these quantitative laws, and through stoichiometry, atomic theory allows for the prediction of mass relationships in chemical reactions.

Quantitative measurements can also be made during the course of a chemical reaction. One of the most revealing measurements to be made during the course of a reaction is the rate of a chemical reaction. By measuring this rate, chemists can formulate hypotheses about the specific sequence of atomic and molecular events occurring during the course of a chemical reaction. The study of chemical reaction rate is called **chemical kinetics**, and the sequence of atomic and molecular events occurring during a chemical reaction is referred to as the **reaction mechanism**.

Although the details associated with the formulation of a mechanistic hypothesis for a chemical reaction are beyond the scope of this text, the reaction mechanism is an important concept for beginning chemistry students to understand. To this end, it might be useful to consider an analogy. Suppose that you are observing a line of cars entering and leaving a tunnel. You are allowed to observe the tunnel entrance and exit, but you cannot make any observations inside of the tunnel. During your observation, you notice the rate of flow of cars into or out of the tunnel seems to be affected by the color of the cars. An increase in the concentration of red cars always slows down the traffic flow, an increase in the concentration of blue cars always speeds up the traffic flow, and an increase in the concentration of any other car color does not affect the traffic flow. From these observations it is possible to formulate hypotheses about the condition inside the tunnel. For example, the traffic flow observations could be explained by the hypothesis that the tunnel contains an express lane for blue cars and that all red cars are being stopped at a police checkpoint. Of course, there are many other possible explanations, but other explanations will suggest experiments to distinguish between competing hypotheses. In an analogous manner the rate at which reactants disappear or products appear during the course of a chemical reaction allow for the formulation of competing mechanistic hypotheses that can be subjected to further testing.

By applying the techniques of chemical kinetics, chemists obtain a theoretical view of what is going on during the chemical reaction process. By way of example, consider the following chemical equation that represents the natural decomposition of ozone (O_3):

Chemical Equation 5.D
$$2\,O_3 \rightarrow 3\,O_2$$

Ozone is a triatomic molecule of the element oxygen that forms in the upper atmosphere by the action of ultraviolet radiation on diatomic molecular oxygen (O_2). The upper atmospheric ozone plays a crucial role in protecting the surface of the earth from exposure to high levels of ultraviolet radiation, and the destruction of this ozone layer by atmospheric pollutants is a current environmental concern. Equation 5.D represents the normal decomposition of ozone. Kinetic studies suggest that this reaction proceeds by the following mechanism:

Reaction Mechanism 5.A
Step 1: $O_3 \rightleftharpoons O_2 + O$
Step 2: $O + O_3 \rightarrow 2\,O_2$
Net Reaction: $2\,O_3 \rightarrow 3\,O_2$

Reaction Mechanism 5.A indicates that the normal decomposition of ozone occurs in two steps. In the first step, an ozone molecule decomposes to form a diatomic oxygen molecule and an oxygen atom. In the second step, the oxygen atom reacts with another ozone molecule to form two oxygen diatomic molecules. Note that the net reaction that results from this two-step mechanism is the chemical reaction represented by Chemical Equation 5.D. The double arrow in step 1 of Reaction Mechanism 5.A indicates that this step is a dynamic equilibrium. In a dynamic equilibrium, the forward and reverse reactions take place at an equal rate. The concept of a dynamic equilibrium is discussed in greater detail in the next section.

Atmospheric pollutants can alter the rate at which ozone decomposes by altering the mechanism of the reaction. By way of example, kinetic studies suggest that nitric oxide (NO) might cause atmospheric ozone to decompose according to the following faster reaction mechanism:

Reaction Mechanism 5.B
Step 1: $O_3 \rightleftharpoons O_2 + O$
Step 2: $NO + O_3 \rightarrow O_2 + NO_2$
Step 3: $O + NO_2 \rightarrow NO + O_2$
Net Reaction: $2\,O_3 \rightarrow 3\,O_2$

Notice that the atmospheric pollutant nitric oxide (NO) is consumed in step 1 of this mechanism, and it is produced in step 3. A chemical substance that alters the rate of a chemical reaction without changing the net reaction (net stoichiometry) is called a **catalyst**. In Reaction Mechanism 5.B, nitric oxide (NO) acts as a catalyst to speed up the decomposition of ozone in the upper atmosphere. Ironically, this same pollutant can catalyze the production of ozone in the lower atmosphere where the ozone itself becomes a pollutant.

Synthetic catalysts are immensely important to the chemical industry where they are used to control synthetic reactions. Household chemistry problems 11 and 12 at the end of Chapter 4 demonstrated chemical catalysis.

5.12 Chemical Equilibrium

Step one in both Reaction Mechanisms 5.A and 5.B involve a dynamic equilibrium. During a chemical **dynamic equilibrium**, both forward and reverse chemical reactions are taking place at equal rates. This condition is noted in a chemical equation by replacing the single arrow with a double arrow. An example of an industrially important dynamic equilibrium is the formation of the gas ammonia (NH_3) from nitrogen and hydrogen:

Chemical Equation 5.E
$$N_2 + 3\,H_2 \rightleftharpoons 2\,NH_3$$

The chemical reaction described by Chemical Equation 5.E represents the industrial fixation of atmospheric nitrogen. Diatomic molecular nitrogen does not react readily with other substances. The formation of chemical compounds from molecular nitrogen is called nitrogen fixation, and natural nitrogen fixation processes are an essential part of nature's nitrogen cycle. The industrial process that produces ammonia from nitrogen and hydrogen is called the Haber process, after the German chemist Fritz Haber, who developed the technology associated with the industrialization of this important equilibrium reaction.

When a chemical reaction reaches its dynamic equilibrium point, the reaction mixture contains both reactant and product substances. The concentration of these substances at the equilibrium point is controlled by the law of mass action. Although the details of this quantitative law are beyond the scope of this text, it should be noted that the equilibrium concentrations for a given chemical reaction are related to a constant that is dependent on temperature and pressure. A useful qualitative rule that describes the behavior of an equilibrium system was discovered by Henri Louis Le Chatelier in 1884. **Le Chatelier's principle** states that if a stress is placed on a system that is in a state of dynamic equilibrium, the system will experience a readjustment in the equilibrium that tends to diminish the original stress. For example, in the Haber process, the removal of ammonia (NH_3) from the reaction container would tend to shift the reaction to the right in order to produce more ammonia. In the industrialization of this reaction, Haber actually investigated the stresses of temperature and pressure; nevertheless, the removal of ammonia serves to illustrate Le Chatelier's principle.

Chemical reactions that take place in closed systems should be represented by equilibrium equations. In order to determine the quantity of product substances present at equilibrium, the equation gravimetric ratio techniques introduced in Section 5.8 must be coupled with the law of mass action. Fortunately, an understanding of equilibrium principles allows many industrially important reactions to be driven to completion. In equilibrium reactions that cannot be driven to completion, the recycling of valuable reactant compounds allows for their complete conversion into product compounds. In either case, the use of equation gravimetric ratios to compute product quantities is an important aspect of industrial chemistry.

EXAMPLE XV

Percent Yield of a Chemical Reaction

Stoichiometry can be used to calculate the mass of a product that should be produced by a chemical reaction. For various reasons including equilibrium problems, the amount of product actually isolated is usually less than the theoretical amount predicted by stoichiometry. During the study of any chemical reaction process, chemists record this product loss by calculating the percent yield, which is defined as the percent value of the following ratio:

$$\frac{\text{Mass of Product Isolated}}{\text{Mass of Product Predicted by Stoichiometry}}$$

Consider the following equation:

$$N_2 + 3\,H_2 \rightleftharpoons 2\,NH_3$$

If the reaction of 38.1 grams of hydrogen (H_2) with an excess of nitrogen (N_2) results in the isolation of 178 grams of ammonia (NH_3) what is the percent yield of the reaction? Atomic Masses: N = 14.0 H = 1.01 amu

SOLUTION: 83.2%

Since nitrogen is present in excess, the theoretical mass of ammonia will be determined by the amount of hydrogen. Hydrogen is referred to as the limiting reagent. Using the approach illustrated in Example X, the theoretical mass of ammonia can be calculated from the gravimetric ratio $3(H_2)/2(NH_3)$:

$$\frac{6.06 \text{ amu } H_2}{34.06 \text{ amu } NH_3} = \frac{38.1 \text{ g } H_2}{X \text{ g } NH_3}$$

THEORY FACT

$$X = 214 \text{ g } NH_3 \text{ (Theoretical Yield)}$$

From the definition of percent yield given in the problem, the following result can be obtained:

$$\frac{178 \text{ g } NH_3 \text{ Isolated}}{214 \text{ g } NH_3 \text{ Predicted by Stoichiometry}} (100) = 83.2\,\%$$

5.13 Do Atoms Really Exist?

In Chapter 4, the atom was presented as an abstraction that adequately rationalized the facts associated with the chemical reaction process. In Chapter 5, the atomic/molecular theory was used as a useful computational tool. Using primitive molecular "pictures" (empirical formulas), the nineteenth century chemist began the systematic synthesis of natural and totally new chemical compounds. Using formulas and equations, chemists predicted mass behavior during the chemical reaction process. And using the principles of chemical kinetics and chemical equilibrium, chemists speculated about the behavior of atoms and molecules during the chemical reaction process. Throughout all of this, a question persisted. Were atoms simply useful abstractions, or did they really exist?

Well, do atoms really exist? This is a question that we must answer for you in a very personal way. As chemists, we want you to know that we have come an awfully long way since 1803. And although no one has ever seen an atom, in the sense that human beings normally see, the evidence for their existence has become quite impressive. In recent years, a technique called scanning tunneling microscopy has come very close to providing atomic pictures. As chemists, we must say that atoms do, indeed, exist. BUT! There is a philosopher that lives within us, and the philosopher warns that we must proceed with caution where such lofty matters are concerned. There is no conflict here. The chemist within us is a pragmatic scientist, and the philosopher is an idealistic thinker. The two never argue, but they do have some very interesting discussions.

"Philosophy is like a mother who gave birth to and endowed all other sciences. Therefore one should not scorn her in her nakedness and poverty, but should hope, rather, that part of her Don Quixote ideal will live on in her children so that they do not sink into philistinism." – Albert Einstein, German/American Physicist

Chapter Five
Performance Objectives

P.O. 5.0

Review all of the boldfaced terminology in this chapter, and make certain that you understand the use of each term.

atomic mass
chemical kinetics
dynamic equilibrium
gravimetric ratio
molecular mass
symbol, chemical

catalyst
constant composition
empirical formula
Le Chatelier's principle
reaction mechanism
transmutation

chemical equation
diatomic molecule
formula, chemical
mole
stoichiometry

P.O. 5.1

You must be able to identify an element name given its symbol, and you must be able to identify an element symbol given its name.

EXAMPLE:
Which of the following compounds contains the element tin?

a) $TiCl_4$ b) $ZnCl_2$ c) $SnCl_4$ d) $AgCl$

SOLUTION: c

The answer is c. The selection of the correct answer requires the rote memorization of the boldfaced elements and symbols listed in Figure 3.7 in the text book.

ADDITIONAL EXAMPLE:
Which of the following compounds contains the element gold?

a) $AgCl$ b) $HgCl_2$ c) GaN d) $Gd_2(SO_4)_3$ e) none of these

ANSWER: e

P.O. 5.2

You must be able to read a chemical formula.

EXAMPLE:
How many atoms of oxygen are represented in each of the following chemical formulas?

$$Ca(C_2H_3O_2)_2 \qquad C_6H_4(COOH)_2$$

SOLUTION: Each formula represents 4 oxygen atoms.

Textbook Reference: Section 5.3

ADDITIONAL EXAMPLE:
How many atoms of hydrogen are represented by the following chemical formula?

$$C_6H_8(OH)_6$$

ANSWER: 14

136

P.O. 5.3

You must be able to calculate a formula mass (molecular mass) given the formula of a compound.

EXAMPLE:
Calculate the formula mass for each of the following compounds:

$$Ca(C_2H_3O_2)_2 \qquad C_6H_4(COOH)_2$$

Atomic Masses: Ca = 40.1   C = 12.0   H = 1.0   O = 16.0 amu

SOLUTION: 158 and 166 amu respectively

Textbook Reference: Section 5.3

ADDITIONAL EXAMPLE:
The mineral topaz has the formula $Al_2SiO_4F_2$. What is the "molecular mass" of topaz?
Atomic Masses: Al = 27.0   Si = 28.1   O = 16.0   F = 19.0 amu

ANSWER: 184.amu

P.O. 5.4

You must be able to read and then balance a chemical equation.

EXAMPLE:
Balance the following chemical equations:

$$K + H_2O \rightarrow KOH + H_2$$
$$C + O_2 \rightarrow CO$$

SOLUTION:

$$2\,K + 2\,H_2O \rightarrow 2\,KOH + H_2$$
$$2\,C + O_2 \rightarrow 2\,CO$$

Textbook Reference: Section 5.4

ADDITIONAL EXAMPLE A:
Balance the following equation:

$$CH_4 + O_2 \rightarrow CO_2 + H_2O$$

ANSWER: $CH_4 + 2\,O_2 \rightarrow CO_2 + 2\,H_2O$

ADDITIONAL EXAMPLE B:
When hydrogen peroxide, H_2O_2, is poured onto an open wound, it decomposes into water and oxygen gas, O_2. Write a balanced chemical equation for the decomposition of hydrogen peroxide. After the equation is balanced, how many molecules of water appear in the equation?

ANSWER: 2 molecules of water

P.O. 5.5

You must be able to calculate the atomic mass of an element given its percent composition in chemical combination with an element of known atomic mass. This problem type also demands that the formula of the compound be known. (The logical methods that nineteenth century chemists used to deduce simple chemical formulas are discussed in Chapter 7.)

EXAMPLE:
An unknown element X forms a chemical compound with oxygen that has the formula XO_2. Quantitative analysis of this compound indicates that it is 27.3% X. If the atomic mass of oxygen is known to be 16.0 amu, calculate the atomic mass of the unknown element.

SOLUTION: 12.0 amu

Textbook Reference: Section 5.6

ADDITIONAL EXAMPLE:
An unknown element, X, forms a compound with the element oxygen with the formula XO. The compound has a definite composition of 36.0% X and 64.0% oxygen. If the atomic mass of oxygen is 16.0 amu, what is the atomic mass of X?

ANSWER: 9.00 amu

P.O. 5.6

You must be able to calculate the number of moles of an element or compound given its mass.

EXAMPLE:
How many moles of calcium are contained in a 125 gram sample of calcium? The atomic weight of calcium is 40.1

SOLUTION: 3.12 moles of calcium

Textbook Reference: 5.7

ADDITIONAL EXAMPLE:
How many moles of water (H_2O) are contained in a 26.3 gram sample of water?
Atomic Masses: H = 1.01   O = 16.0 amu

ANSWER: 1.46 moles of water

Note that you must first calculate the formula mass of water (18.0 amu).

P.O. 5.7

You must be able to compare numbers of atoms or molecules given masses of different elements or compounds.

EXAMPLE:
Which of the following contains the greater number of atoms?

(i) 130 grams of silver (Ag)   (ii) 100 grams copper (Cu)

Atomic Masses: Ag = 108   Cu = 63.5 amu

SOLUTION: (ii)

130 grams of Ag equals 1.20 moles of Ag while 100 grams of Cu equals 1.57 moles of Cu. Since a mole is a standard pile, the greater number of moles contains the greater number of atoms.

ADDITIONAL EXAMPLE:

Which of the following contains the greater number of molecules?

        (i) 25.0 grams of water (H_2O)   (ii) 45.0 grams of sodium oxide (Na_2O)

Atomic Masses: H = 1.01   O = 16.0   Na = 23.0

SOLUTION: (i)

First calculate the formula mass of each molecule: H_2O = 18.0, Na_2O = 62.0. Then calculate the number of moles corresponding to each mass: 25.0 grams of water equals 1.39 moles and 45.0 grams of Na_2O equals 0.726 moles of Na_2O.

P.O. 5.8

You must be able to calculate the empirical formula of a chemical compound given its constant composition.

EXAMPLE:

What is the empirical formula of a chemical compound that has the following constant composition:

$$C = 13.6\% \qquad F = 86.4\%$$

Atomic Masses: C = 12.0   F = 19.0 amu

SOLUTION: CF_4

Textbook Reference: Section 5.7

ADDITIONAL EXAMPLE:

What is the empirical formula of a chemical compound that has the following constant composition:

$$C = 80.0\% \qquad H = 20.0\%$$

Atomic Masses: C = 12.0   H = 1.0 amu

ANSWER: CH_3

P.O. 5.9

You must be able to calculate the % composition of a chemical compound given the formula of the compound.

EXAMPLE:

Calculate the percentage of aluminum in the chemical compound aluminum oxide, Al_2O_3.
Atomic Masses: Al = 27.0   O = 16.0 amu

SOLUTION: 52.9% Al

Textbook Reference: Section 5.8

ADDITIONAL EXAMPLE:
Calculate the percentage of carbon in the compound formaldehyde, CH_2O.
Atomic Masses: C = 12.0   H = 1.0   O = 16.0 amu

ANSWER: 40.0% C

P.O. 5.10

You must be able to solve a gravimetric ratio problem that involves a chemical formula. This is a generalization of the previous problem type.

EXAMPLE:
Calculate the weight of aluminum required to produce 20.0 tons of aluminum sulfate, $Al_2(SO_4)_3$.
Atomic Masses: Al = 27.0   S = 32.0   O = 16.0 amu

SOLUTION: 3.16 tons Al

Textbook Reference: Section 5.8

ADDITIONAL EXAMPLE:
Calculate the weight of hydrogen that could in theory be obtained by the decomposition of 20.0 tons of ethane, C_2H_6. Atomic Masses: C = 12.0   H = 1.0 amu

ANSWER: 4.00 tons of hydrogen

P.O. 5.11

You must be able to solve a gravimetric ratio problem that involves a chemical equation.

EXAMPLE:
Calculate the mass of oxygen that would be produced by the decomposition of 25.0 grams potassium chlorate, $KClO_3$, according to the following chemical equation:

$$2 \, KClO_3 \rightarrow 2 \, KCl + 3 \, O_2$$

Atomic Masses: K = 39.1   Cl = 35.5   O = 16.0 amu

SOLUTION: 9.79 g of O_2

Textbook Reference: Section 5.8

ADDITIONAL EXAMPLE:
Copper carbonate decomposes according to the following equation:

$$CuCO_3 \rightarrow CuO + CO_2$$

What mass of copper oxide, CuO, can be produced by the decomposition of 85.0 grams of copper carbonate, $CuCO_3$?

        Atomic Masses: Cu = 63.6   C = 12.0   O = 16.0 amu

ANSWER: 54.7 grams of CuO

P.O. 5.12

You must be able to calculate the percent yield of a chemical reaction given the mass of the product isolated, the mass of the starting material, and the balanced chemical equation.

EXAMPLE:
The following balanced chemical equation represents the chemical reaction that takes place during the synthesis of acetylene (C_2H_2) from calcium carbide (CaC_2) and water:

$$CaC_2 + 2\ H_2O \rightarrow Ca(OH)_2 + C_2H_2$$

If the reaction of 166 grams of calcium carbide with an excess of the other reactant results in the isolation of 58.0 grams of acetylene, what is the percent yield of the reaction based on the calcium carbide consumed?

        Atomic Masses: Ca = 40.1   C = 12.0   H = 1.00   O = 16.0

SOLUTION: 86.1%

Textbook Reference: Section 5.12

ADDITIONAL EXAMPLE:
The following balanced chemical equation represents the chemical reaction that takes place during the synthesis of chloroform ($CHCl_3$) from methane (CH_4) and chlorine (Cl_2):

$$CH_4 + 3\ Cl_2 \rightarrow 3\ HCl + CHCl_3$$

If the reaction of 40.6 grams chlorine with an excess of the other reactant results in the isolation of 15.6 grams of chloroform, what is the percent yield of the reaction based on the chlorine consumed?

        Atomic Masses: H = 1.01   C = 12.0   Cl = 35.5 amu

ANSWER: 68.5%

Chapter Five
Problems[8]

Elements and Symbols

1. Which of the following compounds contains the element silver?

    a) SiO_2 b) $AgNO_3$ c) H_2S d) NaBr   e) none of these

    STUDENT SOLUTION:

2. Which of the following compounds contains the element lead?

    a) LiCl   b) P_2O_5 c) PbI_2 d) $HgCl_2$ e) none of these

    STUDENT SOLUTION:

3. Which of the following compounds contains the element Na?

    a) nitrogen dioxide   b) nickel chloride   c) silver chloride
    d) sodium chloride   e) none of these

    STUDENT SOLUTION:

4. Which of the following compounds contains the element Fe?

    a) iron oxide   b) phosphorus pentoxide   c) sodium fluoride
    d) lead iodide   e) none of these

    STUDENT SOLUTION:

---

8. In solving these problems, do not allow rounding off to become a major problem. Use the atomic masses of the elements as they are recorded in Figure 3.7. If you do round off an answer, attempt to adhere to the guideline of Chapter 2 (footnote 6), but don't worry if your answers are not rounded off correctly. Remember, the most important thing is to understand the chemical logic of a particular problem.

Chemical Formulas

5. The chemical compound glycerine is a by-product of soap manufacturing. Its formula is represented below. How many atoms of each element are contained in a molecule of glycerine?

$$\text{Glycerine: } C_3H_5(OH)_3$$

STUDENT SOLUTION:

6. How many atoms of each element are there in one molecule of $C_6H_3(NO_2)(COOH)_2$?

STUDENT SOLUTION:

7. How many atoms of hydrogen are represented by each of the following formulas?

a) $C_6H_4(COOCH_3)_2$ b) $CH_3(CH_2)_3OH$
c) $CH_3CH(NH_2)CH_3$ d) $C_6H_3(CH_3)_3$
e) $CH_3CH_2CH_2COOH$

STUDENT SOLUTION:

Molecular Mass

8. What is the molecular mass of glycerine?

$$\text{Glycerine: } C_3H_5(OH)_3$$

STUDENT SOLUTION:

9. What is the molecular mass of $C_6H_3(NO_2)(COOH)_2$?

STUDENT SOLUTION:

10. Sodium carbonate, Na_2CO_3 is sold commercially as washing soda. What is the formula mass (molecular mass) of this compound?

    STUDENT SOLUTION:

11. Calculate the formula mass (molecular mass) of each of the following compounds:

    a) C_3H_6O b) $CaCO_3$ c) $Al_2(CO_3)_3$ d) $C_6H_4(CH_3)_2$ e) $CH_3(CH_2)_2CH_3$

    f) $Mg_3(PO_4)_2$ g) $(NH_4)_2CO_3$ h) $C_6H_3(CH_3)_2(COOH)_2$

    STUDENT SOLUTION:

Reading and Balancing Chemical Equations

12. Complete the following analogy:

    FORMULA:COMPOUND = _____:CHEMICAL REACTION

    STUDENT SOLUTION:

13. When hydrogen peroxide, H_2O_2 is poured onto an open wound, it decomposes into water and oxygen gas, O_2. Write a balanced equation for this reaction.

    STUDENT SOLUTION:

14. The element potassium reacts violently with water to produce potassium hydroxide, KOH, and the elemental substance hydrogen. Write a balanced chemical equation for this reaction.

    STUDENT SOLUTION:

15. Balance the following chemical equation:

$$C_2H_6N + O_2 \rightarrow CO_2 + H_2O + NO_2$$

STUDENT SOLUTION:

16. Balance the following chemical equation:

$$C_5H_{10} + O_2 \rightarrow CO_2 + H_2O$$

When this equation is balanced with integer coefficients, what is the total number of water molecules represented on the right hand side of the equation?

STUDENT SOLUTION:

17. At elevated temperatures, the compound zinc sulfide (ZnS) reacts with oxygen to produce zinc oxide (ZnO) and sulfur dioxide (SO_2). Write a balanced chemical equation for this reaction. In writing this equation, remember to represent the oxygen reactant as a diatomic molecule. When this equation is balanced with integer coefficients, what is the total number of zinc sulfide molecules represented on the left hand side of the equation?

STUDENT SOLUTION:

18. Write a balanced equation for the reaction of copper with oxygen to form copper oxide (CuO). In writing this equation, remember to represent the oxygen reactant as a diatomic molecule. When this equation is balanced with integer coefficients, what is the total number of copper atoms represented on the left hand side of the equation?

STUDENT SOLUTION:

Atomic Masses from Percentage Composition

19. An unknown element, X, forms a compound with oxygen with the formula XO_2. The compound has a definite composition of 50.0% oxygen. If the atomic mass of oxygen is 16.0 amu, what is the atomic mass of X?

STUDENT SOLUTION:

20. An unknown element, X, forms a compound with oxygen. The formula of this compound is X_2O_3. The atomic mass of oxygen is known to be 16.0 amu. What is the atomic mass of X if a quantitative analysis of X_2O_3 indicates 47.1% oxygen?

    STUDENT SOLUTION:

21. An unknown metallic element, X, forms a compound with chlorine with the formula XCl_3. One (1.000) gram of this compound contains 0.343 grams of chlorine. If the atomic mass of chlorine is known to be 35.5, calculate the atomic mass of this metallic element.

    STUDENT SOLUTION:

22. An unknown element, X, forms a compound with the element nitrogen. The formula of this compound is X_3N_2. The compound has a definite composition of 72.2% X. If the atomic mass of nitrogen is known to be 14.0 amu, what is the atomic mass of X in amu?

    STUDENT SOLUTION:

23. An unknown element, X, forms a compound with the element oxygen. The formula of this compound is X_2O_3. The compound has a definite composition of 52.9% X. If the atomic mass of oxygen is known to be 16.0 amu, what is the atomic mass of X in amu?

    STUDENT SOLUTION:

Mole Calculations

24. Calculate the number of moles contained in each of the following masses.

    a) 247 grams of carbon   b) 125 grams of sulfur   c) 349 grams of copper

    STUDENT SOLUTION:

25. Calculate the number of moles contained in each of the following masses.

   a. 123 grams of methane (CH_4)
   b. 67.8 grams of calcium carbonate $(CaCO_3)$
   c. 78.9 grams of sodium chloride $(NaCl)$
   d. 15.0 grams of aspirin $(C_9H_8O_4)$

   STUDENT SOLUTION:

26. Determine which substance in each of the following pairs contains the greater number of atoms or molecules.

   a. (i) 57.0 g of aluminum   (ii) 85.0 g of zinc
   b. (i) 100 g of tin   (ii) 100 g of gold
   c. (i) 67.8 g of ammonia gas (NH_3) (ii) 75.6 g of carbon dioxide (CO_2)
   d. (i) 200 g of sulfuric acid (H_2SO_4) (ii) 100 g of carbonic acid (H_2CO_3)

   STUDENT SOLUTION:

27. A binary compound of the elements calcium and chlorine has the formula $CaCl_2$. If 24.1 grams of calcium is reacted with 76.3 grams of chlorine to form this compound, which element is present in stoichiometric excess? (*i.e.* Which element will have some mass left unreacted?) Atomic Masses: Ca = 40.1   Cl = 35.5 amu

   STUDENT SOLUTION:

Empirical Formulas of Chemical Compounds

28. A certain chemical compound gives the following analysis: C = 17.4% and F = 82.6%. What is the empirical formula of this compound?

   STUDENT SOLUTION:

29. Calculate the empirical formula of each of the following compounds:

   a. C = 75.0%   H = 25.0%
   b. S = 40.0%   O = 60.0%
   c. Na = 74.2%   O = 25.8%

   STUDENT SOLUTION:

30. Calculate the empirical formula of the following compounds:

   Compound A: C = 85.6%   H = 14.4%
   Compound B: C = 15.8%   S = 84.2%

   STUDENT SOLUTION:

31. Qualitative analysis indicates that a certain chemical compound contains only the elements nitrogen and oxygen. A quantitative analysis of this compound gives the following analysis: N = 63.6%. What is the empirical formula of this compound?

   STUDENT SOLUTION:

32. Nitrogen and oxygen form a number of different compounds. The quantitative analysis of one of these compounds gives the following analysis: N = 36.8%. What is the empirical formula of this compound?

   STUDENT SOLUTION:

33. Copper metal is obtained from an ore called chalcopyrite, this mineral contains 34.6% Cu, 30.4% Fe, and 35.0% S. Calculate the empirical formula for chalcopyrite.

   STUDENT SOLUTION:

34. The production of iron from iron ore in a blast furnace leads to a byproduct call "slag" which has a composition of 34.5% Ca, 24.2% Si, and 41.3% O. Calculate the empirical formula for slag.

   STUDENT SOLUTION:

35. Many crystalline chemical compounds exist as hydrates. For example, the blue crystals of the compound copper sulfate have the formula: $CuSO_4 \bullet 5H_2O$. This formula indicates that the $CuSO_4$ portion of the compound has five molecules of water chemically bound to it. For historical reasons, chemists retain this rather unorthodox method of writing the chemical formula of a hydrate. If modern notation is used, the formula of the blue copper sulfate hydrate would be written as $CuSO_4(H_2O)_5$. When crystalline hydrates are heated, they often decompose to release water.

The household chemical Epsom salts is a hydrate of the compound magnesium sulfate. The chemical reaction for the dehydration of Epsom salts can be represented by the following incomplete chemical equation:

$$MgSO_4 \bullet XH_2O \rightarrow MgSO_4 + X\,H_2O$$

In this equation, X represents the number of molecules of water that are chemically bound to $MgSO_4$. If the decomposition of 26.0 grams of hydrated magnesium sulfate ($MgSO_4 \bullet XH_2O$) results in the formation of 12.7 grams of anhydrous magnesium sulfate ($MgSO_4$), calculate the value of X required to complete the empirical formula of hydrated magnesium sulfate.

STUDENT SOLUTION:

Percent Composition from Formula

36. Calculate the percentage of carbon in each of the compounds that contain carbon listed in question 11.

STUDENT SOLUTION:

37. Calculate the percentage of aluminum in the compound aluminum sulfide, Al_2S_3.

STUDENT SOLUTION:

38. Calculate the percentage of carbon in the compound ethane, C_2H_6.

STUDENT SOLUTION:

39. The chemical compound anthranilic acid has the formula $C_6H_4(COOH)(NH_2)$. Calculate the percentage of carbon in this chemical compound.

    STUDENT SOLUTION:

40. The chemical compound phthalic acid has the formula $C_6H_4(COOH)_2$. Calculate the percentage of oxygen in this chemical compound.

    STUDENT SOLUTION:

Gravimetric Ratios Involving Chemical Formulas

41. Calculate the mass of calcium that could in theory be obtained from 20.0 metric tons of calcium carbonate, $CaCO_3$.

    STUDENT SOLUTION:

42. In theory, what mass of carbon would be required to produce 25.0 metric tons of each of the compounds listed in question 11 that contain carbon?

    STUDENT SOLUTION:

43. In theory, what mass of carbon would be required to produce 20.0 tons of natural gas, CH_4?

    STUDENT SOLUTION:

44. The chemical compound cinnamic acid has the formula $C_6H_5CHCHCOOH$. In theory, chemists should be able to synthesize this compound from its elements. What mass of this compound in grams could be produced from 125 g of carbon and an excess of the other elements?

    STUDENT SOLUTION:

45. The element aluminum is produced by the decomposition of the chemical compound aluminum oxide, Al_2O_3. How many grams of aluminum can be produced by the decomposition of 185 grams of aluminum oxide?

   STUDENT SOLUTION:

46. Sulfuric acid is manufactured from elemental sulfur. Refer to Figure 5.13. In which year did the United States use approximately 23.5 billion pounds of sulfur in the manufacture of sulfuric acid? Assume that all of the sulfuric acid represented in Figure 5.13 was manufactured directly from sulfur. NOTE: The GNP increased annually from 1970 to 1980.

   STUDENT SOLUTION:

Gravimetric Ratios Involving Equations

47. Sodium nitrate, $NaNO_3$, decomposes according to the following equation:

$$2\ NaNO_3 \rightarrow 2\ NaNO_2 + O_2$$

   What mass of oxygen gas, O_2, can be produced by the decomposition of 125 grams of sodium nitrate?

   STUDENT SOLUTION:

48. What mass of carbon dioxide, CO_2, can be obtained from 1.00 kilogram of blood sugar, $C_6H_{12}O_6$? Assume that all of the carbon in blood sugar is converted to CO_2. HINT: This problem involves a gravimetric ratio of the type 5.D, but because of the above assumption, a complete equation is not required.

   STUDENT SOLUTION:

49. Calculate the mass of sodium, Na, that can be produced by the decomposition of 50 grams of table salt, NaCl. The chemical equation for this decomposition is: $2\ NaCl \rightarrow Cl_2 + 2\ Na$.

   STUDENT SOLUTION:

50. Potassium nitrate, KNO_3, decomposes according to the following equation:

$$2 \, KNO_3 \rightarrow 2 \, KNO_2 + O_2$$

a. What is the percentage of oxygen in the compound KNO_3?

b. What mass of oxygen gas, O_2, can be produced by the decomposition of 404 grams of KNO_3 according to the above equation? HINT: The gravimetric ratio for part a is of the type 5.D. The gravimetric ratio for part b is of the type 5.E.

STUDENT SOLUTION:

51. If 125 grams of potassium reacts according to the reaction described in question 14, how many grams of hydrogen will be produced?

STUDENT SOLUTION:

52. Consider the following balanced chemical equation:

$$2 \, Al + 6 \, HCl \rightarrow 2 \, AlCl_3 + 3 \, H_2$$

According to this equation, how many grams of hydrogen (H_2) are produced by the complete reaction of 22.5 grams of aluminum (Al)?

STUDENT SOLUTION:

53. Consider the following balanced chemical equation:

$$4 \, NH_3 + 7 \, O_2 \rightarrow 3 \, H_2O + NO_2$$

According to this equation, if 125 grams of ammonia (NH_3) are completely reacted, (a) how many grams of oxygen (O_2) are consumed and (b) how many grams of nitrogen dioxide (NO_2) are produced?

STUDENT SOLUTION:

54. Consider the following balanced chemical equation:

$$2 \, C_2H_6 + 7 \, O_2 \rightarrow 4 \, CO_2 + 6 \, H_2O$$

According to this equation, if 35.0 grams of ethane (C_2H_6) are completely reacted, (a) how many grams of oxygen (O_2) are consumed and (b) how many grams of carbon dioxide (CO_2) are produced?

STUDENT SOLUTION:

Problems on Percent Yield

55. The following balanced chemical equation represents the chemical reaction that takes place during the synthesis of ethyl chloride (C_2H_5Cl) from ethane (C_2H_6) and chlorine:

$$C_2H_6 + Cl_2 \rightarrow HCl + C_2H_5Cl$$

If the reaction of 57.0 grams of chlorine with an excess of the other reactant results in the isolation of 44.1 grams of ethyl chloride, what is the percent yield of the reaction based on chlorine consumed?
Atomic Masses: C = 12.0   H = 1.00   Cl = 35.5 amu

STUDENT SOLUTION:

56. The following balanced chemical equation represents the chemical reaction that takes place during the synthesis of oxygen difluoride (OF_2) from fluorine (F_2) and sodium hydroxide:

$$2 \, F_2 + 2 \, NaOH \rightarrow 2 \, NaF + H_2O + OF_2$$

If the reaction of 78.2 grams of sodium hydroxide with an excess of the other reactant results in the isolation of 39.7 grams of oxygen difluoride, what is the percent yield of the reaction based on sodium hydroxide consumed?
Atomic Masses: F = 19.0   Na = 23.0   O = 16.0   H = 1.00 amu

STUDENT SOLUTION:

57. The following balanced chemical equation represents the chemical reaction that takes place during the synthesis of dichlorobenzene ($C_6H_4Cl_2$) from benzene (C_6H_6) and chlorine:

$$C_6H_6 + 2 \, Cl_2 \rightarrow 2 \, HCl + C_6H_4Cl_2$$

If the reaction of 156 grams of chlorine with an excess of the other reactant results in the isolation of 123 grams of dichlorobenzene, what is the percent yield of the reaction based on chlorine consumed?

Atomic Masses: C = 12.0   H = 1.00   Cl = 35.5 amu

STUDENT SOLUTION:

58. The following balanced chemical equation represents the chemical reaction that takes place during the synthesis of boron carbide (B_4C) from carbon (C) and boron oxide:

$$7\ C + 2\ B_2O_3 \rightarrow 6\ CO + B_4C$$

If the reaction of 110 grams of B_2O_3 with an excess of the other reactant results in the isolation of 29.1 grams of boron carbide, what is the percent yield of the reaction based on B_2O_3 consumed?

Atomic Masses: C = 12.0   B = 10.8   O = 16.0 amu

STUDENT SOLUTION:

59. The following balanced chemical equation represents the chemical reaction that takes place during the synthesis of calcium carbide (CaC_2) from carbon (C) and calcium oxide:

$$3\ C + CaO \rightarrow CO + CaC_2$$

If the reaction of 44.7 grams of CaO with an excess of the other reactant results in the isolation of 39.5 grams of calcium carbide, what is the percent yield of the reaction based on CaO consumed?

Atomic Masses: C = 12.0   Ca = 40.1   O = 16.0 amu

STUDENT SOLUTION:

Problems Involving Household Chemistry and Science

60. A common technique in analytical chemistry is to use a chemical reaction of a substance to quantitatively analyze that substance. This experiment explores the application of this technique to the analysis of vitamin C. The experiment requires two different vitamin C-containing juices, antiseptic tincture of iodine, a vitamin C tablet, and a can of spray starch.

Crush a vitamin C tablet and dissolve it in ½ cup of water. The binder that holds the tablet together will not dissolve, but this will not affect the experiment. Place a drop of iodine solution on a 1x1 inch piece of paper towel and spray the spot very briefly with starch. Starch forms a dark blue complex with iodine that can be used to detect very small concentrations of iodine.

Drop the starch/iodine complex-stained paper into the vitamin C solution. The fading of the blue color is due to the following chemical reaction between vitamin C ($C_6H_8O_6$) and iodine (I_2):

$$C_6H_8O_6 + I_2 \rightarrow C_6H_6O_6 + 2\ HI$$

This chemical reaction can be used in the quantitative analysis of vitamin C.

Add 1 oz (30 mL) of orange juice to a glass and spray the juice with starch. Add the iodine solution to the orange juice dropwise with stirring until the blue color remains. The blue color indicates that all of the vitamin C has reacted. Count the number of drops of iodine required to produce this color change. Repeat the analysis with 30 mL of another vitamin C-containing juice. This drop analysis allows a comparison of the relative amounts of vitamin C in the two juices. How could this procedure be modified so that stoichiometric calculations could be used to determine the number of milligrams of vitamin C in a juice sample?

STUDENT SOLUTION:

61. Almost all biological chemical reaction rates are regulated by catalysts called enzymes. The activity of salivary amylase, a digestive enzyme found in saliva, is particularly easy to demonstrate. This enzyme regulates the initial breakdown of starch during the chewing process, and its activity can be demonstrated using the starch/iodine complex illustrated in the previous problem.

Add ¼ teaspoon of spray starch to ½ cup cold water. Dilute this starch solution by adding two tablespoons of the starch solution to one cup of cold water. Pour ¼ cup of diluted starch solution into each of two small glasses. Add two drops of tincture of iodine to each glass. The solutions should turn blue due to the formation of the starch/iodine complex. Add one teaspoon of saliva to one of the glasses containing the starch/iodine complex. Stir this mixture and note the time required for the disappearance of the blue color. Note that the color in the control glass does not fade. Repeat the experiment using warm water. How does temperature affect the rate of the reaction?

STUDENT SOLUTION:

Library Problems

62. The article "Estimation of Avogadro's Number" (King, L. Carroll; Neilsen, E. K. *J. Chem. Educ.* **1958**, 35, 198) describes an experimental procedure for approximating Avogadro's number. Read this article and construct an analogy showing how similar measurements on a pile of real marbles would allow for an estimation of the number of marbles in a pound.

STUDENT SOLUTION:

63. The concept of balancing a chemical equation was introduced in Chapter 5. Since the balancing of a chemical equation must precede any equation stoichiometric calculation, you should not be surprised that the *Journal of Chemical Education* is constantly publishing articles on the subject of equation balancing. While many of these articles are beyond the scope of this text, portions of the following two articles might be of interest to beginning chemistry students: "How Should Equation Balancing be Taught?" (Porter, Spencer K. *J. Chem. Educ.* **1985**, 62, 507) and "Balancing Complex Redox Equations by Inspection (CP)" (Kolb, Doris. *J. Chem. Educ.* **1981**, 58, 642). The method of balancing chemical equations presented in Chapter 5 involves a set of very general instructions. Read the articles by Porter and Kolb. Select one of the specific balancing procedures described in these articles and apply the procedure to the balancing of the following simple equation:

$$H_2S + O_2 \rightarrow H_2O + SO_2$$

STUDENT SOLUTION:

A Special Library Problem
Dedicated to Fifty Percent of the Students Reading This Text

64. You may have noticed that the historical approach taken in this text has focused on the men of chemistry. The reason for this is that during the revolutionary and evolutionary period from 1770 to 1900, chemistry was, indeed, a male dominated profession. Ironically, in ancient times, women played a key role in the chemical profession. As the twentieth century progressed, women slowly began to reenter the profession. At first, progress was slow, but during the early twentieth century, stars such as Marie Curie, Lise Meitner, Irene Joliot-Curie, and Ida Noddack began to shine their light on the chemical profession. Today, chemistry is an exciting profession that is beginning to attract young men and young women in equal numbers. Read the article "Chemistry's Creative Women" (Roscher, Nina Matheny. *J. Chem. Educ.* **1987**, 64, 748). After reading this article, briefly describe the purpose of the American Chemical Society's Garvin Medal? In what year was the medal first awarded? Has a Garvin medalist ever won a Nobel Prize? Has a Garvin medalist ever been elected President of the American Chemical Society?

STUDENT SOLUTION:

Formal Laboratory Exercise

Using Stoichiometry to "See" Molecules
A Quantitative Chemical Analysis

Introduction

Section 5.8 (Example XI) illustrates the use of stoichiometry to calculate the quantity of one of the substances involved in a chemical reaction. The calculation is based on the balanced chemical equation for the reaction being studied, and the example illustrates a common type of stoichiometric prediction. The ability to make material predictions based on balanced chemical equations is fundamental to both process and research chemistry. In addition to allowing predictions based on balanced chemical equations, stoichiometry is also a useful tool to help establish the atomic/molecular identity of products of a chemical reaction. To this end, the monitoring of the "mass balance" of the reactants and products of a chemical reaction that is under investigation is fundamental to the study of chemistry.

As an example, consider the thermal decomposition of the complex copper compound $Cu_2CH_2O_5$. During the heating of this compound, qualitative observations indicate that the compound decomposes into a solid and water. Based on qualitative observation, the equation for this decomposition can therefore be represented as follows:

$$Cu_2CH_2O_5 \rightarrow \text{Solid Product} + H_2O + \text{Gaseous Product (?)}$$

After writing this partial equation, an experienced chemist can begin making educated guesses as to the molecular identity of the solid product. For example, based on the elements in the original copper compound, the solid product is probably CuO or Cu_2CO_4. While this may not be obvious to a beginning chemistry student, experience dictates that the solid product is almost certainly one of these two compounds. Likewise, experience dictates that the complete equation for the decomposition reaction is most likely one of the following two equations:

$$Cu_2CH_2O_5 \rightarrow 2\ CuO + H_2O + CO_2$$

$$Cu_2CH_2O_5 \rightarrow Cu_2CO_4 + H_2O$$

At this point, mass balance and stoichiometry can point to the correct equation. The purpose of formal laboratory exercise is to monitor the mass balance of this chemical reaction and to use the results of this mass balance to identify the molecular identity of the solid and gaseous products.

Procedural Overview

The decomposition of the complex copper compound ($Cu_2CH_2O_5$) is carried out by heating a sample of the compound in a porcelain crucible supported by a ring stand over a Bunsen burner. This apparatus is shown in Figure 5.14. Proper flame adjustment and positioning are essential for the success of this experiment. The use of the Bunsen burner will be discussed by the laboratory instructor at the beginning of the laboratory period, but two safety issues associated with this apparatus should be emphasized in this procedural overview. First, a properly adjusted burner flame can be hard to see. Exercise due care when working with an active burner. Second, heated porcelain cools very slowly. Crucible tongs must be used for the manipulation of the heated crucible and crucible lid. These safety issues will also be discussed by the laboratory instructor.

This experiment involves several crucible heatings followed by cooling and mass determination. The cooling process can be facilitated

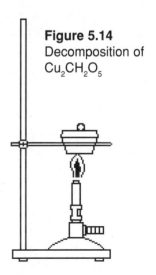

Figure 5.14
Decomposition of
$Cu_2CH_2O_5$

by the following procedure. Turn off the burner and allow the crucible to cool on the ring stand apparatus. After about 5 minutes, use crucible tongs to move the crucible and lid from the ring stand apparatus to a wire gauze. Allow the crucible and lid to cool for an additional 5 minutes on the wire gauze.

The use of the electronic balance was discussed in the formal laboratory exercise associated with chapter two. In the present experiment, all mass determinations are performed by weighing the crucible and its contents. The original mass of complex copper compound is measured by a technique that chemists find quite useful in quantitative experiments. According to this procedure, the mass of the empty crucible is first determined, and then an estimated quantity of complex copper compound is added to the crucible. The crucible and complex copper compound are weighed after this estimated addition, and the mass of the complex copper compound is determined by difference. All of the mass determinations for this experiment are described in detail in the experimental procedure below.

Materials

porcelain crucible and lid, crucible tongs, clay triangle, wire gauze, ring stand and ring, Bunsen burner, laboratory balance

Chemicals

$Cu_2CH_2O_5$

Procedure

1. Determine the mass of an empty, clean and dry crucible and lid. Record this mass in the data table.

2. Measure approximately 2 grams of $Cu_2CH_2O_5$ into a small dry beaker. The reagent table will have a 2 gram sample on display to help you make this estimate.

3. Add the estimated 2 grams of $Cu_2CH_2O_5$ to the crucible and determine the mass of the crucible, lid, and $Cu_2CH_2O_5$ Record the mass of the crucible, lid, and complex copper compound ($Cu_2CH_2O_5$) in the data table.

4. Place the crucible and contents and crucible lid on the clay triangle support as shown in Figure 5.14.

5. Use the Bunsen burner to heat the crucible, lid, and $Cu_2CH_2O_5$ at full heat for 10 minutes.

6. After the 10 minute heating, turn off the burner and allow the crucible to cool for 10 minutes as described in the procedural notes.

7. After the crucible has cooled, determine the mass of the crucible, lid, and solid product. Use the crucible tongs to manipulate the crucible and lid. Record this mass in the data table.

8. Ask the laboratory instructor to check the crucible contents and the data table entries. Based on this check, the instructor may request that the experiment be repeated

9. After the instructor checks the crucible contents, the solid product should be place in the disposal bag at the reagent table. The crucible should be washed well with water.

10. Use the information in the data table to answer the questions in the conclusion and discussion section.

Data Table

mass of empty crucible & lid grams measured	
mass of crucible, lid, & complex copper compound grams measured	
mass of crucible, lid, & solid product grams measured	
mass of complex copper compound reacted grams calculated	
mass of solid product formed grams calculated	

Conclusion and Discussion

1. The following chemical equation describes the chemical reaction that takes place if the complex copper compound, $Cu_2CH_2O_5$, decomposes to form copper oxide, CuO:

    $$Cu_2CH_2O_5 \rightarrow 2\ CuO + H_2O + CO_2$$

    Use the calculation technique illustrated in section 5.8 (Example XI) to calculate the mass of solid product, CuO, that would have been formed from the mass of complex copper compound, $Cu_2CH_2O_5$, in your experiment.

2. The following chemical equation describes the chemical reaction that takes place if the complex copper compound, $Cu_2CH_2O_5$, decomposes to form copper carbonate, Cu_2CO_4:

    $$Cu_2CH_2O_5 \rightarrow Cu_2CO_4 + H_2O$$

    Use the calculation technique illustrated in section 5.8 (Example XI) to calculate the mass of solid product, Cu_2CO_4, that would have been formed from the mass of complex copper compound, $Cu_2CH_2O_5$, in your experiment.

3. According to your calculations above and your experimental results, which of the following equations describes the reaction that took place during the decomposition of the complex copper compound ($Cu_2CH_2O_5$)? Your ability to answer this question demonstrates the use of stoichiometry to "see" atoms and molecules.

    (1) $Cu_2CH_2O_5 \rightarrow 2\ CuO + H_2O + CO_2$

    (2) $Cu_2CH_2O_5 \rightarrow Cu_2CO_4 + H_2O$

Chapter Five Notes

Chapter Six

The Periodic Table or You Too Can Speak Chemistry

6.1 Introduction

By 1794, the French chemist Antoine Lavoisier had accumulated a list of thirty-two chemical elements. Although several of Lavoisier's elements were identified in error, the list of elements continued to grow. As the list of elements grew during the nineteenth century, so also did the feeling among chemists that nature was filled with great confusion. For each new element brought a new set of chemical and physical properties to be dealt with. Was there no natural order to all of this?

There was order indeed, but it could not be perceived until the atomic/molecular theory of John Dalton reached maturity during the mid-nineteenth century. The order, once discovered, was a thing of beauty. The purpose of this chapter is to investigate the natural order of the chemical elements.

6.2 Early Attempts to Organize the Chemical Elements

In the early part of the nineteenth century, the atoms of the elements only possessed one important property that was related to factual information. This property was the unique atomic mass of an element's atom. As the list of chemical elements grew, many chemists looked for a trend or pattern that could be used to relate the chemical and physical properties of the elements to one another. An important aspect of many of the early attempts to find such a trend was the ordering of the elements according to atomic mass. Since trends were easier to memorize than isolated facts, it was hoped that the organization of the elements according to one characteristic would automatically organize the elements according to all characteristics. A blind faith in the inherent orderliness of nature has always been an aspect of the scientific mentality. Hence a search for an elemental trend was entirely in character.

One particularly interesting attempt to organize the elements according to atomic mass was carried out by the English chemist John Alexander Reina Newlands in 1864. Newlands, an amateur musician, assigned each element an ordinal number that represented the mass ranking of its atom. For example, hydrogen, the lightest elements with an atomic mass of 1.01 amu, was assigned the ordinal number one. The element with the next-to-lightest mass, lithium with an atomic mass of 6.94 amu, was assigned the ordinal number 2.[1] This process of assigning ordinal numbers to the various elements according to the mass of their atoms was continued throughout the entire list that existed in 1864. Eventually, the ordinal number mass ranking of a given element became known as the element's **atomic number** (*e.g.* atomic number hydrogen = 1). When Newlands arranged the chemical elements in order of increasing atomic number, an interesting fact caught his musician's eye. This ordering produced a list of elements where the chemical and physical properties of each element differed from the preceding element. However, at element number eight, the trend started to repeat, with element number eight having characteristics similar to element number one, and element number nine having characteristics similar to element number two. To Newlands, this was reminiscent of the musical

---

1. The second lightest element is, in fact, helium, but Newlands was not aware of this fact, since helium was not discovered until 1868. If Newlands had been aware of the existence of the element helium, it actually might have hindered his important discovery.

octave, and it was a rather romantic concept. With respect to the chemical elements, could nature be "singing" a kind of diatonic scale—do, re, me, fa, sol, la, ti, do—H, Li, Be, B, C, N, O, F? The variation and repetition of elemental chemical characteristics suggested to Newlands that the answer to this question was yes. Newlands enunciated his "law of octaves," which was immediately rejected by the scientific establishment.

This was not a case of pigheadedness on the part of the scientific establishment, for Newlands made several scientific errors. First, he overplayed the musical analogy a bit. The scientific establishment has always been conservative (as it must be), and Newlands won no allies by playing up the romance. Second, and more importantly, Newlands became too enamored with his musical analogy. Nature does, indeed, "sing" in octaves, at least for the first twenty-one elements known to Newlands. But then she begins to "sing" a different "tune." Newlands missed the order that exists beyond element twenty-one because he was obsessed with finding his own order. Nature has romance to spare; she needs no assistance from the artificial contrivances of humans. Still, it must be said that Newlands' contribution was significant, and, indeed, history has given him credit where credit is due.

6.3 Additional Atomic Characteristics: Volume and Valence

In order to explain the law of constant composition, John Dalton established the atomic theory of chemical action. The only atomic characteristic required by this theory was the concept of atomic mass. It soon became apparent to chemists that other atomic characteristics could be related to factual observation. For example, a system of relative atomic volumes could be established by measuring the volume of one mole of the solid elements. The calculation of the volume of one mole of an element requires only the density and the atomic mass of the element. This calculation is illustrated in Example I.

EXAMPLE I

Calculation of Atomic Volume

The density of silver is 10.5 g/mL and the density of copper is 8.96. g/mL. Calculate the volume of one mole of each of these elements. Atomic Masses: Cu=63.5 Ag=108 amu.

SOLUTION: Cu = 7.09 mL/mole and Ag = 10.3 mL/mole

$$\text{Density (D)} = \frac{\text{Mass of a sample of matter (M)}}{\text{Volume of a sample of matter (V)}}$$

after algebraic transposition

$$V = \frac{M}{D}$$

For one mole of copper: $\quad V = \dfrac{63.5 \text{ g/mole}}{8.96 \text{ g/mL}} \qquad$ 7.09 mL/mole

For one mole of silver: $\quad V = \dfrac{108 \text{ g/mole}}{10.5 \text{ g/mL}} \qquad$ 10.3 mL/mole

---

Since one mole of each of the elements contains the same number of atoms, the larger molar volume of silver as calculated in Example I can be attributed to the larger volume of individual silver atoms. If it is assumed that atoms are spheres and that these spheres are very closely packed in the solid state, then the relative molar volumes will be a measure of relative atomic volumes.

In 1865, the German chemist Loschmidt used the kinetic molecular theory (Chapter 7) to calculate molecular volumes based on the factual properties of gases. These calculations, which are beyond the scope of this text, ultimately allowed for the estimation of the actual atomic volume of the atoms of various elements. Atomic volume, as a characteristic of atoms related to factual information, was actually preceded by another atomic characteristic. This characteristic, called valence, was related to chemical information. Although both new atomic characteristics played an important role in revealing nature's natural elemental order, valence ultimately proved to be the more useful concept. For this reason, it is appropriate to consider the concept of valence in some detail.

A consistent set of atomic masses allows the chemist to calculate a chemical compound's empirical formula. This calculation, which is illustrated in Chapter 5, makes use of the chemical compound's constant elemental composition. A close look at the empirical formulas of a large group of chemical compounds reveals an interesting pattern. In order to observe this pattern, consider the empirical formulas of a selective group of chemical compounds involving the elements hydrogen (H), sodium (Na), calcium (Ca), aluminum (Al), oxygen (O), and chlorine (Cl). The formulas for these compounds are shown in Figure 6.1. Remember that these formulas are theoretical statements computed from the factual law of constant composition.

$$HCl \qquad\qquad Na_2O$$
$$H_2O \qquad\qquad CaCl_2$$
$$NaCl \qquad\qquad CaO$$
$$AlCl_3$$

Figure 6.1 Formula Listing

At first glance, the pattern inherent in these formulas is not obvious. If we are willing, however, to make three assumptions, then a pattern does begin to emerge:

ASSUMPTION I: An atom of each element is capable of forming a characteristic number of chemical bonds. The number of chemical bonds that an atom of a given element forms is called the element's valence.

ASSUMPTION II: When atoms combine chemically, they tend to use all of their available chemical bonds. Thus an atom is said to saturate its valence.

ASSUMPTION III: The element with the least massive atom, hydrogen, has a valence of one. (*i.e.* A hydrogen atom forms one chemical bond.)

Assumption I defines a new atomic characteristic called valence. According to this definition, an element's **valence** is an integer that describes the number of chemical bonds that its atom can form. Now consider the above list of formulas in light of these three assumptions.

First, since the valence of hydrogen is one (assumption), the formulas HCl and H_2O can be used to deduce the valences of chlorine and oxygen, respectively. Since one chlorine atom bonds to one hydrogen atom (HCl), saturation of valence (assumption) demands that the valence of chlorine be one also. In a similar manner, the ability of an oxygen atom to bond with two hydrogen atoms (H_2O) must be an indication that oxygen is forming two bonds (*i.e.* has a valence of two). On the assumption that the valence is an inherent characteristic of an element's atom, the above logic can now be applied to the formula NaCl. Since the valence of chlorine is one (deduction), the formula NaCl demands that the

Element	Valence
H	1
Cl	1
O	2
Na	1
Ca	2
Al	3

Figure 6.2 Valence Listing

valence of sodium be one also. Notice that the next formula in the list, Na_2O, indicates that a pattern is, indeed, beginning to emerge. This is exactly the formula that this compound must have if the deduced valences and the valence assumptions are correct. If a sodium atom can form one bond (valence one) and an oxygen atom can form two bonds (valence two), then two atoms of sodium would be required to saturate oxygen's valence—hence the formula must be Na_2O. The assumptions lead to an internally consistent pattern; they are, therefore, justifiable. At this point, observe that the remainder of the formulas in the above list can be used to deduce the valences of calcium and aluminum. This valence information is summarized in Figure 6.2.

Since the valence of an element represents the number of chemical bonds that an atom of the element forms during chemical combination, knowledge of valence allows the prediction of the empirical formulas of **binary compounds** (two element compounds). This process was illustrated above for the formula of a binary compound of sodium and oxygen (Na_2O). It is convenient, at this point, to illustrate an algorithm that can be used to predict the formula of a binary compound using the valences of the compound's elements (Fig. 6.3).

In writing the formula of a binary compound, the subscript written for a given element is the valence of the other element. Arrows are used to represent this operation in the above algorithm. Hence, the formula is written Na_2O.

Figure 6.3 Formula Writing Algorithm

If the application of this algorithm results in a formula where the subscripts have a highest common factor, then both of the subscripts should be divided by this factor (*i.e.* reduced to lowest terms). For example, the application of this algorithm to the empirical formula of the compound of calcium is shown in Figure 6.4.

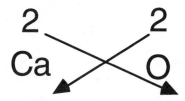

Initially, the algorithm yields the formula Ca_2O_2, but this formula is reduced to CaO. The empirical formula of the binary compound thus reflects the smallest whole number ratio of atoms.

Figure 6.4 Formula Writing Algorithm with Highest Common Factor

---

EXAMPLE II

The Formula Writing Algorithm—Writing a Formula

Use the valences of aluminum and oxygen listed above to predict the formulas of the binary compound of these two elements.

SOLUTION: Al_2O_3

Since the valence of aluminum is three and the valence of oxygen is two, we can write:

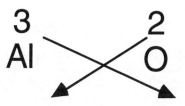

Hence, the formula is written Al_2O_3. Note that the algorithm simply assures that the valence of each element is saturated. Two aluminum atoms form a total of six bonds (2x3), and three oxygen atoms also form a total of six bonds (3x2). A total of six bonds from aluminum to oxygen represents an exact saturation of valence.

☐

Since the valences are derived from empirical formulas, the use of valence to "predict" an empirical formula may seem to be an empty gesture. This is not the case, however, and later in this chapter it should become evident that the skill illustrated in Example II represents a powerful communication device.

The valence concept as presented above was introduced by the English chemist Edward Frankland in 1852. Frankland originally used the term "combining power" to express the number of bonds that an atom can form. He later changed this to the term "atomicity." The modern term valence was suggested by the nineteenth century chemist Friedrich August Kekule. It was derived from the Latin word meaning strength. It should also be noted that Edward Frankland realized that in several cases elements could exhibit more than one valence. For example, Frankland pointed out that the element iron manifested a valence of two in some compounds and valence of three in other compounds. In this textbook, the valence that an element manifests in its most common oxide(s) will be called the element's primary valence(s).

EXAMPLE III

The Formula Writing Algorithm—Determining a Valence

The element iron forms two binary compounds with the element oxygen. The formulas of these compounds are FeO and Fe_2O_3. Assuming that the valence of oxygen is two in both of these compounds, what is the valence of iron in each compound?

SOLUTION: FeO: $Fe = 2$ and Fe_2O_3: $Fe = 3$

By reversing the logic of Example II, we have:

$$FeO \quad Fe \text{ valence} = 2$$
$$Fe_2O_3 \quad Fe \text{ valence} = 3$$

By the year 1865, the chemist's atom, regarded as a theoretical entity, possessed three theoretical characteristics that could be related to factual information:

Atomic Characteristics

Theoretical Characteristic	Factual Information
1. Atomic Mass	Constant Composition
2. Relative Volume	Molar Volume of Solids
Atomic Volume	Physical Properties of Gasses
3. Valence	Constant Composition

In a subsequent section, we will see how these atomic characteristics helped reveal the natural order of the chemical elements.

6.4 The Periodic Table of the Chemical Elements

The discovery of the natural order of the chemical elements was made almost simultaneously by several nineteenth century chemists. The general adoption of this discovery by the scientific community followed most directly from the work of the German chemist, Lothar Meyer, and the Russian chemist, Dmitrii Ivanovich Mendeleev. Although both chemists published periodic tables at about the same time (1869–1870), Mendeleev took the bolder approach in announcing his discovery, with his reward being a lion's share of the credit for the discovery. Basically, both men took the same approach, and that was to extend the method of Newlands. As a guide to whether or not atomic mass ordering resulted in chemical characteristic ordering, Mendeleev monitored, among other things, valence, and Meyer monitored atomic volume and other physical properties.[2] As a result of these approaches, both men saw beyond the octaves of Newlands. Perhaps the easiest way to illustrate the discovery of both men is by means of Meyer's atomic volume correlation. A modern version of Meyer's atomic volume correlation is shown in Figure 6.5.

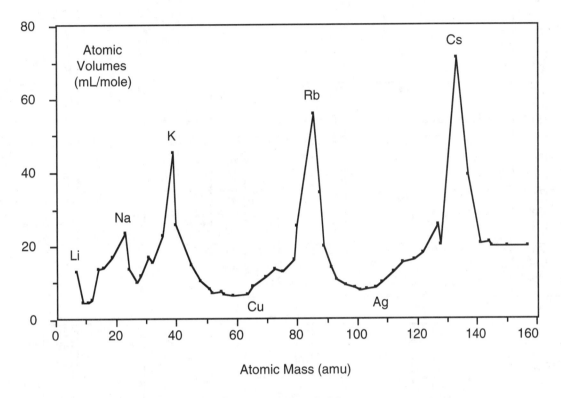

Figure 6.5   Atomic Volume as a Function of Atomic Mass

This graph represents the fluctuation of atomic volume that can be observed when the chemical elements are arranged in order of increasing atomic mass. It is obvious from the graph that the property of atomic volume varies in a regular way. Such regular variation is true of all elemental chemical and physical properties, and this fact is the basis of the modern periodic table of the chemical elements. This table is represented in Figure 6.6. Consider now the logic of its formulation.

In order to understand the logic behind the periodic table, it is instructive to emphasize a point that Newlands missed. In the graph that relates atomic volume to atomic mass (Fig. 6.5), five elements are listed as markers. These elements (Li, Na, K, Rb, Cs) mark the positions in the elemental order where atomic volume maxima occur. If these elements are listed with Newlands' atomic numbers, the error in the law of octaves can be illustrated. For reasons that will become clear shortly, the atomic numbers listed below are *not correct* by modern standards (Fig. 6.7).

---

2.   The first actual atomic volumes were calculated after 1865. Lothar Meyer did not use actual atomic volumes for his study of chemical periodicity; he did the next best thing. He assumed that the volume of one mole of an element as a solid was proportional to the actual atomic volume. After reading Chapter 7, you may want to return to this point and consider the logic of this assumption.

The Periodic Table

IA	IIA	IIIB	IVB	VB	VIB	VIIB	VIIIB			IB	IIB	IIIA	IVA	VA	VIA	VIIA	VIIIA
1 H 1.01																1 H 1.01	2 He 4.00
3 Li 6.94	4 Be 9.01											5 B 10.8	6 C 12.0	7 N 14.0	8 O 16.0	9 F 19.0	10 Ne 20.2
11 Na 23.0	12 Mg 24.3											13 Al 27.0	14 Si 28.1	15 P 31.0	16 S 32.1	17 Cl 35.5	18 Ar 40.0
19 K 39.1	20 Ca 40.1	21 Sc 45.0	22 Ti 47.9	23 V 50.9	24 Cr 52.0	25 Mn 54.9	26 Fe 55.9	27 Co 58.9	28 Ni 58.7	29 Cu 63.6	30 Zn 65.4	31 Ga 69.7	32 Ge 72.6	33 As 74.9	34 Se 79.0	35 Br 79.9	36 Kr 83.8
37 Rb 85.5	38 Sr 87.6	39 Y 88.9	40 Zr 91.2	41 Nb 92.9	42 Mo 95.9	43 Tc (98)	44 Ru 101	45 Rh 103	46 Pd 106	47 Ag 108	48 Cd 112	49 In 115	50 Sn 119	51 Sb 122	52 Te 128	53 I 127	54 Xe 131
55 Cs 133	56 Ba 137	57 La* 139	72 Hf 179	73 Ta 181	74 W 184	75 Re 186	76 Os 190	77 Ir 192	78 Pt 195	79 Au 197	80 Hg 201	81 Tl 204	82 Pb 207	83 Bi 209	84 Po 209	85 At 210	86 Rn 222
87 Fr (223)	88 Ra 226	89 Ac† (227)	104 Rf (261)	105 Db (262)	106 Sg (263)	107 Bh (262)	108 Hs (265)	109 Mt (266)	110 Uun	111 Uuu	112 Uub						

*	58 Ce 140	59 Pr 141	60 Nd 144	61 Pm (145)	62 Sm 150	63 Eu 152	64 Gd 157	65 Tb 159	66 Dy 163	67 Ho 165	68 Er 167	69 Tm 169	70 Yb 173	71 Lu 175
†	90 Th 232	91 Pa (231)	92 U 238	93 Np (237)	94 Pu (244)	95 Am (243)	96 Cm (247)	97 Bk (247)	98 Cf (251)	99 Es (252)	100 Fm (257)	101 Md (258)	102 No (259)	103 Lr (260)

Figure 6.6

Li 2 } 7 (one octave)
Na 9 } 7 (one octave)
K 16 } 7 (one octave)
Rb 30 } 14 (two octaves)
Cs 44 } 14 (two octaves)

Figure 6.7 Newlands' Atomic Numbers for Li Na K Rb Cs

It is important to understand that each element in this list *belongs* in the list. Not only does each element hold a common position in the atomic volume graph, but each element has a common chemistry. For example, each element reacts vigorously with water according to the following chemical equation:

Chemical Equation 6.A

$$2 \text{ M} + 2 \text{ H}_2\text{O} \rightarrow 2 \text{ MOH} + \text{H}_2$$

Any of the symbols (Li, Na, K, Rb, Cs) can replace M in the above equation, and the equation is still valid. Even the formula subscripts will be correct. Further, where characteristic differences do occur (*e.g.* atomic volume), they at least follow a pattern. Newlands was aware of all of this, but he was so enamored with his musical octave theory that he did two things in an attempt to make nature conform to his sense of order. Hence he missed the greater *"truth."* First, in order to make rubidium and cesium at least work out to two octave jumps, he assigned single atomic numbers to several element pairs. For example, atomic number 22 was assigned to both cobalt and nickel. If Newlands had not played this number game, then the atomic numbers of this list of elements would have worked out as follows. Again, for reasons that will become clear shortly, the atomic numbers listed below are still *not correct* by modern standards (Fig. 6.8):

Li 2 } 7 (one octave)
Na 9 } 7 (one octave)
K 16 } 15 (two octaves + 1 note)
Rb 31 }
Cs 48 } 17 (two octaves + 3 notes)

Figure 6.8 Newlands' Atomic Numbers without Multiple Numbering

Clearly, the jumps of 15 and 17 would be a real problem to the theory of octaves. Newlands, therefore, double assigned atomic numbers to make rubidium work out to number 30 and cesium work out to number 44. The second thing that Newlands did was to force the elements copper (Newlands' number 23) and silver (Newlands' number 37) into the list in Figure 6.7. This neatly completed all of the octaves, but chemically copper and silver should have been placed in a separate element grouping. Note the positions of copper and silver that are marked on the atomic volume graph (Fig. 6.5). Newlands played this number game with all of the chemical elements. Hence his elemental groupings ended up being somewhat contrived.

In 1869, Dmitrii Ivanovich Mendeleev published an article in the *Journal of the Russian Chemical Society* in which he suggested an approach similar to Newlands, with one important difference. Mendeleev did not use the concept of atomic number. To be sure, he did arrange the elements in order of increasing atomic mass; he just didn't bother to assign each element an ordinal ranking number.[3] Instead, Mendeleev tabulated the chemical elements, allowing the factual properties of the elements to dictate the elemental groupings. In his table, Mendeleev listed the elements in order of increasing atomic mass. The elements were listed in a horizontal row, but each time an element showed chemical and physical properties that represented the repetition of a cycle

---

3. Mendeleev probably felt that the ordinal numbers had no physical meaning, and they were, therefore, redundant. In the twentieth century, the concept of atomic number was revived because the ordinal numbers do have physical meaning. The periodic table used in this text (Fig. 6.6) anticipates this fact, and therefore, the modern atomic numbers are listed for each element.

(*e.g.* Li, Na, K, Rb, Cs), a new row was started. In this manner, vertical columns recorded elements of common characteristics. As his table developed, Mendeleev allowed chemistry to dictate an element's position. Since he did not use the ordinal number ranking, he could not be unduly influenced by it. This was no trivial point, for it allowed Mendeleev to make three bold predictions that all others had failed to make:

1. Occasionally, an element's characteristics necessitated leaving positions in the tabular array of elements empty. Mendeleev predicted that eventually elements would be discovered to fill the spaces.[4]
2. Occasionally, an element's characteristics necessitated placing it in a position out of atomic mass order. Mendeleev predicted that closer study would reveal that the atomic mass of such elements had been calculated incorrectly.
3. Occasionally, an element placed in the tabular array had a single characteristic that seemed out of place. Mendeleev predicted that closer study would reveal that the single characteristic of such elements had been measured in error.

There is a Yiddish word used to describe the ultimate in brazenness. The word is *chutzpah*, and *chutzpah* is too mild a word to describe Mendeleev's predictions! In the conservative world of natural science, you don't go around saying "I'm right, and everyone else is wrong" and expect to be taken seriously. Mendeleev was not taken seriously—at first.

We are dealing here with one of the greatest discoveries in the field of chemistry, not just in the nineteenth century, but one of the greatest discoveries ever! Given the importance of this discovery, a word about scientific gamesmanship is in order. Mendeleev had a revolutionary insight which he boldly stated. In doing this, he was both courageous and correct. The scientific establishment regarded Mendeleev as a revolutionary kook. Although this view did not require much courage, it was correct. There is no contradiction here; it is just a way of stating that science places a high price on revolution and revolutionaries. It must, for there is precious little time for pursuing false revolution. At least in theory, most scientists appreciate this point, and the adversary system of doing science is not viewed as being closed-minded or cruel. It is simply essential.

If Mendeleev was bold, courageous, and correct, he was also lucky. He made his predictions in 1871, and by 1875, the first prediction was fulfilled. In that year, the French chemist, Paul Emile Lecoq de Boisbaudran, discovered a new element that fit perfectly into the empty position below the element aluminum. Mendeleev had called the missing element by the name "eka-aluminum," inferring one below aluminum (eka: one in Sanskrit). Lecoq called the new element gallium, the modern name for this element.[5] Soon all of Mendeleev's predictions were being fulfilled, although he was a bit off the mark with prediction number two. Some of the atomic masses had, indeed, been calculated incorrectly, but occasionally, a slightly more massive element had to be placed before a less massive one (*e.g.* Co and Ni). With the fulfillment of these predictions came support from the scientific community. The tabular array of elements, the periodic table, was recognized as a masterpiece of organization.

Figure 6.6 represents a modern version of the periodic table. It contains the ninety naturally occurring elements as well as all of the synthetically made elements (Fig. 3.7) discovered prior to 1976. Although Mendeleev did not use ordinal numbers in his table, the use of atomic numbers for the

---

4. The failure to leave gaps for undiscovered elements is one reason that Newlands' atomic numbers did not correlate to the modern atomic numbers. The other reason for this lack of atomic number correlation was the fact that helium, the first member of an entire element grouping, was not discovered until 1868.
5. In 1879, Lecoq discovered another element and broke an old tradition in chemistry by naming the element after a person. The element was named samarium after an obscure Russian mining official named Samarski. The mineral from which Lecoq isolated this element was named after Samarski, and Lecoq simply named the element after the mineral. In so doing, of course, he unknowingly named the element after Samarski. Actually, Lecoq may have knowingly broken this tradition when he named gallium in 1875. Ostensibly, the name "gallium" was selected to honor that region of Europe south and west of the Rhine, Gaul. The New Latin word for this region, *Gallia*, also referred to one of the symbols of the region, the rooster—in French, Lecoq! Paul Emile Lecoq may well have named an element after himself—that's *chutzpah*.

elements was reinstated about 1913 (see Chapter 9). When the atomic number concept was reinstated, the ordinal numbers were assigned on the basis of the periodic table and not on the basis of atomic mass ranking. Hence for elements like cobalt and nickel, atomic number order is not the same as atomic mass order. Using modern atomic numbers for our marker elements, we can see that nature does, indeed, have an elemental rhythm, but she is not "singing" a diatonic scale (Fig. 6.9).

In the next section we will consider the organization of the modern periodic table.

Li	3	} 8
Na	11	} 8
K	19	} 8
Rb	37	} 18
Cs	55	} 18

Figure 6.9 Modern Atomic Numbers

6.5 The Organization of the Modern Periodic Table

As indicated in the last section, the periodic table is a tabular arrangement of the chemical elements according to factual and theoretical characteristics. To the chemist, the table is a beautiful example of the order that exists in nature. It is also a very practical tool. The table makes manageable an immense volume of factual and theoretical minutiae. The purpose of this section is to become familiar with the organization of the modern periodic table (Fig. 6.10).

In becoming familiar with the periodic table, the first thing to note is that the table lists the chemical elements in horizontal rows called **periods** and vertical columns called **families**.

There are seven periods (Arabic numerals) of varying lengths. The first period contains two elements, the second and third contain eight elements, the fourth and fifth contain eighteen elements, and the sixth contains thirty-two elements. The fourteen elements following the asterisk in the above schematic of the periodic table are really a part of period six. They should be listed following the element lanthanum (La) in the sixth period, but because of space limitations, they are listed by footnote. The seventh period also has footnoted elements. These elements should be listed following the element actinium (Ac) in period seven. The seventh period is an incomplete period containing six natural and twenty synthetic elements. Present theory seems to indicate that the seventh period is capable of containing thirty-two elements, thus the elemental rhythm of nature follows the following completed pattern: 2, 8, 8, 18, 18, 32, 32(?). The listing of the chemical elements according to seven periods of these varying lengths automatically arranges the elements in vertical columns called families. As the name implies, all of the elements in a chemical family have similar chemical and physical properties. We will return to this point in the next section. The designation of the various families is a subject of much debate among chemists. In this text, each family is designated with a Roman numeral and a Latin letter (*e.g.* family IA). Over the years, some families have been given actual names. For example, family IA is also called the alkali metal family. These names are of historical interest to chemists, and they will only be used incidentally in this text.

Just as the notes of the diatonic scale change in pitch in a regular way, so also the elements in a given period show a regular change in chemical and physical properties. This regular variation of chemical and physical properties tends to divide the table into regions of elements, with each region containing elements holding some common trait. For example, the chemical elements can be divided into two extreme chemical groups called **metals** and **nonmetals**.[6] In the schematic of the periodic table (above), the metallic elements fall to the left of the step-like heavy line, and the nonmetallic elements fall to the right. Notice that there are many more metallic elements than there are nonmetallic elements. Another kind of grouping that exists on the periodic table is indicated by the family designations *A* and *B*. The *A* elements are called the representative elements. This name is meant to imply that these elements show the regular variation in chemical and physical properties— from metallic to nonmetallic—that is characteristic of the *periodic* nature of the periodic table. The *B* elements are called the **transition** elements. They do not show as great of a regular variation

---

6. The terms metal and nonmetal can be precisely defined. For the purpose of this discussion, however, an intuitive grasp of these terms will suffice.

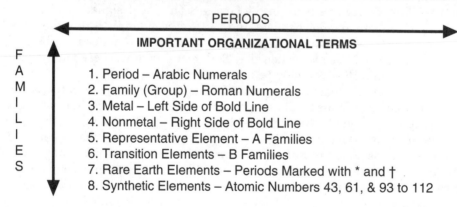

IA	IIA	IIIB	IVB	VB	VIB	VIIB	VIIIB			IB	IIB	IIIA	IVA	VA	VIA	VIIA	VIIIA
1 H 1.01																1 H 1.01	2 He 4.00
3 Li 6.94	4 Be 9.01											5 B 10.8	6 C 12.0	7 N 14.0	8 O 16.0	9 F 19.0	10 Ne 20.2
11 Na 23.0	12 Mg 24.3											13 Al 27.0	14 Si 28.1	15 P 31.0	16 S 32.1	17 Cl 35.5	18 Ar 40.0
19 K 39.1	20 Ca 40.1	21 Sc 45.0	22 Ti 47.9	23 V 50.9	24 Cr 52.0	25 Mn 54.9	26 Fe 55.9	27 Co 58.9	28 Ni 58.7	29 Cu 63.6	30 Zn 65.4	31 Ga 69.7	32 Ge 72.6	33 As 74.9	34 Se 79.0	35 Br 79.9	36 Kr 83.8
37 Rb 85.5	38 Sr 87.6	39 Y 88.9	40 Zr 91.2	41 Nb 92.9	42 Mo 95.9	43 Tc (98)	44 Ru 101	45 Rh 103	46 Pd 106	47 Ag 108	48 Cd 112	49 In 115	50 Sn 119	51 Sb 122	52 Te 128	53 I 127	54 Xe 131
55 Cs 133	56 Ba 137	57 La * 139	72 Hf 179	73 Ta 181	74 W 184	75 Re 186	76 Os 190	77 Ir 192	78 Pt 195	79 Au 197	80 Hg 201	81 Tl 204	82 Pb 207	83 Bi 209	84 Po 209	85 At 210	86 Rn 222
87 Fr (223)	88 Ra 226	89 Ac † (227)	104 Rf (261)	105 Db (262)	106 Sg (263)	107 Bh (262)	108 Hs (265)	109 Mt (266)	110 Uun	111 Uuu	112 Uub						

*	58 Ce 140	59 Pr 141	60 Nd 144	61 Pm (145)	62 Sm 150	63 Eu 152	64 Gd 157	65 Tb 159	66 Dy 163	67 Ho 165	68 Er 167	69 Tm 169	70 Yb 173	71 Lu 175
†	90 Th 232	91 Pa (231)	92 U 238	93 Np (237)	94 Pu (244)	95 Am (243)	96 Cm (247)	97 Bk (247)	98 Cf (251)	99 Es (252)	100 Fm (257)	101 Md (258)	102 No (259)	103 Lr (260)

PERIODS

FAMILIES

IMPORTANT ORGANIZATIONAL TERMS

1. Period – Arabic Numerals
2. Family (Group) – Roman Numerals
3. Metal – Left Side of Bold Line
4. Nonmetal – Right Side of Bold Line
5. Representative Element – A Families
6. Transition Elements – B Families
7. Rare Earth Elements – Periods Marked with * and †
8. Synthetic Elements – Atomic Numbers 43, 61, & 93 to 112

Figure 6.10 The Organization of the Modern Periodic Table

throughout a given period as in the case of the *A* elements. They are, in fact, all metallic in nature. This decrease of regular variation of chemical and physical properties is carried to an extreme in the case of the footnoted elements (asterisk and ampersand). These elements, called the **rare earth** elements, show only a small variation in chemical and physical properties. All of the rare earth elements are metals.

Perhaps we can get a better feel for the various regions of the periodic table by pushing the analogy of Newlands a bit. If we consider only the representative *A* elements, then nature "sings" a low note and a high note to get our attention (period 1), and then she "sings" an eight note scale over and over again. If we add to this the transition *B* elements, then the scale is interrupted after the second note by a variable "hum" gradually increasing in "pitch." Finally, the addition of the rare earth elements results in the variable "hum" being preceded by a monotonous "hum." This may all seem a bit farfetched, but the analogy did play a role in the history of the development of the periodic table, and it may be a useful tool for understanding the organization of the periodic table.

6.6 Chemical Families Exemplified

One of the most important things that the periodic table does for the chemist is to render manageable a huge amount of factual information. In order to see how this works, consider family IA (the alkali metal family) and family VIIA (the halogen family).

The group IA elements (Li, Na, K, Rb, Cs, Fr) derive their family name from the Arabic word for ash.[7] The name alkali refers to the fact that plant ashes are rich in sodium (Na) and potassium (K). This is not a terribly important fact to remember, but it does illustrate the point that, historically, chemical family names are related to chemical fact. Each of the elements in this family share common physical and chemical characteristics. For example, all are soft metals of low density. They all possess low melting points, decreasing in a regular manner from Li (181°C) to Cs (29°C). This latter fact illustrates another important aspect of chemical families—when the physical and chemical properties of the elements within a family vary, they vary in a regular way (Fig. 6.11).

Li	181	°C
Na	98	°C
K	64	°C
Rb	39	°C
Cs	29	°C

Figure 6.11 Melting Points of the Alkali Metals

Loosely defined, the melting point is the temperature where a solid begins to form a liquid. It is an example of a physical property, but the same type of regular family behavior is to be found in the study of chemical properties. By way of example, recall the following family chemical reaction of the alkali metals:

Chemical Equation 6.A

$$2 M + 2 H_2O \rightarrow 2 MOH + H_2$$

Another chemical property that the alkali metals hold in common is a valence of one.

The group VIIA elements (F, Cl, Br, I, At) derive their family name from Swedish words meaning "salt former." Group VIIA is referred to as the halogen family. Again, this name refers to an aspect of family chemistry that will be discussed below. Actually, the halogens are nonmetals that resemble one another more closely than do members of any other family in the periodic table. Once again, where properties vary, they vary in a regular way (Fig. 6.12):

---

7. Although most periodic tables include hydrogen in group IA, hydrogen is truly a unique element. Since hydrogen exhibits properties associated with both groups IA and VIIA, it is listed in this textbook as the first member of both of these groups.

F	-220	°C	(yellow gas at room temperature)
Cl	-110	°C	(green gas at room temperature)
Br	-7	°C	(red liquid at room temperature)
I	114	°C	(black solid at room temperature)

Figure 6.12 Melting Points of the Halogens

The primary valence of the elements in the halogen family is one, and all of these elements occur as diatomic molecules—F_2, Cl_2, Br_2, I_2 (see Chapter 5).[8] A characteristic chemical reaction of the halogen family is the reaction of its elemental members with the elements of group IA. In the reaction shown below, M indicates the symbol of any alkali metal and X indicates the symbol of any halogen:

Chemical Equation 6.B
$$2\ M + X_2 \rightarrow 2\ MX \text{ (Note the Diatomic Formula, } X_2)$$

Since compounds of the general formula MX are called *salts* (*e.g.* NaCl or table salt), the group VIIA elements are appropriately called *salt formers*.

The purpose of the above discussion was to illustrate the principle of family properties. It is important for you to understand this concept, but with the exception of the primary valences of the representative elements (*A* families), it is not necessary to memorize specific family characteristics. The valences of the representative and transition elements are discussed in the next section.

EXAMPLE IV

Variation of Density within a Periodic Table Family

The element germanium (Ge) is a member of family IVA on the periodic table. The element was discovered in 1886 by Clemens Winkler, but its existence was predicted by Mendeleev in 1871. Mendeleev also predicted the density of germanium using his periodic table and the known densities of other family IVA members. Use the periodic table and the known densities of silicon and tin to predict the density of germanium. The density of silicon is 2.62 g/mL, and the density of tin is 7.30 g/mL.

SOLUTION: 4.78 g/mL

Like the other physical characteristics discussed in this section, density varies in a regular way within a periodic table family. Since germanium is between silicon and tin in family IVA, the density of germanium can be approximated as the average of the densities of silicon and tin. The actual density of germanium is 4.96 g/mL.

6.7 The Valences of the Representative and Transition Elements

Each representative family possesses a primary valence as defined in footnote seven. The common primary valences of the representative families form a simple pattern that should be committed to memory (Fig. 6.13):

---

8. The halogens, like many families on the periodic table, exhibit multiple valence. In this text, the term "primary valence" refers to the common oxide valence(s) defined in the first part of this chapter.

Family	Primary Valence
IA	1
IIA	2
IIIA	3
IVA	4
VA	3
VIA	2
VIIA	1
VIIIA	0

Figure 6.13 The Common Primary
Valences of the Representative Elements

The transition elements B also possess primary family valences. However, multiple primary valence is much more prevalent among these elements. For this reason, it is not necessary to commit transition family primary valences to memory. It will be useful, however, to become familiar with the following specific transition element primary valences (Fig. 6.14):

Element	Primary Valence
Fe	2 and 3
Cu	1 and 2
Ag	1
Zn	2
Hg	1 and 2

Figure 6.14 Useful Valences of Transition Elements

An interesting chemical characteristic of transition metals is the tendency to form chemical compounds in which the metals exhibit two types of valence simultaneously. The chemical compound $[Co(NH_3)_6]Cl_2$ illustrates this characteristic. In this compound, the element cobalt is extending two valences to two chlorine atoms and six valences to six ammonia (NH_3) molecules. Since a study of this compound reveals that chlorine and ammonia are bound differently, it is useful to write the formula of the compound so as to emphasize this difference. The formula $[Co(NH_3)_6]Cl_2$ reflects that the cobalt atom is bound to the two atoms of chlorine by relatively strong primary valences, and the cobalt atom is bound to the six molecules of ammonia by six weaker secondary valences. The differences in the relative strengths of these valences is illustrated by the ease of the chemical reaction represented by the following equation:

Chemical Equation 6.C

$$[Co(NH_3)_6]Cl_2 \rightarrow CoCl_2 + 6\,NH_3$$

---

9.   A valence of zero indicates that these elements do not form chemical compounds—*i.e.* they do not form chemical bonds. This is not strictly true, since several of these elements do, indeed, form chemical compounds, albeit reluctantly. Since these elements, which are minor components of the atmosphere, do not react to form chemical compounds under natural conditions, the family primary valence of zero is warranted.

Although the valence of nitrogen in ammonia (NH_3) is three, the valence of each nitrogen atom in $[Co(NH_3)_6]Cl_2$ has expanded to four.

In 1890, the Swiss chemist Alfred Werner developed the theory of primary and secondary valences to explain the behavior of compounds such as $[Co(NH_3)_6]Cl_2$. Werner called these secondary valences coordination valences, and transition metals that exhibit this type of bonding are referred to as **coordination compounds**. The atoms or compounds that bond to transition metals by secondary valences are called **ligands**. The number of ligands bonded to the transition metal is called the **coordination number**.

The development of coordination theory represented a significant development in the chemical atomic theory. Werner received the fourteenth Nobel Prize in chemistry for his work on coordination chemistry. Coordination compounds and the theoretical nature of primary and secondary valences are discussed more completely in Chapters 8 and 9 of this textbook.

6.8 Chemical Nomenclature—Binary Compounds

Among other things, the periodic table is a powerful memory aid (mnemonic). In this section, we will see how the periodic table allows the chemist to "remember" the correct empirical formula for a large number of binary chemical compounds.

Recall that a binary compound is a compound that contains two elements. One of these elements will usually be a metal and the other will be a nonmetal; however, two nonmetals can combine to form a binary compound. We will start with binary compounds consisting of a metal and a nonmetal since these represent the simplest of cases. As indicated earlier in this chapter, all that is required to write the correct empirical formula of a binary compound is the valence of each element in the compound. Since the periodic table organizes, among many other facts, elemental valences, the necessary information can be "read" from the table. Consider the following example:

EXAMPLE V

Formula Writing—Binary Compound A

Write the empirical formula of the compound that forms when the elements aluminum and sulfur react chemically.

SOLUTION: Al_2S_3

Since aluminum is in family IIIA of the periodic table, we can "read" its valence as 3. In a similar manner, since sulfur is a member of family VIA, we can "read" its valence as 2. Now using the algorithm illustrated in Example II, we can write the correct formula of this compound. In writing the formula, the metallic element is usually written first:

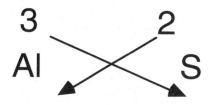

Hence, the formula is written Al_2S_3.

For the purpose of communication, it is convenient to have a system for naming chemical compounds. This system of names is referred to as a nomenclature system. In naming a binary compound, the following rules are used (Al_2S_3 exemplifies rules):

The Nomenclature of Binary Compounds Consisting of a Metal and Nonmetal

1. Write the name of the first element indicated in the formula.

    aluminum

2. Following this, write the name of the second element indicated by the formula, but drop the ending letters.

    aluminum sulf

    NOTE: This rule can be a little tricky for beginners, since it is not always obvious which letters should be dropped. The following list illustrates this principle, and it may be helpful to the beginner:

carbon	*carb*	*nitrogen*	*nitr*
chlorine	*chlor*	*oxygen*	*ox*
iodine	*iod*	*phosphorus*	*phosph*

3. Add the ending "ide" to the word stem formed in rule two.

    aluminum sulfide

In chemical nomenclature, the "ide" ending marks the compound in question as a binary compound.

EXAMPLE VI

Compound Naming—Binary Compound A

Name the following compounds: NaI, Mg_3N_2, CaO.

SOLUTION:
Using the rules above, we arrive at the following names: sodium iodide, magnesium nitride, calcium oxide.

In cases where elements possess two or more common valences, the valence in question is indicated parenthetically. This principle is illustrated below:

FeO	iron (II) oxide
	(II) indicates iron valence = 2
Fe_2O_3	iron (III) oxide
	(III) indicates iron valence = 3

At this point, it is important to note that the principles outlined above allow the empirical formula of a binary compound to be written directly from the name of the binary compound.

EXAMPLE VII

Formula Writing—Binary Compound B

Write the empirical formula for the following compounds: sodium oxide, calcium chloride.

SOLUTION:

Since the name "sodium oxide" indicates a compound of sodium and oxygen, the formula must be Na_2O (see Example II).

By similar reasoning, the name "calcium chloride" indicates a compound of calcium and chlorine. Hence the formula must be $CaCl_2$ (see Example II).

In simple compounds consisting of a metal and nonmetal, the primary valence of each element could be deduced from its position in the periodic table. Thus, a compound of sodium and oxygen is unambiguously assigned a formula of Na_2O.

The situation is different for binary compounds of two nonmetals. Generally, the formulas of these compounds cannot be deduced from the primary valences of the elements. For example, compare the formulas of two gasses that are binary compounds of carbon and hydrogen: CH_4 (methane) and C_2H_2 (ethylene). You would predict CH_4 from primary valences (C = 4, H = 1), but what about C_2H_2? The valences in the carbon atoms of ethylene are said to be unsaturated. The result of unsaturation is that the compound contains "double" bonds or "triple" bonds. We will discuss bonding and structure in more detail in Chapter 8, but it is important to appreciate here that unsaturation leads to combinations of two nonmetals that are not predicted by their primary valence. Consequently, more than one compound often exists from the combination of two nonmetals. As an example, there are five different oxides of nitrogen: N_2O, NO, NO_2, N_2O_3, N_2O_4, and N_2O_5. All of these compounds contain different degrees of unsaturation. The structure of each of these compounds is beyond the scope of this discussion; however, it is important to assign a name to these compounds so that the name unambiguously describes the formula.

In writing the formula of a binary nonmetal compound, the first element in the formula is that which is more to the left or farther down the periodic chart; thus, nitrogen is written first and oxygen follows. In naming the compound, the same rules are applied as above for a nonmetal and metal. However, an additional rule is necessary because the name "nitrogen oxide" does not sufficiently describe which of the five oxides was meant. In order to clear up this ambiguity, we use a Greek prefix before the atom to indicate the number of atoms present. The table below (Fig. 6.15) lists some common Greek prefixes:

Greek Prefix	Number of Atoms
mono	1
di	2
tri	3
tetra	4
penta	5
hexa	6

Figure 6.15   Greek Prefixes

The prefix "mono" is usually omitted for simplicity. Thus, NO_2 is named "nitrogen dioxide" and N_2O_4 is named "dinitrogen tetraoxide." Although this system of nomenclature can be applied to most nonmetal binary compounds, some compounds have retained their older or "trivial" names. For example, NO is nitrogen oxide, but it is also commonly called "nitric oxide." Similarly, N_2O, dinitrogen oxide, is also called "nitrous oxide," and CO is commonly called "carbon monoxide."

EXAMPLE VIII

Compound Naming—Binary Compound B

Name the following compounds: CO_2 and PCl_5.

SOLUTION:

You must first recognize that these are two nonmetals. Using the rules above, the names will be: carbon dioxide and phosphorus pentachloride.

EXAMPLE IX

Formula Writing—Binary Compound C

Write the empirical formula for the following compounds: sulfur dioxide and dinitrogen trioxide.

SOLUTION:

Using the rules above, the formulas will be: SO_2 and N_2O_3.

6.9 Chemical Nomenclature—Polyatomic Compounds

The concepts discussed in the previous section can be extended to polyatomic compounds or compounds more complex than binary compounds. To do this, it is necessary to recognize that certain atom groupings manifest characteristic valences. For example, the atom grouping SO_4 (called the sulfate grouping) acts as a single unit possessing a valence of two. When this grouping bonds to another atom, for example aluminum, the algorithm of Example II applies as follows:

EXAMPLE X

Formula Writing—Polyatomic Compound A

Write the formula of the compound formed between aluminum and the sulfate atom grouping.

SOLUTION:

Since the valence of aluminum is three and the valence of the sulfate grouping is two, we can write:

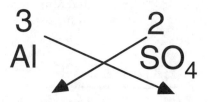

Hence, the formula is written $Al_2(SO_4)_3$. The parentheses are used to represent SO_4 as a single entity. When the subscript outside the parentheses is one, the parentheses are omitted.

EXAMPLE XI

Formula Writing—Polyatomic Compound B

Write the formula of the compound formed between calcium and the sulfate atom grouping.

SOLUTION:

Since the valence of calcium is two and the valence of the sulfate grouping is two, we can write:

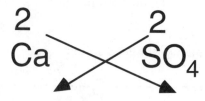

Initially, the algorithm of Example II yields the formula $Ca_2(SO_4)_2$, but this formula is reduced to $CaSO_4$ (*i.e.* 2:2 ratio = 1:1 ratio). Since the subscript outside the parenthesis is one, the parentheses are omitted.

Unfortunately, the valence of an atom grouping can not be "read" from the periodic table. It is, therefore, necessary to become familiar with some of the more common atom groupings (Fig. 6.16).

In writing chemical formulas, all of the atom groupings except the ammonium groupings are treated as nonmetals. The ammonium grouping mimics an alkali metal, and in formula writing it is treated accordingly. It combines with nonmetals, and it is written first in the formula.

The nomenclature of polyatomic compounds is quite simple. In naming polyatomic compounds, the following rules are used (Na_2SO_4 and NH_4Cl examples):

Nomenclature for Polyatomic Compounds

1. Write the name of the first element or atom grouping indicated in the formula.

sodium
ammonium

180

Name	Formula	Valence
Sulfite	SO_3	2
Sulfate	SO_4	2
Nitrite	NO_2	1
Nitrate	NO_3	1
Hypochlorite	ClO	1
Chlorite	ClO_2	1
Chlorate	ClO_3	1
Perchlorate	ClO_4	1
Phosphate	PO_4	3
Bicarbonate	HCO_3	1
Carbonate	CO_3	2
Hydroxide	OH	1
Peroxide	O_2	2
Ammonium	NH_4	1

Figure 6.16 Valence of Important Atom Groupings

2. Following this, write the name of the second element (stem) or atom grouping:

sodium sulfate
ammonium chlor

3. If an element name stem was used in rule 2, add "ide."

ammonium chloride

Notice that when the ammonium grouping forms a compound with a single element, the compound is named like a binary compound.

Again, it is important to note that the principles outlined above allow the empirical formula of a polyatomic compound to be written directly from the name of the polyatomic compound.

EXAMPLE XII

Formula Writing—Polyatomic Compound C

Write the empirical formula for the following compounds: sodium nitrate, ammonium sulfate.

SOLUTION:
Since the name "sodium nitrate" indicates a compound of sodium and the atom grouping nitrate, the formula must be $NaNO_3$ (see Example X).

By similar reasoning, the name "ammonium sulfate" indicates a compound of the atom grouping ammonium and the atom grouping sulfate. Hence, the formula must be $(NH_4)_2SO_4$ (see Example X).

———————————————————————□———————————————————————

These examples require a knowledge of atom groupings (name, formula, and valence). Although there are many atom groupings, only the list of fourteen presented above will find use in this text.

As a final note, it is important to distinguish between atom groupings and compounds. Atom groupings need to combine with other elements (or atom groupings) to form compounds. Thus, the sulfite atom grouping (SO_3) is very much different from the compound sulfur trioxide (SO_3) even though they have the same formula.

6.10 Chemical Periods Exemplified

The regular variation of chemical properties within a period of the periodic table merits closer investigation. Excluding the first period, which is rather anomalous, the elements within a period show a regular variation of characteristics from metal to nonmetal. Although the physical differences between metals and nonmetals are fairly obvious, the chemical differences are a bit more subtle. Chemists find it useful to contrast the chemical differences between metals and nonmetals by focusing on the chemistry of the oxides of the elements. Oxygen is the most abundant element on the surface of the earth. It is also a highly reactive element that makes up about twenty percent of the atmosphere. Contrasting the chemical characteristics of the chemical elements by focusing on oxide chemistry is a very logical thing for chemists to do.

One characteristic of representative metal oxides and many nonmetal oxides is that they react chemically with water to produce solutions that have distinctive properties. For example, both sodium oxide and calcium oxide produce solutions that have a bitter taste. Solutions of these metal oxides also feel slippery when rubbed between two fingers. Dilute solutions of sulfur dioxide and carbon dioxide, however, have a sour taste, and these nonmetal oxide solutions do not produce the slippery feel associated with the metal oxide solutions. Because the distinction between these solutions is of fundamental importance in chemistry, it is convenient to define contrasting terms to describe these solutions. Metal oxide solutions are said to be basic, and nonmetal oxide solutions are said to be acidic. Acidic solutions react chemically with basic solutions to produce solutions that possess a salt taste. The reaction of a simple acidic solution with a basic solution is called **neutralization**. The reaction of a metal oxide with a nonmetal oxide produces a chemical compound that is called a **salt**. This process is illustrated by the following chemical equation:

Chemical Equation 6.D

$$CaO \quad + \quad SO_3 \quad \rightarrow \quad CaSO_4$$

basic   acidic   salt

oxide   oxide

Since taste and feel are not always safe investigative techniques, it is not surprising that chemists developed other methods to distinguish metal oxide solutions and nonmetal oxide solutions. Perhaps one of the most useful methods involves the contrasting effects these solutions have on certain types of chemical dyes. For example, litmus, a coloring matter derived from lichens, is blue in metal oxide solutions and red in nonmetal oxide solutions. Litmus is one of many colored acid/base sensitive compounds called acid/base indicators. An acid/base indicator can be used to distinguish between acidic and basic solutions. It can also be used to establish an operational definition of acid and base. Operationally, an **acid** is any chemical compound that turns litmus red and a **base** is any compound that turns litmus blue. Other indicators could have been used to establish this definition, but the litmus indicator illustrates the concept of an operational definition. Since chemical compounds other than metal and nonmetal oxides are capable of changing the color of litmus red or blue, this operational definition expands the concept of acid and base. For example, since a water solution of hydrogen chloride (HCl) tastes sour and turns litmus red, this solution is an acidic solution even though it did not result from a nonmetal oxide.

Operational definitions are very useful in the study of natural science, but they can be intellectually frustrating. Why are water solutions of nonmetal oxides acidic? What is the atomic/molecular origin of acidity? The answers to these questions do not follow from the operational definition of acid and base. These questions can be answered, but the answers require a more detailed analysis of the valence concept.

6.11 Oxidation State—A Useful Concept Related to Valence

What is the origin of valence? Attempts to answer this question have dominated the field of theoretical chemistry for more than a century. Although the modern theory of chemical bonding is presented in Chapter 9, it is instructive at this point to examine a concept that is related to the origins of chemical bonding.

From the very beginning of the atomic theory, it was recognized that the chemical bond was most likely an electrical force of attraction. During the early part of the nineteenth century, the Swedish chemist Jons Jacob Berzelius developed an electrical theory of chemical bonding that attributed chemical bonding to the attraction of positive (electropositive) metal atoms and negative (electronegative) nonmetal atoms. The assignment of a positive charge to bound metal atoms and a negative charge to bound nonmetal atoms was consistent with electrochemical investigations carried out by the English natural scientist Michael Faraday (Chapter 9). As the valence concept developed, chemists found it convenient to extend the Berzelius theory by formalizing a set of rules that assigned chemically bound and unbound atoms an apparent electrical charge. Although the rules were somewhat arbitrary, they were developed to be consistent with the emerging valence concept and chemical fact. For example, unbound atoms were viewed as being electrically neutral, bound metal atoms were viewed as being electropositive, and bound nonmetal atoms were viewed as being electronegative. Further, in many simple binary compounds, the value of the apparent charge was equated to the signed (+ or –) value of the valence. In sodium chloride, for example, the sodium atom was assigned an apparent charge of +1 and the chlorine atom was assigned an apparent charge of –1. The apparent charge assigned to an atom according to these rules was called the **oxidation state** of the atom.

The rules for assigning oxidation state to bound and unbound atoms are summarized below. In this list, each rule is followed by a rationale that is consistent with an electrical view of the valence concept.

Oxidation State Rules

1. The oxidation state of all unbound elements is 0.
   Rationale: Pure elements are electrically neutral.
2. The oxidation state of all bound family IA elements is +1.
   Rationale: Family IA elements exhibit only a valence of one.
3. The oxidation state of all bound family IIA elements is +2.
   Rationale: Family IIA elements exhibit only a valence of two.
4. The oxidation state of bound oxygen is –2
   Exception: Oxidation state of oxygen peroxide is –1.
   Rationale: Oxygen usually exhibits a valence of two.
5. The oxidation state of bound hydrogen is –1 when bonded to metals and +1 when bonded to nonmetals.
   Rationale: Hydrogen usually exhibits a valence of one.
6. The sum of the oxidation states of all of the atoms in a molecule must equal 0.
   Rationale: Pure compounds are electrically neutral.
7. The sum of the oxidation states of all of the atoms in a nonmetal atom grouping must equal the negative value of the valence of the grouping.
   Rationale: A given nonmetal atom grouping only exhibits one valence.

8. The sum of the oxidation states of all of the atoms in a metal atom grouping must equal the positive value of the valence of the grouping.
Rationale: A given metal atom grouping only exhibits one valence.

These rules allow the assignment of an apparent charge or oxidation state to all of the atoms in many common chemical compounds. In some compounds, electrochemical evidence suggests that the oxidation state represents an actual electrical charge on a bound atom. For example, the fact that a water solution of sodium chloride (NaCl) conducts an electric current supports the idea that the bound sodium and chlorine atoms posses actual positive and negative charges which can take on an independent existence in water solution. Since an electrically charged atom is called an **ion**, compounds that contain charged atoms are called **ionic compounds**. On the other hand, the fact that a water solution of the compound acetone (C_3H_6O) does not conduct an electric current implies that the oxidation state represents only a relative distribution of electrical charge within the molecule of a **nonionic compound**.

In the next section, the concept of oxidation state will be used to establish a chemical reaction classification scheme. This scheme will include acid/base reactions, and the scheme will suggest the atomic/molecular origin of acids and bases.

EXAMPLE XIII

Determination of Oxidation State

What is the oxidation state of each atom in the compound potassium dichromate. The formula of potassium dichromate is $K_2Cr_2O_7$.

SOLUTION: K=+1 Cr=+6 O=−2

The oxidation state of potassium is +1 by rule two. The oxidation state of oxygen is −2 by rule four. Rule six allows the computation of the oxidation state of chromium as follows:

potassium atoms = +1   oxygen atoms = −2   chromium atoms = X

$$2(+1) + 7(-2) + 2(X) = 0$$
$$X = +6$$

6.12 A Classification System for Chemical Reactions

Although there are only ninety naturally occurring elements, the number of chemical reactions involving these elements and their compounds is huge. In order to unify the study of the chemical reaction process, it is common for chemists to classify chemical reactions according to various criteria. Several classification systems are used routinely in the study of chemistry, and the selection of a particular system depends on the nature of the chemical investigation. For example, there are at least three different systems that are used to classify acid/base reactions. These systems vary in the generality of the acid/base definition. Each of these systems has a utility that is related to specific study objectives. In the study of basic chemistry, it is convenient to establish the following simple chemical reaction classification system based on the concept of oxidation state.

Classification of Chemical Reactions

1. Redox Reactions: A **redox reaction** (reduction/oxidation) changes the oxidation state of one or more of the reactant elements. Elements that become more positive are said to experience **oxidation**, and elements that become more negative are said to experience **reduction**.

Chemical Equation 6.E
$$CH_4 + 2\ O_2 \rightarrow CO_2 + 2\ H_2O$$

Carbon changes from −4 to +4 (*i.e.* oxidized).
Oxygen changes from 0 to −2 (*i.e.* reduced).

2. Addition Reactions: An **addition reaction** combines two molecules, atom groupings, or molecular fragments to form a single molecule or atom grouping without a change in oxidation state.

Chemical Equation 6.F
$$CuCl_2 + 6\ NH_3 \rightarrow [Cu(NH_3)_6]Cl_2$$

3. Displacement Reactions: A **displacement reaction** involves the replacement of part of a molecule with another molecule, atom grouping, or molecular fragment. This reaction takes place without a change in oxidation state.

Chemical Equation 6.G
$$[Cu(H_2O)_6]Cl_2 + 6\ NH_3 \rightarrow [Cu(NH_3)_6]Cl_2 + 6\ H_2O$$

4. Acid/Base Reactions: An **acid/base reaction** is a displacement reaction that involves the transfer of hydrogen with an oxidation state of +1 (a hydrogen ion). The molecule, atom grouping, or molecular fragment that donates the hydrogen ion is called an **acid**. The molecule, atom grouping, or molecular fragment that accepts the hydrogen ion is called a **base**. This reaction takes place without a change in oxidation state.

Chemical Equation 6.H
$$HCl + NaOH \rightarrow NaCl + H_2O$$
acid   base

Equation 6.H represents a specific acid base neutralization reaction between HCl and NaOH. A useful generalization of this equation for hydroxide bases is represented in Equation 6.H′ below:

Chemical Equation 6.H′
$$HX + MOH \rightarrow MX + H_2O$$
acid   base

In this chemical equation, X represents any atom grouping attached to a transferable hydrogen (acidic hydrogen) and M represents any metal or metal like atom grouping (*e.g.* ammonium).

Although the rules for assigning oxidation state are somewhat arbitrary, they are based on a logical extension of the valence concept and on electrochemical fact. The foregoing classification system is therefore a useful tool for the study of chemical fact and theory. For example, the electric charge transfer implied by the oxidation state changes in redox reactions is very real, and redox reactions are responsible for the generation of electric current in batteries and fuel cells. Further, the addition and displacement reactions include the coordination chemistry of the transition metals that was discussed in Section 6.7. Finally, the acid/base reaction provides an understanding of the operational definitions of acid and base that were presented in Section 6.10. Evidently, acid/base indicators are capable of detecting hydrogen ion transfers.

6.13 Some Additional Notes on Acid/Base Chemistry

As suggested in Section 6.11, there is evidence that charged atoms or ions can take on an independent existence in water solution. If the acid/base reaction represented by Chemical Equation 6.H is written so as to emphasize the presence of ions, the fundamental nature of acid/base reactions becomes more apparent:

Chemical Equation 6.I
$$H^+(aq) + Cl^-(aq) + Na^+(aq) + OH^-(aq) \rightarrow Na^+(aq) + Cl^-(aq) + H_2O$$

This equation implies that a water solution of HCl contains H^+ and Cl^- ions, and that a water solution of NaOH contains Na^+ and OH^- ions. The parenthetical "aq" indicates that the ions exist in aqueous or water solution. The equation also implies that the product water molecule is nonionic. Notice that the equation shows that the Na^+ and the Cl^- ions do not change during the reaction. The ions that do not change are called **spectator ions**, and they can be removed from the equation. If these ions are removed from this equation, then the equation that represents this acid/base reaction reduces to a very simple form:

Chemical Equation 6.J
$$H^+(aq) + OH^-(aq) \rightleftharpoons H_2O$$

The double arrow in Equation 6.J indicates that the reaction is a dynamic equilibrium process. In any water solution, H^+ and OH^- ions are constantly forming water, and water is constantly reforming these ions. Actually, most chemical reactions should be represented as dynamic equilibrium processes, but for simplicity this point is often ignored. Hydrogen ion equilibrium reactions like the one represented in Equation 6.J can have a powerful effect on any chemical reaction taking place in water solution. Since a significant amount of chemistry on this planet takes place in water solution, acid/base chemistry is of fundamental importance.

The fact that the chemical reaction represented by Equation 6.J is a dynamic equilibrium means that a water solution always contains both H^+ and OH^- ions. Although these ions are not usually present in equal concentrations, the concentration of the H^+ ions can be used as a measure of relative acidity or basicity. Since the nature of the equilibrium expressed by Equation 6.J ensures that H^+ ion concentration in water will usually be quite low, chemists find it convenient to measure this concentration on a logarithmic scale called the **pH** scale. In this textbook, the pH scale will simply be presented as a relative scale that can be used to measure acid concentration. Typically the pH of a water solution can vary between the values of 0 and 14. A pH of 7 corresponds to an equal concentration of H^+ and OH^- ions—a neutral solution. A pH value lower than 7 indicates an excess of H^+ ions—an acidic solution. And a pH value higher than 7 indicates an excess of OH^- ions—a basic solution.

Acid/base indicators are colored compounds that are sensitive to H^+ or OH^- ion concentration. One very useful acid/base indicator called Yamada's universal indicator can be used to measure pH over a fairly wide range. As this indicator is subjected to pH changes from 2 to 12, the indicator color changes from red to violet. The following approximate pH/color chart for Yamada's universal indicator is easy to remember because the order of the colors is the same as the order of the colors in the spectrum. A familiarity with this color chart will be useful in following chemical lecture demonstrations associated with the study of acid/base chemistry (Fig. 6.17).

Red	Orange	Yellow	Green	Blue	Indigo	Violet
pH 2	4	6	7	8	10	12
Acidic			Neutral			Basic

Figure 6.17   pH/Color Chart for Yamada's Universal Indicator

6.14 Chemical Nomenclature—Acids

When binary or polyatomic hydrogen compounds act as acids, it is customary to change the names of these compounds to reflect the acidic character of the compounds. For binary compounds, the prefix *hydro* replaces the word *hydrogen*, the suffix *ic* replaces the suffix *ide*, and the word *acid* is added to the name of the compound. In water solution, the binary compound hydrogen chloride (HCl) acts as an acid, hence to emphasize these acidic characteristics, the name of the compound is changed to hydrochloric acid. For polyatomic compounds, the word *hydrogen* is dropped, the suffix *ic* replaced the suffix *ate* (or *ous* replaces *ite*), and the word *acid* is added to the name of the compound. Applying these rules to HNO_3 and HNO_2 yields the name nitric acid for HNO_3 and nitrous acid for HNO_2.

6.15 Some Specific Chemical Reaction Types

The classification of chemical reactions according to a system of four classes is of some utility in the prediction of the products to be expected from the reaction of a given set of reactant substances. The classification system is not the complete solution to this problem however. In addition to learning how to classify chemical reactions, chemists must also learn how to recognize some specific chemical reaction types which are members of the four general classes. Predicting the products of chemical reactions by analogy to prototype reactions is a powerful organizational tool. The approach is a particularly useful tool for predicting the products of chemical reactions based on the three dimensional structure of reactant molecules. This molecular structure issue is discussed in more detail in Chapter 8. In the present chapter, it is instructive to apply this reaction product prediction technique to five simple reaction types.

In the list below, five common reaction types are identified. Each reaction type is named, classified, and exemplified. Each example equation serves as a prototype equation that should allow the prediction of the products of the incomplete chemical equation that follows the example equation.

EXAMPLE XIV

Prediction of Reaction Products by Analogy

Simple Elemental Combination (Redox Class)

Example:
$$3\,Mg + N_2 \rightarrow Mg_3N_2$$
Comment:
Although a given pair of elements may not react, if a reaction does take place, the only product to be expected is the binary compound.

187

Predict:
$$Mg + O_2 \rightarrow$$

Simple Combustion (Redox Class)

Example:
$$CH_4S + 3\ O_2 \rightarrow CO_2 + 2\ H_2O + SO_2$$
Comment:
Note that the oxide of each element in the compound is formed.
Predict:
$$C_2H_6 + O_2 \rightarrow$$

Indirect Combustion (Redox Class)

Example:
$$CH_4 + 4\ CuO \rightarrow CO_2 + 2\ H_2O + 4\ Cu$$
Comment:
Although a given pair of compounds may not react, if a reaction does take place, the original oxide gives up oxygen to form the oxides of the other elements.
Predict:
$$C_2H_6 + Fe_2O_3 \rightarrow$$

Acid/Base Anhydride (Addition Class)

Examples:
$$SO_2 + H_2O \rightarrow H_2SO_3$$
$$MgO + H_2O \rightarrow Mg(OH)_2$$
Comment:
If nonmetal oxides react with water, they tend to form acids as defined on page above. If metal oxides react with water, they tend to form hydroxide bases as defined above.
Predict:
$$CO_2 + H_2O \rightarrow$$
$$Na_2O + H_2O \rightarrow$$

Single Substitution (Redox Class)

Example:
$$2\ Fe + 3\ CuCl_2 \rightarrow 2\ FeCl_3 + 3\ Cu$$
Comment:
Although a given element and compound may not react, if a reaction does take place, the unbound element replaces the bound element.
Predict:
$$Zn + AgNO_3 \rightarrow$$

SOLUTION: MgO; $CO_2 + H_2O$; $Fe + CO_2 + H_2O$; H_2CO_3; $NaOH$; $Ag + Zn(NO_3)_2$

Each set of reaction products is predicted by analogy to the prototype example.

6.16 You Too Can Speak Chemistry

Chemistry invades all aspects of our daily lives, and because chemistry is a highly organized discipline, our daily interactions with chemistry are comprehensible—even to nonchemists. At this point in your study of chemistry, your chemical communication skills are already quite formidable. To illustrate this point, consider an example of the discipline of chemistry interfacing with the discipline of history. As the example is discussed, keep in mind that your chemical communication skills are probably better than you think.

As I write, I have in front of me a delightful little book titled *Flags of Five Nations*.[10] The book is a history of an area of coastal Georgia known as the Golden Isles of Guale. This area of Georgia is about sixty miles south of Savanna, the oldest city in Georgia. At one point in the book, the author describes the construction of plantation buildings in the town of Brunswick, Georgia:

"barns and other plantation buildings were made of 'tabby,' an all but indestructible mixture of sand, oyster shells, water, and lime from burned shell."

What was this all but indestructible building material called "tabby?" Let us see how a little chemistry can give a more complete view of this simple historical fact.

The first point to consider in understanding how "tabby" functions as a building material is the chemical composition of oyster shells. Chemically, oyster shells are composed of the same chemical compound that makes up the building material marble. This chemical compound is calcium carbonate. Since you can translate the term calcium carbonate directly to an atomic/molecular viewpoint, you are capable of the following understanding:

$$\text{Oyster Shells} = \text{Calcium Carbonate} = CaCO_3$$

The next point to understand is that the oyster shells are not really burned, rather they are heated to a very high temperature in a lime kiln. Chemically, the *calcium carbonate*, when heated, decomposes to form the powdery solid *calcium oxide* and the gaseous compound *carbon dioxide*. Since you can translate each of the italicized terms, you are capable of writing a chemical equation for the first step in the formation of "tabby."

Chemical Equation 6.K
$$CaCO_3 \rightarrow CaO + CO_2$$

The next stage in the formation of a "tabby" construction is to make a slush by mixing the powdery solid calcium oxide with water. Chemically, when calcium oxide is reacted with water, the product of the reaction is calcium hydroxide. Hence, you can write:

Chemical Equation 6.L
$$CaO + H_2O \rightarrow Ca(OH)_2$$

In the final stage of "tabby" construction, the calcium hydroxide slush is mixed with sand and unheated oyster shells. This mixture is poured into molds (a wall mold, for example) where it hardens like cement. Chemically, the calcium hydroxide in the slush reacts with carbon dioxide in the air. This results in the formation of calcium carbonate and water. Hence, you can write:

Chemical Equation 6.M
$$Ca(OH)_2 + CO_2 \rightarrow CaCO_3 + H_2O$$

This last reaction reveals the strategy of the whole process. The building material calcium carbonate (marble, limestone, oyster shells) is rendered pourable by chemical transformations 6.K and 6.L. This pourable mixture is mixed with sand and shells to give it bulk. It is then poured into building molds and allowed to set. Notice that the setting of "tabby" in reaction 6.M simply involves the chemical reformation of the building material calcium carbonate.

The chemical view of "tabby" construction is not better than the historical view, rather both points of view form a more complete picture. This is the true meaning of liberal arts—an appreciation of the unity of all of human knowledge.

"All the sounds of the earth are like music." – Oscar Hammerstein II, American Composer

---

10.   B. Vanstory, *Flags of Five Nations*, Fort Frederica Association, St. Simons Island, Georgia (1971).

Chapter Six
Performance Objectives

P.O. 6.0

Review all of the boldfaced terminology in this chapter, and make certain that you understand the use of each term.

acid	acid-base reaction	addition reaction
atomic number	base	binary compounds
coordination compounds	coordination number	displacement reaction
families	ion	ionic compounds
ligands	metals	neutralization
nonionic compound	nonmetals	oxidation
oxidation state	periods	pH
rare earth	redox reaction	reduction
salt	spectator ions	transition
valence		

P.O. 6.1

You must be able to identify an element name given its symbol, and you must be able to identify an element symbol given its name.

EXAMPLE:
Which of the following compounds contains the element tin?

a) $TiCl_4$ b) $ZnCl_2$ c) $SnCl_4$ d) AgCl

SOLUTION: c

The answer is c. The selection of the correct answer requires the rote memorization of the boldfaced elements and symbols listed in Figure 3.7 in the textbook.

ADDITIONAL EXAMPLE:
Which of the following compounds contains the element gold?

a) AgCl   b) $HgCl_2$ c) GaN   d) $Gd_2(SO_4)_3$ e) none of these

ANSWER: e

P.O. 6.2

You must be able to use the periodic table to determine the primary valence of an element.

EXAMPLE:
What is the primary valence of aluminum?

SOLUTION: 3

Textbook Reference: Section 6.7

ADDITIONAL EXAMPLE:
What is the primary valence of zinc?

ANSWER: 2

P.O. 6.3

You must demonstrate an understanding of the organizational strategy of the periodic table.

EXAMPLE:
The chemical compound sodium hydroxide, NaOH, can be used to remove carbon dioxide, CO_2, from the atmosphere of a closed chamber (*e.g.* space capsule) by reacting according to the following chemical equation:

$$2\ NaOH + CO_2 \rightarrow Na_2CO_3 + H_2O$$

Write the chemical equation for the reaction of potassium hydroxide with carbon dioxide.

SOLUTION: $2\ KOH + CO_2 \rightarrow K_2CO_3 + H_2O$

Textbook Reference: Sections 6.6 and 6.10

ADDITIONAL EXAMPLE:
The element chlorine, Cl_2, reacts with water according to the following chemical equation:

$$Cl_2 + H_2O \rightarrow HCl + HClO$$

Write the chemical equation for the reaction of the element bromine, Br_2, with water.

ANSWER: $Br_2 + H_2O \rightarrow HBr + HBrO$

ADDITIONAL EXAMPLE:
Based on the calculations in Example I, predict if the molar volume of gold (Au) would be higher or lower than silver (Ag).

ANSWER: higher

ADDITIONAL EXAMPLE:
When the product of the following chemical reaction is dissolved in water, will the resulting solution be basic, neutral, or acidic?

$$S + O_2 \rightarrow ?$$

ANSWER: acidic
Textbook Reference: Section 6.10

ADDITIONAL EXAMPLE:
When the product of the following chemical reaction is dissolved in water, will the resulting solution have a pH greater than seven, equal to seven, or less than seven?

$$Mg + O_2 \rightarrow ?$$

ANSWER: greater than 7
Textbook Reference: Sections 6.10 and 6.13

P.O. 6.4

You must be able to use the primary valence of elements to write the formula of a binary (2 element) compound consisting of a metal and nonmetal given the name of the compound. This skill implies the ability to name the compound if you are given the formula.

EXAMPLE:
What is the chemical formula of sodium sulfide?

SOLUTION: Na_2S

Textbook Reference: Section 6.8

ADDITIONAL EXAMPLE:
What is the name of the compound with the formula: ZnS?

ANSWER: zinc sulfide

P.O. 6.5

You must be able to write the formula of a binary compound consisting of two nonmetals given the name of the compound. This skill implies the ability to name the compound if you are given the formula.

EXAMPLE:
What is the chemical formula of dinitrogen pentaoxide?

ANSWER: N_2O_5

ADDITIONAL EXAMPLE:
Name the compound with the formula: PCl_3?

ANSWER: phosphorus trichloride

P.O. 6.6

You must be able to use the primary valence of elements to write the formula of a compound containing a common atom grouping given the name of the compound. The atom grouping valences must either be memorized or written on exam note card. This skill implies the ability to name the compound if you are given the formula.

EXAMPLE:
What is the chemical formula of sodium bicarbonate?

SOLUTION: $NaHCO_3$

Textbook Reference: Section 6.9

ADDITIONAL EXAMPLE:
What is the name of the compound with the chemical formula $Ca_3(PO_4)_2$?

ANSWER: calcium phosphate

P.O. 6.7

You must be able to write the chemical equation for a chemical reaction given a narrative description of the reaction.

EXAMPLE:
Aluminum reacts chemically with hydrogen chloride to produce aluminum chloride and hydrogen gas. Write a balanced chemical equation for this reaction.

SOLUTION: $2 \text{ Al} + 6 \text{ HCl} \rightarrow 2 \text{ AlCl}_3 + 3 \text{ H}_2$

Textbook Reference: Section 6.10

ADDITIONAL EXAMPLE:
Calcium carbonate reacts chemically with carbon dioxide and water to produce calcium bicarbonate. Write a balanced chemical equation for this reaction.

ANSWER: $\text{CaCO}_3 + \text{CO}_2 + \text{H}_2\text{O} \rightarrow \text{Ca(HCO}_3)_2$

P.O. 6.8

You must be able to assign the oxidation state to all of the atoms in a molecule.

EXAMPLE:
What is the oxidation state of all of the atoms in KMnO_4?

SOLUTION: $O = -2$, $K = +1$, $Mn = +7$

Textbook Reference: Section 6.11

ADDITIONAL EXAMPLE:
What is the oxidation state of all of the atoms in CH_4O?

ANSWER: $O = -2$, $H = +1$, $C = -2$

P.O. 6.9

You must be able to classify chemical reactions according to the system presented in Section 6.12.

EXAMPLE:
Classify each of the following reactions:

1. $2 \text{ H}_2 + \text{O}_2 \rightarrow 2 \text{ H}_2\text{O}$
2. $\text{H}_2\text{SO}_4 + 2 \text{ KOH} \rightarrow \text{K}_2\text{SO}_4 + 2 \text{ H}_2\text{O}$

SOLUTION: 1 = redox   2 = acid/base

Textbook Reference: Section 6.12

ADDITIONAL EXAMPLE:
   Classify each of the following reactions:

   1. $AgNO_3 + NaCl \rightarrow AgCl + NaNO_3$
   2. $Ni + 4\ CO \rightarrow Ni(CO)_4$

ANSWER: 1 = displacement   2 = addition

P.O. 6.10

   You must be able to predict the products of chemical reactions according to the system presented in Section 6.15.

EXAMPLE:
   Predict the product(s) to be expected from the chemical reaction of the following reactants:

$$C_8H_{18} + O_2 \rightarrow ?$$

SOLUTION: $CO_2 + H_2O$

   The reactants place this reaction in the type identified as simple combustion (redox class). The products are predicted by analogy to the prototype example reaction.

   Textbook Reference: Section 6.15

ADDITIONAL EXAMPLE:
   Predict the product(s) to be expected from the chemical reaction of the following reactants:

$$C_2H_6 + Ag_2O \rightarrow ?$$

ANSWER: $Ag + CO_2 + H_2O$

Problems

Primary Valences of Elements

1. If the valence of sulfur is two, what is the valence of iron in each of the following compounds: FeS, Fe_2S_3?

    STUDENT SOLUTION:

2. If the valence of chlorine is one, what is the valence of copper in each of the following compounds: Cu_2Cl_2, $CuCl_2$?

    STUDENT SOLUTION:

3. An unknown metal element, X, forms a simple metal oxide with the formula X_2O_3. If this element is a representative element ("A" family), which of the following families on the periodic table is most likely to contain this unknown metal as a member?

    a) IIIA   b) VIIA   c) IA   d) IIA   e) none of these

    STUDENT SOLUTION:

4. Two chemical formulas for compounds of the unknown elements X and Y are shown below:

    $$H_2X \qquad\qquad YX_2$$

    If the valence of hydrogen is known to be one, what is the valence of the unknown element Y that is consistent with the given formulas?

    a) 1   b) 2   c) 3   d) 4   e) none of these

    STUDENT SOLUTION:

5. When plants are burned, their ashes contain potash, P_2O_5. What is the valence of the element phosphorus (P) in potash?

    STUDENT SOLUTION:

Organization of the Periodic Table

6. Calcium reacts chemically with water according to the following equation:

$$Ca + 2\,H_2O \rightarrow Ca(OH)_2 + H_2$$

Magnesium also reacts chemically with water. Write the chemical equation that represents the reaction of magnesium and water.

    STUDENT SOLUTION:

7. Based on the trends listed in Figures 6.11 and 6.12, predict which element in each of the following pairs has the lowest melting point:

    a) Ca or Ba   b) P or As

    STUDENT SOLUTION:

8. The elements zinc (Zn), cadmium (Cd), and mercury (Hg) are members of family IIB on the periodic table. Mercury is the only metal that is a liquid at room temperature. Based on the position of these elements on the periodic table, which of the following statements is the most likely correct?

    a) Zinc has a higher melting point than cadmium.
    b) Cadmium has a higher melting point that zinc.
    c) Cadmium and zinc have about the same melting point.

    STUDENT SOLUTION:

9. When high-sulfur coals are burned in power plants, sulfur dioxide gas (SO_2) is formed. Sulfur dioxide is a major air pollutant and cause of acid rain. However, SO_2 can be removed from the smoke stack by reaction with calcium oxide (CaO) to form calcium sulfite ($CaSO_3$), a solid which can be removed from the stack by an electrostatic precipitator.

$$CaO + SO_2 \rightarrow CaSO_3$$

Write the chemical equation for the reaction of magnesium oxide (MgO) with SO_2.

STUDENT SOLUTION:

10. Based on the trend listed in Figure 6.11, would you expect Francium (Fr) to be a liquid, solid, or gas at room temperature (25°C)?

STUDENT SOLUTION:

11. Based on their position in the periodic table, which of the following elements are metals and which are nonmetals?

N, Cr, Ni, O, P, Cl, Ga, S

STUDENT SOLUTION:

12. In the following chemical reaction: $X + O_2 \rightarrow$?, where X is one of the elements listed below, if the product is dissolved in water, will the resulting solution be basic, neutral, or acidic?
Mg, C

STUDENT SOLUTION:

13. When the product of each of the following chemical reactions is dissolved in water, which reaction product will produce an acidic solution?
Reaction A: $Na + O_2 \rightarrow$?
Reaction B : $CS_2 + O_2 \rightarrow$?

STUDENT SOLUTION:

14. In the following chemical reaction: $X + O_2 \rightarrow$?, where X is one of the elements listed below, if the product is dissolved in water, will the resulting solution have a pH greater than seven, equal to seven, or less than seven?

S, Ca

STUDENT SOLUTION:

15. The element 115 has not been discovered. If it should be discovered someday, can you predict its properties? To which family would it belong? Would it be a metal or nonmetal; a solid, liquid, or gas?

STUDENT SOLUTION:

Naming Compounds

16. Name each of the following compounds: Na_3PO_4, $Ca(ClO)_2$, $ZnCl_2$, $NaClO_3$, H_2S, $CaSO_4$, SiO_2, CCl_4.

STUDENT SOLUTION:

17. Give the empirical formula for each of the following compounds: aluminum sulfate, lithium oxide, sodium sulfate, copper (II) chloride, sodium hydroxide, zinc oxide, silicon tetrachloride, carbon disulfide

STUDENT SOLUTION:

18. Which of the following is the correct formula for zinc phosphate?

a) Zn_2PO_4 b) $Zn(PO_4)_2$ c) $Zn_3(PO_4)_2$ d) $ZnPO_4$ e) none of these

STUDENT SOLUTION:

19. Which of the following is the correct formula for mercury (II) oxide?

a) Hg_2O_3 b) HgO c) Hg_2O d) HgO_2 e) none of these

STUDENT SOLUTION:

20. Which of the following is the correct formula for magnesium carbonate?

a) $Mg_2(CO_3)_3$ b) $MgCO_3$ c) Mg_2CO_3 d) $Mg(CO_3)_2$ e) none of these

STUDENT SOLUTION:

21. Which of the following is the correct formula for calcium peroxide?

a) Ca_2O_2 b) CaO c) Ca_2O d) CaO_2 e) none of these

STUDENT SOLUTION:

22. Which of the following is the correct formula for iron (III) oxide?

a) Fe_2O b) Fe_2O_3 c) Fe_3O_2 d) FeO e) none of these

STUDENT SOLUTION:

23. Which of the following is the correct formula for aluminum chloride?

a) Al_3Cl_2 b) $AlCl$ c) Al_2Cl d) Al_3Cl e) none of these

STUDENT SOLUTION:

24. Which of the following is the correct name for $NaClO_2$?

   a) sodium hypochlorite   b) sodium chlorite   c) sodium chlorate
   d) sodium perchlorate   e) none of these

   STUDENT SOLUTION:

25. Which of the following is the correct name for $CaCO_3$?

   a) calcium carbonate   b) calcium bicarbonate   c) calcium hypocarbonate
   d) calcium percarbonate   e) none of these

   STUDENT SOLUTION:

Chemical Equations From Narrative Descriptions

26. When heated, potassium nitrate decomposes into potassium nitrite and oxygen gas. Write a balanced chemical equation for this reaction.

   STUDENT SOLUTION:

27. Ammonia, NH_3, can be produced by heating ammonium chloride with calcium oxide. Calcium chloride and water are also products of this reaction. Write a balanced equation for this reaction.

   STUDENT SOLUTION:

28. Carbon dioxide reacts chemically with sodium hydroxide to form the compound sodium carbonate and the compound water. Write a balanced chemical equation for this chemical reaction.

   STUDENT SOLUTION:

29. Sodium oxide reacts chemically with water to produce the compound sodium hydroxide. Write a balanced chemical equation for this chemical reaction.

   STUDENT SOLUTION:

30. The element silver will react with a mixture of hydrogen sulfide and oxygen. Silver sulfide and water are the products of this reaction. Write a balanced chemical equation for the reaction of silver with hydrogen sulfide and oxygen. When this equation is balanced with integer coefficients, what is the total number of water molecules represented on the right side of the equation?

a) 4   b) 1   c) 2   d) 3   e) none of these

STUDENT SOLUTION:

31. The element oxygen can be prepared by heating potassium chlorate. When potassium chlorate is heated to produce oxygen, potassium chloride is also formed as a product. Write a balanced chemical equation for this reaction. When this equation is balanced with integer coefficients, what is the coefficient in front of oxygen on the right hand side of the equation?

a) 2   b) 3   c) 4   d) 1   e) none of these

STUDENT SOLUTION:

32. Write a balanced chemical equation for the reaction of copper with oxygen to form copper (II) oxide. When this equation is balanced with integer coefficients, what is the total number of copper atoms represented on the left hand side of the equation?

a) 2   b) 3   c) 4   d) 1   e) none of these

STUDENT SOLUTION:

33. Nitrogen dioxide is a major air pollutant that produces a red-brown haze you often see in urban areas where there is significant air pollution. It causes eye irritation and respiratory problems. It can react with sunlight to produce other pollutants, or it can react with water to form acid rain. Nitrogen dioxide is formed by two reactions. The first occurs at high temperature, for example, in an internal combustion engine, by reaction of atmospheric (molecular) nitrogen with atmospheric oxygen to produce nitric oxide (nitrogen oxide). When nitric oxide is released into the atmosphere it reacts with atmospheric oxygen to produce nitrogen dioxide. Write the balanced chemical equations for these two reactions.

STUDENT SOLUTION:

Oxidation States

34. What is the oxidation state of all of the atoms in each of the following molecules?

    LiF, CaC_2, $KMnO_4$, H_2SO_4, CuO, CuO_2, N_2O, NO

    STUDENT SOLUTION:

35. The oxidation state of the phosphorus atom in the phosphate atom grouping is:

    a) –5   b) +5   c) +8   d) –3   e) none of these

    STUDENT SOLUTION:

36. The oxidation of the nitrogen atom in the nitrite atom grouping is:

    a) –1   b) –3   c) +1   d) +4   e) none of these

    STUDENT SOLUTION:

Classification of Chemical Reactions

37. Classify the following reactions as a redox, addition, displacement or acid/base reaction:

    a. $Na_2SO_3 + CaCl_2 \rightarrow CaSO_3 + 2\ NaCl$
    b. $Cu_2S + O_2 \rightarrow 2\ Cu + SO_2$
    c. $AgCl + 2NH_3 \rightarrow [Ag(NH_3)_2]Cl$
    d. $Pb(NO_3)_2 + 2\ NaI \rightarrow 2\ NaNO_3 + PbI_2$
    e. $NaHCO_3 + HCl \rightarrow H_2CO_3 + NaCl$
    f. $H_3PO_4 + 3\ NaOH \rightarrow Na_3PO_4 + 3\ H_2O$
    g. $2\ Al_2O_3 \rightarrow 4\ Al + 3\ O_2$
    h. $H_2O + BF_3 \rightarrow H_2OBF_3$
    i. $2\ Na_3PO_4 + 3\ Mg(ClO_3)_2 \rightarrow Mg_3(PO_4)_2 + 6\ NaClO_3$
    j. $I_2 + 2\ Na_2S_2O_3 \rightarrow 2\ NaI + Na_2S_4O_6$
    k. $2\ HNO_3 + Mg(OH)_2 \rightarrow Mg(NO_3)_2 + 2\ H_2O$

    STUDENT SOLUTION:

Predicting Chemical Reaction Products by Analogy

38. Predict the products of each of the following reactions by analogy to the appropriate prototype reaction in textbook section 6.15:

   a. $Na + S \rightarrow$
   b. $CH_4O + O_2 \rightarrow$
   c. $CH_4 + PbO \rightarrow$
   d. $SO_3 + H_2O \rightarrow$
   e. $CaO + H_2O \rightarrow$
   f. $Al + HCl \rightarrow$
   g. $Cl_2 + KI \rightarrow$

   STUDENT SOLUTION:

Problems Involving Household Chemistry and Science

39. The article "Edible Acid-Base Indicators" (Mebane, Robert C.; Rybolt, Thomas R. *J. Chem. Educ.* **1985**, 62, 285) suggests a number of acid/base indicators that can be found in the kitchen. The article describes the preparation of these indicators, and it lists the pH/color relationships for each indicator. After reading this article, attempt to use several of these indicators to determine the approximate pH of one regular aspirin tablet dissolved in one cup of water (approximately 250 mL) and one buffered aspirin tablet dissolved in one cup of water.

   STUDENT SOLUTION:

40. As indicated in Section 6.7, it is common for transition metals to form primary and secondary valences. This aspect of transition metal chemistry can be demonstrated with some simple household materials. Obtain three one ounce disposable plastic medicine cups. Place a shiny penny in the bottom of each cup. Add one teaspoon of vinegar to one of the plastic cups. Add one teaspoon of 3% hydrogen peroxide to a second cup. To the third cup add one teaspoon of vinegar and one teaspoon of 3% hydrogen peroxide. After about fifteen minutes you will notice a significant change in the cup containing the penny, vinegar, and hydrogen peroxide. There is actually quite a bit of interesting chemistry going on in this simple experiment.

   Although vinegar (acetic acid) does not react with copper, it does react with copper oxide. The reaction between copper oxide (basic metal oxide) and acetic acid is a logical extension of the reaction represented by Chemical Equation 6.D. While a mixture of copper and oxygen is not extremely reactive, a copper containing penny will eventually become coated with black copper (II) oxide. This oxide coating can be removed by dilute acetic acid. When copper oxide reacts with acetic acid, copper (II) acetate, $Cu(C_2H_3O_2)_2$, is formed. Copper (II) acetate is an ionic compound, and in water solution, the the Cu^{+2} ion forms six secondary bonds to water molecules. The resulting blue-green coordination compound has the formula $[Cu(H_2O)_6](C_2H_3O_2)_2$.

   A 3% hydrogen peroxide solution decomposes to form oxygen and water, but as observed above, a 3% hydrogen peroxide solution does not react with a shiny copper penny in fifteen minutes. The formation of the blue-green solution in the cup containing a mixture of hydrogen peroxide and acetic acid suggests that the acetic acid is involved in some type of catalytic activity since the formation of black copper (II) oxide on the surface of the penny is quite evident. The acetic acid then reacts with the copper (II) oxide as described above.

There is, indeed, a lot of chemistry going on in the third cup. Write balanced chemical equations for each of the following net chemical reactions taking place in the third cup: 1) the decomposition of hydrogen peroxide to form water and oxygen; 2) the reaction of copper and oxygen to form copper (II) oxide; 3) the the reaction of copper (II) oxide with acetic acid to form copper (II) acetate and water; 4) the reaction of water and copper (II) acetate to form the blue-green coordination compound.

Repeat the above experiment using a nickel in place of a penny. What do the results of this experiment suggest about the chemical composition of a nickel?

STUDENT SOLUTION:

Library Problems

41. The title of this chapter is "The Periodic Table or You Too Can Speak Chemistry." Read the article "Acid Rain Effects on Stone Monuments" (Charola, A. Elena. *J. Chem. Educ.* **1987**, 64, 436). After reading the article, demonstrate your ability to read, write, and speak chemistry by answering the following questions with verbal descriptions and chemical equations:

    a. Why is "natural" rain acidic?
    b. What are the origins of industrial acid rain?
    c. How does acid rain cause the dissolution of calcite?
    d. What is the secondary effect that causes mechanical damage to calcareous stones after dissolution?

    STUDENT SOLUTION:

42. The alcohol breathalyzer that is used to measure levels of alcoholic intoxication involves some interesting redox chemistry. The article "Oxidation-Reduction in Blood Analysis: Demonstrating the Reaction in a Breathalyzer" (Anderson, John M. *J. Chem. Educ.* **1990**, 67, 263) gives a brief description of the breathalyzer and breathalyzer chemistry. A follow-up letter to the editor of the Journal (*J. Chem. Educ.* **1992**, 69, 258) presents a more complete view of the redox chemistry associated with the breathalyzer. After reading this article and follow-up letter, briefly describe how a breathalyzer works.

    STUDENT SOLUTION:

Formal Laboratory Exercise

Using Acid/Base Chemistry as an Analytical Tool
A Quantitative Chemical Analysis

Introduction

Equation 6.H' in Section 6.12 represents the generalized neutralization reaction for an acid and a hydroxide base.

Chemical Equation 6.H'
$$HX + MOH \rightarrow MX + H_2O$$
$$\text{acid} \quad \text{base}$$

Because acid/base indicators can detect the point of neutralization in this type of reaction, these indicators are useful tools in a wide range of analyses that involve acidic or basic substances. For example, the drug aspirin ($C_9H_8O_4$) is an acid. Only one of the eight hydrogen atoms in the aspirin molecule is acidic according to the definition in Section 6.12, and the formula for aspirin can be written as $HC_9H_7O_4$ to emphasize this point. Since aspirin is an acid, it reacts with hydroxide bases according to the generalized reaction represented by Equation 6.H'. The specific equation for the reaction with the base sodium hydroxide is shown below:

$$HC_9H_7O_4 + NaOH \rightarrow NaC_9H_7O_4 + H_2O$$
$$\text{aspirin (acid)} \quad \text{base}$$

If the acid/base indicator phenolphthalein is added to a water solution of aspirin, then the indicator will assume its colorless acidic state. After the addition of a sufficient quantity of sodium hydroxide to neutralize the aspirin, the phenolphthalein indicator will assume its red basic state. This fact allows the aspirin neutralization reaction to be used as an analytical method to determine the amount of aspirin in an unknown sample. The purpose of this formal laboratory exercise is to use this analytical method to determine the amount of aspirin in a multi-component analgesic tablet.

The first step in this type of analysis is to determine the quantity of sodium hydroxide solution required to neutralize a known mass of pure aspirin. This sodium hydroxide standardization procedure leads to a simple stoichiometric equivalence ratio for the neutralization reaction which can be expressed as follows:

$$\frac{\text{grams of aspirin}}{\text{milliliters of NaOH}} = \text{constant for a given NaOH solution}$$

If this same sodium hydroxide solution is then used to neutralize the aspirin in an unknown sample, then the following use of linear reasoning allows for the calculation of the mass of aspirin in the unknown sample.

$$\frac{\text{grams of aspirin}}{\text{milliliters of NaOH}} = \frac{\text{grams of aspirin'}}{\text{milliliters of NaOH'}}$$

The procedure that is used to determine when the neutralization reaction is complete is described in the procedural overview.

Procedural Overview

This acid/base analysis requires the controlled addition of a solution of sodium hydroxide (base) to an measured quantity of aspirin (acid) until the acid/base indicator color change shows that neutralization of the acid is complete. The controlled addition of a liquid reagent to a chemical reaction container is carried out by using a volume measuring tool called a "buret" in a process referred to as titration. A typical titration apparatus is shown in Figure 6.18.

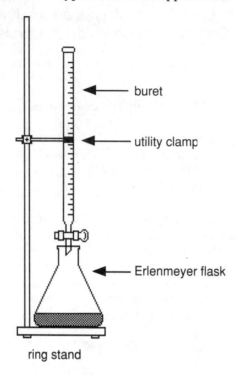

Figure 6.18 Titration Apparatus

The buret is a precision instrument for delivering controlled volumes of liquid. Before using the buret in this experiment, it should be rinsed with distilled water followed by a rinse with about 10 mL of the sodium hydroxide solution to be used in the experiment. The rinse solutions are drained through the tip of the buret into a beaker by opening the stopcock of the buret. The ringstand, utility clamp, Erlenmeyer flask, and buret are assembled as shown in Figure 6.18. After closing the buret stopcock, the buret is filled with sodium hydroxide solution using a funnel to transfer the solution. The buret is filled so that the sodium hydroxide solution level is well above the zero calibration line. The buret stopcock is then opened to drain the sodium hydroxide solution until the liquid is below the zero calibration line (0.00 mL). It is not necessary to stop exactly at the zero line. After draining below the zero calibration line, there should be no air bubbles in the buret tip.

The volume of liquid delivered by a buret is measured by the difference between the initial buret reading and the final buret reading after the controlled addition of the liquid to the reaction. When placed in a buret, most liquids form a curved surface called the "meniscus." The bottom of the meniscus is read to obtain the volume. It is also necessary to estimate the distance that the meniscus lies between the graduation marks. In the Figure 6.19 below, the volume is read as 3.15 mL by estimating to the nearest 5/10 of the smallest division on the scale.

During the titration, the sodium hydroxide solution is slowly added to a solution of the aspirin (acid) and phenolphthalein indicator in the Erlenmeyer flask. As long as the flask contains unreacted aspirin, each small addition of sodium hydroxide solution will produce a pink phenolphthalein indicator color that will disappear as the flask is swirled. As the neutralization point (endpoint) is reached, the pink color fades more slowly, and the sodium hydroxide solution should be added more slowly by dropwise addition. The endpoint occurs when the pink color remains for 30 seconds. Since swirling the flask can splash solution onto the tip of the buret and onto the sides of the flask, this splashed solution should be washed into the flask just before the endpoint. This can be accomplished by using a distilled water wash bottle to rinse the tip of buret and the sides of the flask. If the endpoint is passed, the indicator turns red, and the titration must be repeated.

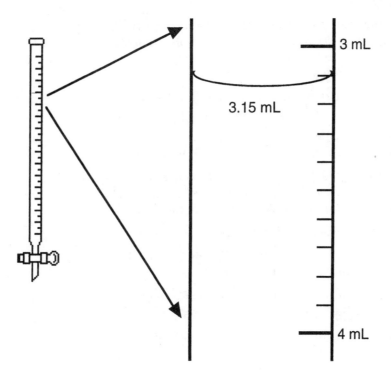

Figure 6.19   Reading a Buret

Materials

hotplate, ringstand, utility clamp, buret, 250 mL Erlenmeyer flask, 150 mL beaker, funnel, wash bottle, 6x4 depression well plate

Chemicals (Test solution concentrations are listed in Appendix III.)

pure aspirin standard, aspirin tablets, starch, povidone, calcium carbonate, ethanol, sodium hydroxide solution (0.1 M), phenolphthalein solution, iodine solution

Procedure

Standardization of Sodium Hydroxide Solution with Pure Aspirin:

1. Using a piece of weighing paper, determine the exact mass of about 0.3 grams of pure aspirin. Quantitatively transfer this aspirin to a 250 mL Erlenmeyer flask and record the exact mass of the aspirin in the data table.

2. Add 5 mL distilled water and 5 mL of ethanol to the flask containing the aspirin.

3. Gently heat the flask containing the aspirin, water, and ethanol on a hotplate for two to three minutes. Do not boil. After heating, add 50 mL of distilled water and 3 drops of phenolphthalein indicator solution to the contents of the flask.

4. Obtain about 60 mL of sodium hydroxide solution from the reagent table. Fill the buret as described in the procedural notes.

5. Titrate the aspirin solution in the Erlenmeyer flask as described in the procedural notes. Record the initial and final sodium hydroxide volume levels in the data table.

6. Ask the laboratory instructor to check the end point color of your titration and your data table entries. Based on this check, the instructor may request that the standardization be repeated.

7. Discard the contents of the Erlenmeyer flask down the drain, wash the flask thoroughly with water, and proceed to the aspirin tablet analysis in the next section of the procedure.

Analysis of Aspirin Tablet:

1. Determine the mass of an aspirin containing tablet and transfer the tablet to a 250 mL Erlenmeyer flask. Record the exact mass of the aspirin tablet in the data table.

2. Add 5 mL of distilled water to the flask containing the aspirin-containing tablet. Shake gently to disperse the tablet, then add 5 mL of ethanol.

3. Gently heat the flask containing the tablet, water, and ethanol on a hotplate for two to three minutes. Do not boil. After heating, add 50 mL of distilled water and 3 drops of phenolphthalein indicator solution to the contents of the flask.

4. Fill the buret as described in the procedural notes.

5. Titrate the aspirin solution in the Erlenmeyer flask as described in the procedural notes. Record the initial and final sodium hydroxide volume levels in the data table.

6. Ask the laboratory instructor to check the end point color of your titration and your data table entries. Based on this check, the instructor may request that the titration be repeated.

7. Discard the contents of the Erlenmeyer flask down the drain and wash the flask thoroughly with water. Discard the contents of the buret down the drain and wash the buret thoroughly with water.

8. Use the information in the data table to answer the questions in the conclusion and discussion section.

Data Table

Standardization of Sodium Hydroxide Solution with Pure Aspirin:

mass of pure aspirin grams measured	
initial buret reading milliliters measured	
final buret reading milliliters measured	
volume NaOH added from buret milliliters calculated	

Analysis of Aspirin Tablet:

mass of aspirin tablet grams measured	
initial buret reading milliliters measured	
final buret reading milliliters measured	
volume NaOH added from buret milliliters calculated	

Conclusion and Discussion

1. Since the standardization and analysis were carried out with the same sodium hydroxide solution, the following relationship can be used to calculate the mass of aspirin in the aspirin containing tablet:

$$\frac{\text{grams of aspirin}}{\text{milliliters of NaOH}} = \frac{\text{grams of aspirin'}}{\text{milliliters of NaOH'}}$$

    Use this relationship and information in the data table to calculate the mass of aspirin in the aspirin-containing tablet.

2. Calculate the percentage of aspirin in the aspirin-containing tablet.

3. In addition to aspirin, the aspirin-containing tablet contains either starch, providone, or calcium carbonate. Small solid samples of starch, providone, and calcium carbonate can be distinguished by the addition of iodine solution. Using the 6x4 24 depression well plate, add 1 or 2 drops of iodine to match head size quantities of pure starch, pure providone, and pure calcium carbonate. Perform this iodine solution spot test on an aspirin containing table. Which of these three substances is in the aspirin-containing tablet?

Chapter Six Notes

Chapter Seven

Energy Relationships in Chemical Reactions

7.1 Introduction

In Chapters 4 and 5, the central focus was on material changes that took place during the chemical reaction process. Using atomic/molecular theory, chemists become the masters of these material changes. But altered matter is not the only useful product of the chemical reaction process.

Almost all chemical reactions are accompanied by energy changes. In some cases, the reactions release energy, and in other cases, the reactions consume energy. Energy flow in chemical reactions can be measured quantitatively, and the results of these measurements can be incorporated into chemical equations. The resulting chemical equations are called thermochemical equations, and the purpose of this chapter is to investigate the process of energy flow during a chemical reaction as expressed by the thermochemical equation. This investigation will involve an expansion of atomic/molecular theory that incorporates the concept of energy into the theory.

7.2 Consider a Barrel of Petroleum

There is perhaps no better way to illustrate the material and energy value of chemical reactions than to consider a barrel of petroleum. Petroleum is a basic raw material that can be used to synthesize a wide range of chemical substances—pharmaceuticals, fibers, plastics, detergents, and dyes, to name a few. Petroleum can also be processed to produce chemical substances that are used primarily to produce energy—gasoline, kerosene, and fuel oil. Figure 7.1 shows how society typically uses a barrel (42 gallons) of petroleum. In this diagram, the material sector is represented in terms of synthetic fibers (shirts) and the energy sector is represented in terms of gasoline. These examples are selected because most chemistry students are familiar with the market value of these petrochemicals.

A consideration of the cost of a barrel of petroleum, the cost of 36 gallons of gasoline, and the cost of 100 synthetic fiber shirts suggests that there are some interesting marketplace dynamics associated with the fate of a barrel of petroleum. These dynamics are made more interesting by the fact that the two competing sectors split this valuable resource quite unequally. Petroleum or crude oil is a complex mixture of chemical compounds containing carbon and hydrogen (hydrocarbons). In its native state it is black-yellow "gunk" of very little use to society. In the hands of the chemical profession, petroleum becomes a material of great value—a material that has a profound influence on global politics.

In this chapter, an investigation of chemical energy will focus on basic principles, and these principles will be illustrated by a variety of chemical and physical systems. In the next chapter of this textbook, the chemical creativity associated with the application of atomic/molecular theory to the material and energy sectors of the petroleum industry will be explored. This study will necessitate an expansion of the theory to include the concept of molecular structure. The study of the petroleum industry will also apply the concepts of molecular energy developed in the present chapter. The study of chemical energy begins with a simple question. What is energy?

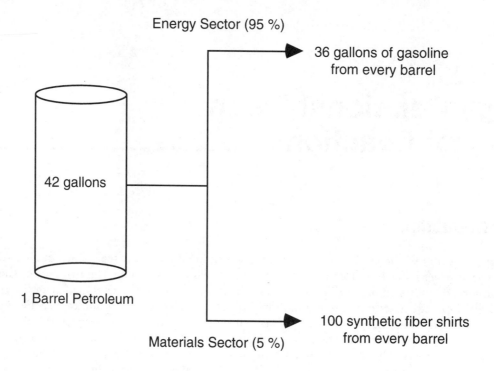

Energy Sector (95 %)

36 gallons of gasoline
from every barrel

42 gallons

1 Barrel Petroleum

100 synthetic fiber shirts
from every barrel

Materials Sector (5 %)

Figure 7.1   The Fate of a Barrel of Petroleum

7.3 Energy

The concept of energy can be somewhat confusing. This confusion is caused by the fact that energy as a physical reality can manifest itself in so many different ways. Heat, light, sound, and electricity are examples of the different ways in which energy can show itself. In the early days of chemistry, this multiplicity of expression created some problems. Lavoisier, for example, listed heat (French: Calorique) as one of the natural chemical elements. The best way to deal with the confusion created by energy's many forms is to attempt to classify the various forms of energy. Such classification requires a specific definition.

Energy is usually defined as the ability to do work. This definition is fine as a qualitative starting point, but it is through quantitative study that nature reveals its innermost secrets. It is appropriate, therefore, to state a quantitative definition of energy. This is accomplished by focusing attention on the word *work* and attempting to quantify this familiar human experience. One of the most common types of physical work is lifting. The following equation definition for work is fully compatible with the human experience of lifting:

DEFINITION: Energy is the ability to do work. **Work** is defined quantitatively as follows:

$$\textbf{Expression 7.A}$$
$$\text{The Definition of Work}$$

$$W = F_w S$$

W = Work
F_w = Weight of Object Lifted
S = Distance Object Lifted

EXAMPLE: The simple act of human lifting is an example of work.

Using the English system of measurement, the Expression 7.A indicates that lifting a 50 pound weight 1 foot would require 50 foot-pounds (50 pound × 1 foot) of work. In a similar manner, the equation indicates that it would require 100 foot-pounds to lift the same weight 2 feet (50 pounds × 2 feet). These arithmetic results are perfectly compatible with human experience. A person lifting a 50 pound weight 2 feet would "feel" that the task required twice as much work as lifting a 50 pound weight 1 foot. Notice that Expression 7.A is nothing more than a statement of common sense.

In the example presented above, English units of measurement are used to simplify the explanation. It is appropriate at this point to introduce the metric units of measurement that will be employed whenever Expression 7.A is used in this text:

$$S = \text{Distance} = \text{meters (abbreviated m)}$$
$$F_w = \text{Weight} = \text{newtons (abbreviated N)}$$
$$W = \text{Work} = \text{newton-meters (abbreviated N} \bullet \text{m)}$$
$$1 \text{ newton-meter} = 1 \text{ joule (abbreviated J)}$$

The best way to gain an understanding of these metric units is by considering an example involving a familiar quantity of work:

EXAMPLE I

The Work Equation

A male 6 feet tall and of average build will weigh about 160 pounds. Expressed in the metric system, this weight is almost exactly 712 newtons. A 712 newton man lifts his own weight 0.500 meters while doing one pull-up. How many joules of work will the man have done after completing ten pull-ups?

SOLUTION: 3560 J

$$W = F_w S$$
$$W = (712 \text{ N})(0.500 \text{ m})$$
$$W = 356 \text{ N} \bullet \text{m} = 356 \text{ J}$$

Ten pull-ups would require ten times this amount of work, 3560 J.

Since most people have attempted pull-ups, Example I provides a touchstone for the new metric units of weight and work.

Although the **joule** (J) is the unit of work accepted by the Système International (SI), it is not the usual energy unit used in commerce. For this reason, chemists often use other units of energy. One such unit is the calorie (abbreviated cal), and since it is a unit familiar to many nonscientists, the calorie will be used frequently in this text. The kilocalorie (1000 cal abbreviated kcal) is also called the nutritional Calorie (abbreviated Cal). This can be somewhat confusing, since the only difference in the spelling of **calorie** and **Calorie** is the capital "C" used in spelling the nutritional Calorie. It is the nutritional Calorie that most nonscientists are familiar with, and it is important to remember that 1 Calorie = 1000 calories. Since it is instructive to use two different units for measuring work (calorie and joule), the application of conversion factors is necessary in certain situations.

EXAMPLE II

Conversion of Energy Units

In order to do ten pull-ups, a 160 pound (712 newton) man does 3560 joules of work (Example I). How many calories of work is this? How many nutritional Calories?

CONVERSION FACTORS:

$$1 \text{ cal} = 4.18 \text{ J}$$
$$1 \text{ kcal} = 1000 \text{ cal}$$

SOLUTION: 852 cal or 0.852 Cal

As illustrated in Chapter 2, all conversion factors may be written as constant ratios. These ratios are subject to the arithmetic of linear reasoning:

$$\frac{X}{Y} = \frac{X'}{Y'}$$

Conversion A: $\dfrac{4.18 \text{ J}}{1.00 \text{ cal}} = \text{Constant}$

$$\frac{4.18 \text{ J}}{1.00 \text{ cal}} = \frac{3560 \text{ J}}{X \text{ cal}}$$

X = 852 cal

Conversion B: $\dfrac{1000 \text{ cal}}{1.00 \text{ kcal}} = \text{Constant}$

$$\frac{1000 \text{ cal}}{1.00 \text{ kcal}} = \frac{852 \text{ cal}}{X \text{ kcal}}$$

X = 0.852 kcal
Since 1 kcal = 1 nutritional Calorie (abbreviated Cal):
X = 0.852 Cal

For the arithmetically sophisticated person, this latter conversion is trivial. It involves only decimal point "pushing." Such ease of conversion is one of the strengths of the metric system. Although the approach taken above involves doing the problem the long way, it does present certain advantages to the arithmetically "rusty." Of course, both conversion factor problems may also be solved by the factor label method described in Example V of Chapter 2.

The introduction of Expression 7.A attempted to make use of the familiarity that most people have with the act of lifting. Weight, however, is only one example of a more general phenomenon called "force."

DEFINITION: A **Force** is a push or a pull. An unopposed force changes the motion of a mass. Force may be defined quantitatively in terms of this latter effect:

214

Expression 7.B
The Definition of Force

$$F = ma$$

F = Force Producing Acceleration (newtons, N)
m = Mass Being Accelerated (kilograms, kg)
a = Acceleration (meters/second², m/s²)

EXAMPLE: A force must be applied in order to stop a moving mass.

Students who have had a physics course will recognize Expression 7.B as Sir Isaac Newton's second law of motion. Newton's first law of motion states that the constant velocity of a mass will not change unless acted upon by a force. Newton's second law quantitatively describes the change in velocity produced by an unopposed force.

EXAMPLE III

The Force Equation

During free fall, the force called "weight" accelerates all masses on the planet at a constant gravitational acceleration rate of 9.80 m/s². When a mass is at rest on the surface of the planet, the weight force is opposed by an equal but opposite force applied by the surface of the earth. This latter situation is actually a specific example of Newton's third law of motion—every action produces an equal but opposite reaction. Use the force equation to calculate the weight of a 75.0 kg person experiencing a constant gravitational acceleration of 9.80 m/s².

SOLUTION: 735 N

$$F = ma$$
$$F = (75.0 \text{ kg})(9.80 \text{ m/s}^2)$$
$$F = 735 \text{ kg} \bullet \text{m/s}^2$$
$$F = 735 \text{ N}$$

Notice that this equation implies that $1 \text{ N} = 1 \text{ kg} \bullet \text{m/s}^2$

If the current view of nature held by the scientific community is correct, a force must have one of four origins:

Nature's Forces

1. Gravitational Force (Weight)
2. Electromagnetic Force[1]
3. Atomic Nuclear Strong Force
4. Atomic Nuclear Weak Force

All human beings are familiar with gravitational force, and most have experienced electromagnetic force—the attractions and repulsions experienced by static electric charges and magnets. The latter two nuclear forces are not well known by nonscientists. The important point here is not to discuss the

---

1.   Electricity and magnetism are opposite sides of the same coin, hence the term electromagnetic.

details of the four basic forces of nature, but rather to extend the meaning of Expression 7.A to include all of nature's forces. Hence work is done whenever any push or pull (force) is exerted over a distance:

Expression 7.C
A General Definition of Work

$$W = FS$$

W = Work (joules, J)
F = Any Force (newtons, N)
S = Distance Through Which Force Is Exerted (meters, m)

Expression 7.C applies equally well to lifting a weight (gravitational force) or stretching a rubber band (electromagnetic force).[2]

7.4 Classification of Energy

The specific definition of energy presented in the last section establishes exactly what it is that the various forms of energy have in common. At least in theory, all forms of energy represent the ability to lift a weight.[3] Hence, electric current is recognized as a form of energy because it is possible to imagine electric current turning an electric motor, which in turn lifts a weight. By the same type of reasoning, a woman standing at the top of a step ladder is recognized as possessing energy because it is possible to imagine her jumping from the ladder onto a teeter-totter, which in turn lifts a weight. In both of these examples, the familiar human experience of lifting is being used as a common denominator for evaluating different forms of energy. Because of Expression 7.A, this common denominator is quantitative, and hence all forms of energy can be expressed in joules.

As the various forms of energy are identified, it becomes evident that all forms of energy can be placed in one of two major categories. All energy is either kinetic energy or potential energy.

DEFINITION: Kinetic energy is the energy possessed by any mass in motion. Quantitatively, the kinetic energy of an object is expressed by the following equation:

Expression 7.D

Quantitative Definition of Kinetic Energy

$$K.E. = \frac{1}{2}mv^2$$

K.E. = Kinetic Energy of Object in Motion (joules, J)
m = Mass of Object in Motion (kilograms, kg)
v = Velocity of Object in Motion (meters/second, m/s)

EXAMPLE: A moving ball possesses kinetic energy (Fig. 7.2):

---

2. Although it may not be obvious that the force involved in stretching a rubber band is electrical in nature, this is indeed the case. The electrical nature of chemical bonding was introduced in Chapter 6, and it is discussed in more detail in chapter nine.
3. In practice, nature places severe restrictions on the "spending" of energy. These restrictions are considered in the last section of this chapter.

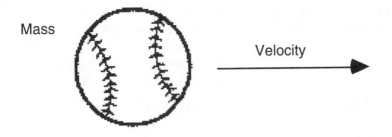

Mass

Velocity

A moving ball posesses kinetic enegy because
it has mass and velocity.

Figure 7.2   An Example of Kinetic Energy

DEFINITION: Potential energy is the energy stored in a system whenever work is done on the
system. Quantitatively, this work is expressed by the general definition of work:

Expression 7.C
Quantitative Definition of Potential Energy

$$W = FS$$

W = Work (joules, J)
F = Any Force (newtons, N)
S = Distance Through Which Force Is Exerted (meters, m)

EXAMPLE: When a stationary weight is suspended above the ground, the earth and the weight form
a system that stores potential energy equal to the work required to lift the weight to its stationary
position (Fig. 7.3).

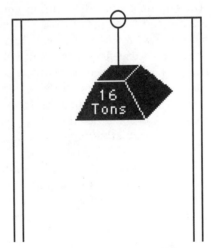

Figure 7.3   An Example of
Potential Energy

    A given object can possess both kinetic and potential energy. For example, a baseball in flight
between the pitcher's mound and home plate possesses kinetic energy because it is a *mass* in *motion*.
It also possesses potential energy because it is a *force* (*weight*) that has been moved (lifted) through a
distance (*i.e.* the distance from the ball to the ground). In the above example, the italicized words

217

represent markers that can be used to identify the presence of a particular form of energy. These italicized quantities are the quantities contained in Expressions 7.C and 7.D. A change in any of these quantities will always be accompanied by a corresponding energy transformation.

7.5 The First Law of Thermodynamics

The energy of the universe is constantly undergoing change. These changes or energy transformations are subject to natural laws called the laws of thermodynamics. There are three laws of thermodynamics, and each will be discussed briefly in this chapter. At this point, it is appropriate to introduce the first law of thermodynamics.

In the early part of the nineteenth century, physicists (and some chemists) were involved with the task of energy classification described in the previous section. The concepts of kinetic and potential energy emerged from the *Principia*, the late seventeenth century work in which Sir Isaac Newton developed a precise theory of motion.[4] But how did such things as heat and light fit into this classification scheme? If the answer to this question is not obvious to you, you must realize that the answer was far from obvious to the natural scientists of the early nineteenth century. About 1842, a major breakthrough occurred. This breakthrough was the discovery of the first law of thermodynamics, first enunciated by the German physician Julius Robert Mayer. This law was more commonly known as the law of conservation of energy.

The First Law of Thermodynamics
Energy cannot be created or destroyed;
it can only be changed in form (transformed).

The first law of thermodynamics was the product of many different nineteenth century minds, and it was also the product of the pragmatic side of science.[5] During the late eighteenth and early nineteenth centuries, steam engines were becoming of increasing industrial importance. The performance of these engines had to be evaluated, and this was accomplished by measuring the engine's "duty"—the number of feet the engine could lift a million pounds of water, using one bushel of coal. The "duty" of an engine was thus defined in terms of work (Equation 7.A), and it was not long before the heat of the engine was associated with the work also. Finally, by 1847, the British physicist James Prescott Joule had performed a quantitative experiment in which he determined the work equivalent of heat. It was Joule's value of 4.18 J/1.00 cal that was used in Example II.

We have returned to an important philosophical theme. There is a practical side to pure science that must provide society with a service of value. In the present case, we see that this situation is a two-way street. The first law of thermodynamics is a bookkeeping principle of pure science, but its origins can be found in the practical world of industrial machines. Science begets technology and technology begets science.

With the invention and development of the steam engine, energy became a major item of commerce. While the total amount of work that an engine could do as a function of fuel consumed was an important measurement, an equally important measurement related to how rapidly the engine could do work. This latter concern led to the development of the concept of power.

DEFINITION: Power is the rate at which work is done. In a more general sense, power measures the rate of the flow of energy:

---

4.   Interestingly, the concept of energy is not mentioned in the *Principia*. Energy as a human concept, however, logically followed from the work of Sir Isaac Newton.
5.   Since this is a chemistry text, it seems appropriate to list the major chemical contributors to this law. In this spirit, the German chemists Karl Friedrich Mohr and Justus Liebig must be mentioned.

Expression 7.E
Quantitative Definition of Power

$$P = \frac{W}{t}$$

P = Power (watts, W)
W = Work (joules, J)
t = Time (second, s)

EXAMPLE: A 100-watt light bulb consumes energy at the rate of 100 joules per second.

EXAMPLE IV

The Power and Kinetic Energy Equations

What power must the engine develop to accelerate a 70.0 kg mass from a velocity of 0 m/s to a velocity of 10.0 m/s in 2.00 seconds?

SOLUTION: 1750 watts

The initial kinetic energy of the mass is zero. The kinetic energy of the mass after 2.00 seconds is given by Expression 7.D:

$$K.E. = \frac{1}{2}mv^2$$

$$K.E. = \frac{1}{2}(70.0 \text{ kg})(10.0 \text{ m/s})^2$$

$$K.E. = 3500 \text{ kg} \bullet \text{m}^2/\text{s}^2$$

$$K.E. = 3500 \text{ J } (1 \text{ J} = 1 \text{ kg} \bullet \text{m}^2/\text{s}^2)$$

Since this work is accomplished in 2.00 seconds, the power is calculated from Expression 7.E as follows:

$$P = \frac{W}{t}$$

$$P = \frac{3500 \text{ J}}{2.00 \text{ s}}$$

$$P = 1750 \text{ J/s}$$

$$P = 1750 \text{ W } (1 \text{ J/s} = 1 \text{ W})$$

Expression 7.E should be familiar to readers who have paid electrical energy bills. When electric power companies record information related to Expression 7.E on an electrical energy bill, they use equation units that are suited to home energy usage—power is measured in kilowatts (kW), time is measured in hours (h), work is measured in kilowatt•hours (kW•h).

7.6 Heat

How did the first law of thermodynamics help solve the energy classification problem? Consider the case of heat.

Heat is something that enters into and exits from matter. But what, exactly, is heat? If we have no knowledge of the law of energy conservation, this is not an easy question to answer. The best that we can do is state what heat does to matter when it enters. For most substances, the addition of **heat** has one of two observable effects (assuming no chemical change):

1. The addition of heat to matter causes the **temperature** of the matter to increase, OR
2. The addition of heat to matter causes a **phase change** to take place (*i.e.* solid to liquid or liquid to gas).

Both of these situations present somewhat of a mystery, since they both seem to involve the disappearance of heat into matter with the concomitant appearance of a new characteristic (*i.e.* a new temperature or a new phase of matter). What happened to the heat?

Historically, the problem was further complicated by the fact that no one really understood what temperature was. The only thing that the early nineteenth century scientist could do was define temperature in terms of how it was measured. Hot objects caused liquid mercury in a thin glass tube to expand, and cold objects caused the liquid mercury to contract. If the mercury level in such a glass tube (thermometer) was marked while inserted in a standard hot object and then marked again while inserted in a standard cold object, the resulting thermometer could be used to measure the degree of temperature relative to these two standards. In 1742, the Swedish astronomer Aners Celsius suggested boiling water and freezing water as the hot and cold standards. This system was eventually accepted as the metric standard. (Fig. 7.4).[6]

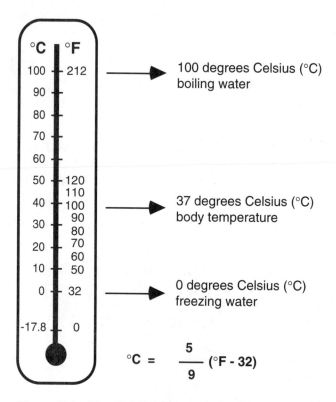

$$°C = \frac{5}{9}(°F - 32)$$

Figure 7.4   The Celsius Temperature Scale

---

6.   The Celsius scale is not the temperature scale used by the Système International (SI). It is, however, the temperature scale that will be used in this text.

This operational definition only defined temperature in terms of how it was measured, and it did not really explain the relationship between temperature and heat. With the advent of the first law of thermodynamics, however, the cloud of mystery began to disappear.

Once it is recognized that heat is a form of energy and that energy cannot be destroyed, it becomes clear that heat cannot disappear when it passes into matter. As energy, heat can only be deposited in matter. Further, it must be deposited as kinetic and/or potential energy. But what is it inside of the matter that is moving and experiencing force? Could it be that the molecular world is really a dynamic world of energetic molecules, with temperature change and phase change simply being real world reflections of changes in the kinetic and potential energy of molecules? By the end of the nineteenth century, the physicist's answer to this latter question was a definite "yes." The molecule, which was so successful in explaining chemical change, was to become equally successful in explaining physical change.

7.7 Energy and the Molecular World

In the early part of the nineteenth century, the molecule was a theoretical concept that explained mass relationships in chemical reactions. Although the application of this concept to chemical change met with great success, this was not the case with physical change. For example, John Dalton's attempts to use the molecular concept to explain the physical behavior of gases were a dismal failure. It was not until the latter part of the nineteenth century that a satisfactory molecular theory of physical change emerged. What follows is a brief qualitative outline of this kinetic molecular theory:

The Kinetic Molecular Theory of Solids, Liquids, and Gases

1. Molecules exist. (This may have been obvious to chemists in the early part of the nineteenth century, but in the world of physics, the molecule as a viable theoretical entity was far from obvious.)
2. Molecules are in a constant state of motion, and hence they possess an inherent kinetic energy.
3. Collisions between molecules result in no net loss of kinetic energy.
4. Molecules attract one another. Hence any large collection of molecules possesses an inherent potential energy.

   NOTE: It must be remembered that the atoms of a molecule are held together by strong attractions called *chemical bonds*. Hence, the inherent potential energy of a large collection of molecules has two origins (Fig. 7.5):

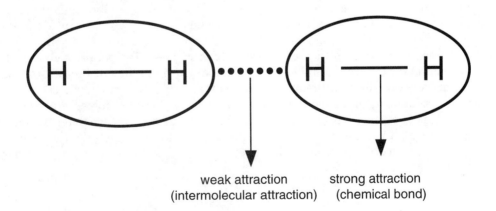

weak attraction | strong attraction
(intermolecular attraction) | (chemical bond)

Figure 7.5 Forces Related to Molecular Potential Energy

NOTE: In the diagram above, the chemical bond between the two hydrogen atoms represents a force acting at a distance. The chemical bond, therefore, holds stored or potential energy. In a similar manner, the weak intermolecular attraction between the H_2 molecules holds a smaller amount of potential energy.[7]

5. Average molecular kinetic energy is directly proportional to the Kelvin temperature of a large collection of molecules.

   NOTE: The Kelvin temperature scale (°K) has the same degree size as the Celsius temperature scale (°C), but it has a different zero point that is 273.15 degrees lower than the Celsius zero point (°K = °C + 273.15). The Kelvin temperature scale moves beyond a purely operational definition of temperature. It actually defines temperature in terms of the following theoretical expression:

   $$\text{Average Molecular Kinetic Energy} \propto \text{Kelvin Temperature}$$

   The experimental details associated with measuring the Kelvin zero point of temperature are beyond the scope of this text. The important thing to be appreciated about item five is that a temperature increase is a real world reflection of an increase in the average kinetic energy of molecules.

6. Phase changes represent changes in the average distances between molecules. Hence a phase change corresponds to a change in molecular potential energy.
7. The molecules of a solid move by vibrating about fixed ordered positions, while the molecules of a liquid or gas move more freely in a random manner.

These seven points represent a brief qualitative summary of a theory that correlates and explains a vast amount of factual information. It is instructive to go through a short thought experiment that correlates the aspects of the theory to factual information.

Consider adding heat to a pure solid substance that is at an initial temperature below its melting point and under the influence of normal atmospheric pressure. As heat is added to the solid, it is observed that the temperature of the solid increases. This is a fact that most people expect. Now attempt to visualize the events taking place at the molecular level. Initially, the molecules of the solid substance are vibrating about fixed ordered positions. As the heat energy (symbolized by the letter Q) enters the substance, the molecules begin to vibrate faster. This has the effect of increasing the average kinetic and potential energy of the molecules in accordance with the law of energy conservation. The increase in kinetic energy is caused by the increase in the motion of the molecules. In this case, since a vibrating system must also store energy, the average potential energy of the molecules increases along with the average kinetic energy of the molecules (Fig. 7.6).

As additional heat energy is added to the solid, the average vibrational kinetic energy of the molecules continues to increase. Factually, it is observed that the temperature of the solid continues to increase. Finally, the melting point of the solid is reached, and at this point, an interesting observation is made—the addition of heat no longer causes a temperature change. It is observed that the solid melts to form a liquid, but the temperature remains at the constant **melting point**. If these events are visualized at the molecular level, the liquefaction process corresponds to the molecules randomizing and moving further apart. Since there are forces of attraction between the molecules, this increase in distance means an increase in the average potential energy of the molecules. During the liquefaction process, the average kinetic energy of the molecules does not change. The motion of the molecules, however, changes from vibration to translation or motion in a straight line (Fig. 7.7).

---

7. The nineteenth century chemist did not fully understand the nature of these attractions. Since all forces are subject to Expression 7.C, this lack of understanding did not prevent application of the kinetic molecular theory. Understanding the basic nature of chemical bonds has been a central theme of twentieth century chemistry.

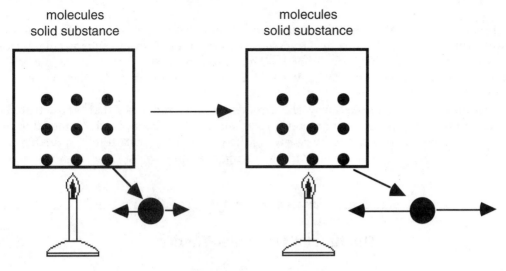

molecules
solid substance

molecules
solid substance

Addition of heat increases molecular vibration
in a solid substance.

Figure 7.6 The Addition of Heat to Matter (Temperature Change)

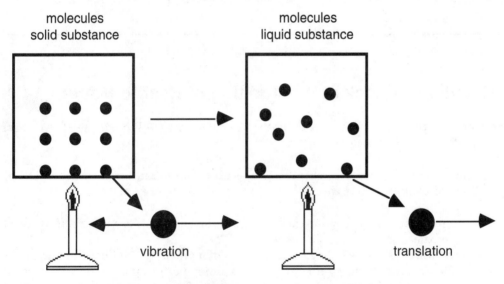

molecules
solid substance

molecules
liquid substance

vibration

translation

Addition of heat causes increased molecular distance.
Molecular vibration changes to molecular translation.

Figure 7.7 The Addition of Heat to Matter (Phase Change)

After the solid has totally melted, the addition of heat will again cause the temperature to increase. During this period of temperature increase, the thermal energy that enters the liquid is again stored as molecular kinetic and potential energy, with the kinetic energy increase reflected as a measurable temperature increase. Although the molecules of the liquid no longer vibrate about fixed positions, some of the added heat energy is stored as molecular potential energy because of internal vibrations within complex molecules. At this point, the scenario of the previous paragraph is repeated. This time, however, the phase change—which again takes place at constant temperature—is from liquid to gas. The constant temperature in this case is called the **boiling point** of the substance. Once again, the heat energy that is absorbed during this phase change serves to increase the average distance between the molecules (*i.e.* increase the average potential energy of the molecules). During this final phase change, the gas that is formed fills the entire container. Evidently, the molecular separation in the gaseous state is quite large, with the molecules occupying the space by virtue of their motion.

The dynamic picture of the molecular world of solids, liquids, and gases that is illustrated above can help us understand many physical characteristics of matter. For example, the removal of heat from a pure substance can be understood by reversing the logic of the illustration. To further illustrate the use of this theory to explain fact, consider the following example:

EXAMPLE V

The Kinetic Molecular Theory

When water evaporates from your skin, your skin feels cold. This is a fact. Explain this fact in terms of the kinetic molecular theory.

SOLUTION:
In order for the liquid-to-gas phase change to take place, the molecules of the liquid must move further apart. This represents an increase in the average potential energy of the water molecules. Your skin supplies this energy to the water in the form of heat, hence the cooling effect.

7.8 A Quantitative Look at the Addition of Heat to Matter

We have seen that the addition of heat to most pure substances will have one of two observable effects (assuming no chemical change):

1. The addition of heat to matter causes the temperature of the matter to increase, OR
2. The addition of heat to matter causes a phase change to take place.

We have also considered a kinetic molecular rationalization for both of these effects. Consider now the quantitative aspects of these two observable effects.

For all pure substances, a phase change is always accompanied by the liberation or consumption of energy. Further, the amount of energy liberated or consumed per gram of material is a constant for a phase change of a given substance at a given temperature and pressure. All pure substances have a constant **heat of vaporization** and a constant **heat of melting**. For example, the evaporation (liquid to gas) of one gram of water consumes 540 calories. In chemistry, any process that consumes energy is said to be **endothermic**. Hence for water we can write:

An Endothermic Process
1 gram of water liquid → 1 gram of water gas

↑

consumes 540 calories

If the above process is reversed, and one gram of gaseous water is condensed (gas to liquid), then the direction of the energy flow is reversed. In chemistry, any process that liberates energy is said to be **exothermic**. Hence for water we can write:

An Exothermic Process

1 gram of water gas → 1 gram of water liquid

↓

liberates 540 calories

Since the heat consumed or liberated in these processes did not result in a temperature change for the water, early scientists tended to view the heat as "hidden." The heat required to evaporate one gram of a substance was, therefore, called the latent heat (hidden heat) of vaporization. Pure substances also have constant heats of melting. For water, the heat of melting is 80 calories per gram.

Expression 7.F
Heat Producing a Phase Change

$$Q = P_c m$$

Q = Heat Added to (or Subtracted from) Substance (cal)
P_c = Constant Heat of Phase Change (cal/g)
m = Mass of Substance (g)

——— EXAMPLE VI ———

Heat Producing a Phase Change

How much heat is required to melt a 25.0 gram ice cube? The constant heat of melting for water is 80.0 cal/g.

SOLUTION: 2000 cal

$$Q = P_c m$$
$$Q = (80.0 \text{ cal/g})(25.0 \text{ g})$$
$$Q = 2000 \text{ cal}$$

If the addition of heat to a substance does not produce a phase change, then it will produce a temperature increase. Just as each substance requires a unique quantity of heat to produce a phase change in one gram of material, so also each substance requires a unique quantity of heat to raise the temperature of one gram of material one degree Celsius. This unique quantity of heat is called the substance's specific heat. Since it requires 1.00 calorie to raise the temperature of one gram of water one degree Celsius, the specific heat of water is 1.00 calorie per gram degree Celsius (abbreviated cal/g•°C). Each substance has its own unique specific heat.[8] For example, the specific heat of the element nickel is 0.104 cal/g•°C.

How many calories would be required to raise the temperature of 50.0 grams of water from 25.0 degrees Celsius to 30.0 degrees Celsius? Since it requires 1.00 calorie to raise the temperature of each gram of water by one degree Celsius, it would require 50.0 calories to raise the temperature of 50.0 grams of water by one degree Celsius. An increase in the temperature from 25.0 to 30.0 degrees Celsius, however, represents a net increase of 5.0 degrees Celsius. Hence the total amount of heat

---

8.   The specific heat of a substance is actually a function of temperature, pressure, and physical phase. For a given phase of matter, the change in the specific heat of a substance over a relatively small temperature range is not very large.

required is 250 calories (50.0 cal/°C X 5.0°C). Notice that the logic of the solution to this problem is summarized by a simple equation:

Expression 7.G
Heat Producing a Temperature Change

$$Q = cm\Delta T$$

Q = Heat Added to (or Subtracted from) Substance (cal)
c = Specific Heat of Substance (cal/g•°C)
m = Mass of Substance (g)
ΔT = Change in Temperature That Substance Experiences (°C)

EXAMPLE VII

Heat Producing a Temperature Change

How much heat is required to raise the temperature of 25.0 grams of nickel from 25.00°C to 29.00°C. The specific heat of nickel is 0.104 cal/g•°C.

SOLUTION: 10.0 cal

$$Q = cm\Delta T$$
$$Q = (0.104 \text{ cal/g•°C})(25.0 \text{ g})(4.00°C)$$
$$Q = 10.4 \text{ cal}$$

7.9 The Measurement of Heat Flow During a Chemical Reaction

The heat flow that occurs during a chemical reaction is measured with a device called a **calorimeter**. In most calorimeters, the heat to be measured is transferred to or from water in such a way that a phase change does not occur. By observing the temperature change of a known mass of water, Expression 7.G can be used to calculate the heat transferred. In the case of combustion reactions, a specially designed calorimeter called a bomb calorimeter (Fig. 7.8) is used. Its use exemplifies calorimetric measurement in general.

Consider the combustion of methanol, CH_4O, which is represented by the following chemical equation:

Chemical Equation 7.A
$$2 \text{ } CH_4O + 3 \text{ } O_2 \rightarrow 2 \text{ } CO_2 + 4 \text{ } H_2O$$

If a known mass of methanol is placed inside of the bomb along with a large excess of oxygen, then the ignition wire when heated electrically will cause the complete combustion of the methanol. If the combustion is carried out while the bomb is immersed in the water contained in the calorimeter, then the heat evolved by the combustion reaction will be transferred to the water. Knowledge of the mass of the water and the temperature change allows Expression 7.G to be used to calculate the heat transferred. Actually, some of the heat is transferred to the calorimeter itself, but this fact can be ignored without destroying the logic of the experimental approach. In practice, the heat transferred to the calorimeter and the heat transferred to the water must be calculated.

—— EXAMPLE VIII ——

Calorimeter Calculation

A sample of methanol, CH_4O, weighing 1.00 gram is burned in a bomb calorimeter (Fig. 7.8). The calorimeter contains 1780 grams of water (specific heat = 1.00 cal/g•°C). After combustion, the temperature of the water rises from 25.00°C to 28.83°C. Is the chemical reaction endothermic or exothermic? How many calories of heat are exchanged during the chemical reaction?

SOLUTION: The reaction is exothermic because the heat evolved by the chemical reaction increases the temperature of the water; 6.82 kcal of heat is exchanged.

Since the heat transferred to the water causes a temperature increase, Expression 7.G is applicable:

$$Q = cm\Delta T$$
$$Q = (1.00 \text{ cal/g•°C})(1780 \text{ g})(3.83°C)$$
$$Q = 6820 \text{ cal (rounded off)}$$
$$Q = 6.82 \text{ kcal}$$

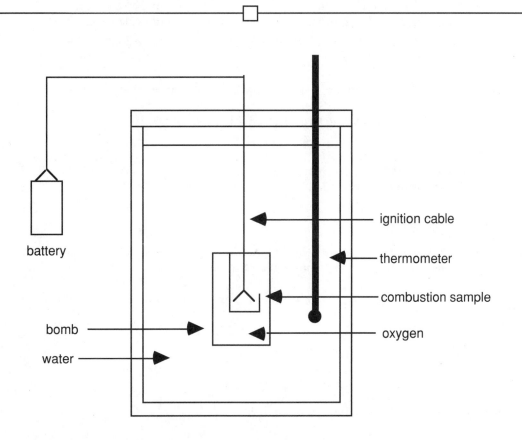

Figure 7.8 A Bomb Calorimeter

7.10 The Thermochemical Equation

The example above (Example VIII) illustrates how the heat of a combustion reaction is measured. The result of this calculation indicates that 6.82 kcal of heat is released when 1.00 gram of methanol is burned. This result can be used to make Chemical Equation 7.A a more complete description of the burning of methanol. Recall from Chapter 5, that a chemical equation always implies a set of atomic/molecular combining masses. Hence Chemical Equation 7.A can be written as follows:

Chemical Equation 7.A

$$2\ CH_4O\ +\ 3\ O_2\ \rightarrow\ 2\ CO_2\ +\ 4\ H_2O$$

$$\begin{array}{cccc} 2(32.0) & 3(32.0) & 2(44.0) & 4(18.0) \\ \text{amu} & \text{amu} & \text{amu} & \text{amu} \end{array}$$

In order to incorporate energy information into this chemical equation in a meaningful way, chemists agree to rewrite the equation with a new set of implied combining masses—the mole masses:

Chemical Equation 7.B

$$2\ CH_4O\ (l)\ +\ 3\ O_2\ (g)\ \rightarrow\ 2\ CO_2\ (g)\ +\ 4\ H_2O\ (l)$$

$$\begin{array}{cccc} 2(32.0) & 3(32.0) & 2(44.0) & 4(18.0) \\ \text{grams} & \text{grams} & \text{grams} & \text{grams} \end{array}$$

In Chemical Equation 7.B, the implied combining masses are the atomic/molecular combining masses expressed as grams. As discussed in Chapter 5, this represents an extension of the mole concept to include chemical compounds. The parenthetical letters indicate the phase of each substance in the reaction—(s) solid, (l) liquid, and (g) gas, and the reason for their inclusion in the equation will be discussed later. The obvious energy value to incorporate into Chemical Equation 7.B is the amount of heat that the combustion of 64.0 grams (2 × 32.0) of methanol would release (*i.e.* 2 moles of methanol). This can be calculated from the result of Example VIII by the application of linear reasoning.

EXAMPLE IX

Chemical Energy and Linear Reasoning

If the combustion of 1.00 gram of methanol releases 6.82 kcal, how much heat will be released by the combustion of 64.0 grams of methanol?

SOLUTION: 436 kcal

It seems reasonable to assume that doubling the quantity of methanol burned would double the amount of heat released. This logic is demanded by the law of energy conservation. Hence:

$$Q \propto \text{Mass of Methanol}$$

OR

$$\frac{Q}{\text{Mass of Methanol}} = \text{Constant}$$

Linear reasoning now allows the following calculation:

$$\frac{6.82\ \text{kcal}}{1.00\ \text{g methanol}} = \frac{X}{64.0\ \text{g methanol}}$$

$$X = 436\ \text{kcal}$$

If the energy value computed in Example IX is listed with Chemical Equation 7.B, the result is a **thermochemical equation** in which the energy value is related to each of the combining masses implied by the equation:

Chemical Equation 7.C
An Exothermic Thermochemical Equation

$$2 \text{ CH}_4\text{O (l)} + 3 \text{ O}_2\text{(g)} \rightarrow 2 \text{ CO}_2\text{(g)} + 4 \text{ H}_2\text{O (l)} \qquad Q = -436 \text{ kcal}$$
$$\begin{array}{cccc} 2(32.0) & 3(32.0) & 2(44.0) & 4(18.0) \\ \text{grams} & \text{grams} & \text{grams} & \text{grams} \end{array}$$

The minus sign is incorporated in the thermochemical equation to indicate that the reaction is exothermic (liberates energy). Endothermic reactions (consume energy) are indicated by the similar use of a plus sign.

Chemical Equation 7.D
An Endothermic Thermochemical Equation

$$\text{Br}_2\text{ (l)} + \text{Cl}_2\text{ (g)} \rightarrow 2 \text{ BrCl (g)} \qquad Q = +6.27 \text{ kcal}$$
$$\begin{array}{ccc} 160 & 71.0 & 231 \\ \text{grams} & \text{grams} & \text{grams} \end{array}$$

Since the energy liberated or consumed by a reaction is linearly related to the combining mass of each component of the reaction, the following set of constant ratios can be constructed from any thermochemical equation:

Expression 7.H
Constant Thermochemical Ratios

$$\frac{Q \text{ Liberated or Consumed by Reaction}}{\text{Combining Mass of Any Reaction Component}} = \text{Constant}$$

EXAMPLE X

Energy/Mass Ratios in a Thermochemical Equation

Express all of the energy/mass constant ratios implied by the thermochemical Equation 7.D.

SOLUTION:
Using Expression 7.H above:

$$6.27 \text{ kcal}/(1 \times \text{Br}_2) = 6.27 \text{ kcal}/160 \text{ g Br}_2$$

$$6.27 \text{ kcal}/(1 \times \text{Cl}_2) = 6.27 \text{ kcal}/71.0 \text{ g Cl}_2$$

$$6.27 \text{ kcal}/(2 \times \text{BrCl}) = 6.27 \text{ kcal}/231 \text{ g BrCl}$$

These constant ratios are subject to the arithmetic of linear reasoning. The application of this arithmetic allows the chemist to make predictions about energy flow during the chemical reaction process. In the next section, we will consider a practical application of this type of prediction.

7.11 An Example of Applied Thermochemistry

The thermochemical equation represents an extension of the atomic/molecular theory of chemical change. This extension incorporates energy information into the chemical equation. In the same way the chemical equation allows the chemist to make predictions about the mass behavior during chemical reactions, the thermochemical equation allows the chemist to make predictions about energy flow. In a society that has come to value the energy produced by chemical reactions with the same relish as the materials produced by chemical reactions, there are many examples of the utility of thermochemical prediction. This section considers one such example from the area of nutrition.

The result of Example II in this chapter indicates that ten pull-ups done by an average male are equivalent to less than one nutritional Calorie of work. If you consider that an apple contains the energy equivalent of about 70 nutritional Calories, one nutritional Calorie for every ten pull-ups seems quite low. In the language of a weight watcher, it would seem that more than 700 pull-ups are required to "burn" off an apple. This statement assumes that human beings convert the Caloric value of food into work with 100 percent efficiency, and this is far from the truth. How many nutritional Calories does an average man or woman require in order to do a normal day's work?

To understand how the chemist can help answer this question, it is essential to understand the biochemical fate of the food we eat. Food is eaten to provide essential nutrients for body maintenance, and it is eaten for its fuel value. This latter function is evaluated in terms of the Caloric value of food, a measurement that is actually made with a bomb calorimeter (Fig. 7.8). To a reasonable approximation, the food that the human body uses as fuel is processed to form blood sugar (glucose), $C_6H_{12}O_6$, which is then biologically "burned" according to the following thermochemical equation representing the process of **respiration**:

Chemical Equation 7.E
Biological "Burning" or Respiration

$$C_6H_{12}O_6\,(aq) + 6\ O_2\,(g) \rightarrow 6\ CO_2\,(g) + 6\ H_2O\,(l) \qquad\qquad Q = -686\ kcal$$
Exothermic

This chemical reaction takes place in cells throughout the body, and the value of 686 kcal corresponds to the biological pathway of this reaction, with the oxygen being supplied via the circulatory system. The waste carbon dioxide and water are exhaled, and the oxygen is inhaled during the breathing process. For simplicity, the parentheticals *aq, g,* and *l* are being used somewhat erroneously in this equation to indicate glucose respiration in cellular "water" or "aqueous" solution. If Chemical Equation 7.E represents the major source of biochemical energy, then the energy produced is linearly related to the carbon dioxide exhaled. One method for determining the Caloric output of normal activity is to collect and weigh the carbon dioxide exhaled during normal activity.

EXAMPLE XI

A Practical Use of a Thermochemical Equation

During one hour of normal daily activity, a male is found to exhale 32.4 grams of *carbon dioxide*. Assuming that Chemical Equation 7.E represents the major source of biochemical energy and that the man's level of activity remains constant throughout a 24-hour day, calculate the man's *Caloric* output during a normal day.

SOLUTION: 2020 Cal (kcal)

The first step in a problem of this type is to identify the energy/mass ratio that applies to the wording of the problem. The italicized words in this problem indicate that the appropriate ratio is 686 kcal/$(6 \times CO_2)$ = 686 kcal/264 g CO_2.

NOTE: In calculations of this type, it is the magnitude of the energy that is important. Hence, to avoid confusion, the minus sign indicating an exothermic reaction is omitted when writing the energy/mass ratio.

The second step of the problem is to use linear reasoning to make the thermochemical equation predict a useful fact:

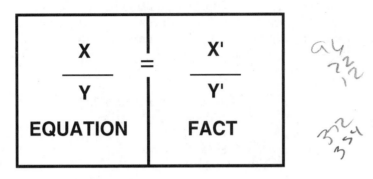

$$\frac{686 \text{ kcal}}{264 \text{ g } CO_2} = \frac{X \text{ kcal}}{32.4 \text{ } CO_2}$$

EQUATION = FACT

X = 84.2 kcal per hour (rounded off)

(84.2 kcal/hr) × (24 hr/day) = 2020 kcal/day (rounded off)

2020 nutritional Calories

Actually, the average Caloric intake for an American male is about 2500 Calories per day. If this figure is compared to an average of about 6000 Calories per day for an American male at the turn of the century, the term sedentary lifestyle takes on new meaning. For many of us, the discovery of a more sedentary lifestyle requires but a glance in the mirror. It does not require an act of chemical clairvoyance!

In human muscle tissue, the biological pathway of the respiration reaction provides only 263 kilocalories of force times distance (lifting) work for every mole of glucose "burned." This means that 423 kcal/mole are liberated as body heat. This corresponds to about 38% efficiency. When this efficiency is compared to the miserable performance of an automobile engine (about 5%), it becomes clear that life energy processes are really quite amazing. The miserable performance of the internal combustion engine and other heat engines is explored in more detail in section 7.14.

Chemical Equation 7.E represents a bit of chemistry that unites almost all life forms on this planet. Trees, bacteria, mice, and humans all use glucose, $C_6H_{12}O_6$, as biological fuel. Since glucose is this planet's universal biological fuel, there must be some regeneration mechanism. There is, of course, and this regeneration mechanism is called **photosynthesis**. It takes place in green plants under the influence of chlorophyll, and its thermochemical equation is shown below:

Chemical Equation 7.F
Photosynthesis

$6 \text{ } CO_2 \text{ (g)} + 6 \text{ } H_2O \text{ (l)} \rightarrow C_6H_{12}O_6 \text{ (aq)} + 6 \text{ } O_2 \text{ (g)}$ 　　　 Q = +686 kcal
　　　　　　　　　　　　　　　　　　　　　　　　　　　　　　　 Endothermic

Since the net equation for photosynthesis is simply the reverse of the net equation for respiration, the thermochemical equation for photosynthesis indicates that it is an endothermic process. During natural photosynthesis, this energy input is supplied by solar energy. Note that the magnitudes of the energies represented in Chemical Equations 7.E and 7.F are the same, but that the signs of the energy flows are opposite. This condition holds for all reversible chemical reactions, and it is a consequence of the law of conservation of energy.

To at least some extent, almost all life forms on this planet also share another bit of chemistry. If deprived of oxygen, most life forms can still use glucose for energy by means of **anaerobic glycolysis**. The process of yeast anaerobic glycolysis is called **fermentation**, and the thermochemical equation for this oxygen starvation energy production is described in Chemical Equation 7.G. Human anaerobic glycolysis is described in Equation 7.H.

Chemical Equation 7.G
Yeast Fermentation

$$C_6H_{12}O_6 \text{(aq)} \rightarrow 2\ C_2H_6O \text{ (aq)} + 2\ CO_2 \text{(g)} \qquad Q = -26.0 \text{ kcal}$$
$$\text{ethyl alcohol} \qquad\qquad\qquad \text{Exothermic}$$

Chemical Equation 7.H
Human Anaerobic Glycolysis

$$C_6H_{12}O_6 \text{(aq)} \rightarrow 2\ C_3H_6O_3 \text{(aq)} \qquad Q = -26.0 \text{ kcal}$$
$$\text{lactic acid} \qquad\qquad \text{Exothermic}$$

Twenty-six kcal is a poor substitute for 686 kcal, but at least life goes on. In the case of yeast, the products of this reaction are carbon dioxide, CO_2, and ethyl alcohol, C_2H_6O. The equation represents the process of fermentation, and it illustrates nicely a dichotomy that exists in chemistry. When the brewer chokes off the poor yeast cell by sealing the brewing vat, the brewer is the chemical materialist interested in a material product (actually a waste product) of the reaction. But chemical reactions also involve energy, and to the little yeast cell this is the whole ball game—life itself. Here is the true pragmatism of chemistry. It is the provider of material, energy, and life itself.[9]

7.12 The Atomic/Molecular Origin of Chemical Energy

In writing a thermochemical equation, factual information is appended to a normal chemical equation. This act alone does not constitute an extension of atomic/molecular theory. When an attempt is made to identify the atomic/molecular origin of this energy, however, the original theory is being extended. What is the origin of the energy that flows during the chemical reaction process?

In order to answer this question, consider again Chemical Equation 7.D.

Chemical Equation 7.D
An Endothermic Thermochemical Equation

$$Br_2 \text{ (l)} + Cl_2 \text{ (g)} \rightarrow 2\ BrCl \text{ (g)} \qquad Q = +6.27 \text{ kcal}$$

| 160 grams | 71.0 grams | 231 grams |

---

9. The use of microorganisms to do our material chemistry has come a long way. Fermentation to produce ethyl alcohol is perhaps the oldest example of industrial microbiological chemistry. In recent years, in a feat that staggers the imagination, genetically engineered bacteria have been used to synthesize human insulin!

Whenever a thermochemical equation such as this is recorded, it is recorded at a constant reaction temperature—usually 25 degrees Celsius. How this is accomplished technically is not important. It is important to realize that the reaction represented by Chemical Equation 7.D consumes 6.27 kcal when the reactants and products are at the same temperature (25°C). Since the reaction is held at a constant temperature, the kinetic molecular theory dictates that the average kinetic energy of the molecules remains constant. If this is the case, the origin of the chemical energy must be identified with changes in the inherent potential energy of the reacting molecules. Recall that the inherent potential energy of the molecules has two origins—chemical bonds between atoms within a molecule and weak attractions between molecules (Fig. 7.9).

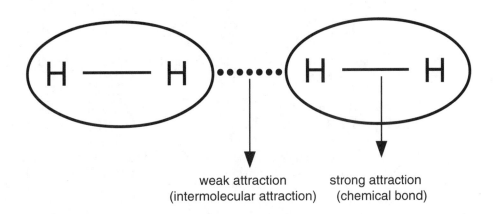

weak attraction   strong attraction
(intermolecular attraction)   (chemical bond)

Figure 7.9 The Source of Chemical Energy

Chemical Equation 7.D involves changes in both types of bonding. The reactants are a liquid element (Br_2) and a gaseous element (Cl_2). The product is a gaseous compound (BrCl). Clearly, the weak attractions that hold the molecules of a liquid close together must be overcome during this chemical reaction. This is the reason that a thermochemical equation always indicates the phase of each reacting substance. That Chemical Equation 7.D also involves an alteration of chemical bonds between atoms is made clear by the following representation of the equation for this reaction:

Chemical Equation 7.D
An Endothermic Thermochemical Equation Showing Bond Changes

$$Br–Br \ (l) + Cl–Cl \ (g) \rightarrow Br–Cl \ (g) + Br–Cl \ (g)$$

The atomic/molecular origin of chemical energy can actually be investigated quantitatively by using the concept of **bond dissociation energy**. The bond dissociation energy for a diatomic molecule is simply the energy required to break the molecule into two atoms. For many diatomic molecules this energy can be measured experimentally, and then expressed as an appropriate thermochemical equation. Equations 7.I, 7.J, and 7.K show the thermochemical equations for the bond dissociation of three diatomic molecules:

Chemical Equations 7.I, 7.J, & 7.K

7.I	$H_2 \ (g) \rightarrow 2 \ H \ (g)$	Q = +104.0 kcal
7.J	$Cl_2 \ (g) \rightarrow 2 \ Cl \ (g)$	Q = +58.0 kcal
7.K	$HCl \ (g) \rightarrow H \ (g) + Cl \ (g)$	Q = +103.2 kcal

Notice that the reactions represented by these equations are all endothermic, and that bond dissociation energies have positive values. Bond dissociation energies for diatomic and more complex molecules can be used to calculate energy transfers for other chemical reactions. These calculations are based on three rules that apply to the manipulation of all thermochemical equations. All of these rules are a consequence of the law of conservation of energy.

Thermochemical Equation Rules

1. Reversing the direction of a thermochemical equation reverses the sign of the energy of the equation:

$$H_2 (g) \rightarrow 2 H (g) \qquad\qquad Q = +104.0 \text{ kcal}$$
$$2 H (g) \rightarrow H_2 (g) \qquad\qquad Q = -104.0 \text{ kcal}$$

2. Increasing the stoichiometric coefficients of a thermochemical equation by an integer multiple increases the energy by the same multiple:

$$HCl (g) \rightarrow H (g) + Cl (g) \qquad\qquad Q = +103.2 \text{ kcal}$$
$$2 HCl (g) \rightarrow 2 H (g) + 2 Cl (g) \qquad\qquad Q = +206.4 \text{ kcal}$$

3. Adding two or more thermochemical equations to form a net thermochemical equation results in a net thermochemical equation with an energy that is the algebraic sum of the added equations:

$$H_2 (g) \rightarrow 2 H (g) \qquad\qquad Q = +104.0 \text{ kcal}$$
$$Cl_2 (g) \rightarrow 2 Cl (g) \qquad\qquad Q = +58.0 \text{ kcal}$$
$$2 H (g) + 2 Cl (g) \rightarrow 2 HCl (g) \qquad\qquad Q = -206.4 \text{ kcal}$$
$$H_2 (g) + Cl_2 (g) \rightarrow 2 HCl (g) \qquad\qquad Q = -44.4 \text{ kcal}$$

Rule three in this list is called **Hess' law** of constant heat summation after Germain Henri Hess the nineteenth century chemist who enunciated this form of the law of conservation of energy. In the example above, Hess' law is being used to calculate the energy of a chemical reaction from bond dissociation energies, thus tracing the origin of this energy to the chemical bonds of the reactant and product molecules. Hess' law is of general utility, and it applies to the summation of all thermochemical equations.

EXAMPLE XII

Hess' Law of Constant Heat Summation

One of the uses of Hess' law is the calculation of reaction heat values for chemical reactions that are difficult or impossible to run in calorimeters. The following two thermochemical equations represent complete combustion reactions that are very easy to run in a bomb calorimeter:

$$(1) \quad C (s) + O_2 (g) \rightarrow CO_2 (g) \qquad Q = -94.1 \text{ kcal}$$
$$(2) \quad 2 CO (g) + O_2 (g) \rightarrow 2 CO_2 (g) \qquad Q = -135.4 \text{ kcal}$$

The heat value for the thermochemical equation that represents the incomplete combustion of carbon to form carbon monoxide cannot be determined by direct calorimetric methods. Use Hess' law to determine the heat of this reaction from thermochemical equations (1) and (2).

$$2 C (s) + O_2 (g) \rightarrow 2 CO (g) \qquad Q = ? \text{ kcal}$$

234

SOLUTION: –52.8 kcal

The following thermochemical manipulations of equations (1) and (2) allow for the calculation of the missing reaction heat value:

$$2\ C\ (s) + 2\ O_2\ (g) \rightarrow 2\ CO_2\ (g) \qquad Q = -188.2\ \text{kcal}$$
$$2\ CO_2\ (g) \rightarrow 2\ CO\ (g) + O_2\ (g) \qquad Q = +135.4\ \text{kcal}$$
$$2\ C\ (s) + O_2\ (g) \rightarrow 2\ CO\ (g) \qquad Q = -52.8\ \text{kcal}$$

———————————————□———————————————

The attempt to explain the molecular origin of chemical energy represents a milestone in the development of the atomic/molecular theory of chemical change. The problem with the chemical equation, however, is that it is a statement about individual molecules. Chemical reactions, on the other hand, take place between large collections of molecules. In an attempt to explain the factual energy value appended to the thermochemical equation, the chemist turns to a physical theory that deals with large collections of molecules, the kinetic molecular theory of solids, liquids, and gases. Once the behavior of large numbers of molecules is understood, the chemical equation becomes a much more meaningful expression.

7.13 Statistics: A Mathematical Tool for the Kinetic Theory

The qualitative statement of the kinetic molecular theory given in Section 7.7 makes reference to the average characteristics of molecules. For example, the Kelvin temperature of a gas is said to be directly proportional to the average kinetic energy of the gas molecules. At the quantitative level, statements such as this present some real problems. What is the meaning of the word "average" in a molecular world controlled by random factors?

The quantitative understanding of the behavior of a large number of molecules moving in a random way requires the mathematical tool of statistics. Since the study of statistics is of great importance in many areas of human study, it will be useful at this point to consider the application of statistics to the kinetic molecular theory. Consider the following nonmolecular example.

Imagine an immense room with 1,000 people standing in the middle. Each individual desires to reach the west wall of the room. In order to accomplish this, however, each individual must follow a very specific rule: Flip a coin. If the coin falls heads, take one step east. If the coin falls tails, take one step west.

Although it is impossible to predict if a given individual will ever reach the west wall, the fate of all 1,000 individuals can be predicted with a fair degree of certainty. This point is illustrated by the bar graphs below which summarize the results after zero, one, and two coin flips, respectively (Fig. 7.10).

The trend that seems to be developing here will continue. With a really immense number of individuals in a truly immense room, the bar graph will assume the shape of a smooth curve called a **Gaussian distribution** (Fig. 7.11).

This distribution of randomly controlled movements indicates that most individuals will end up in the center region of the room; some will make progress toward the west wall; and some, alas, will make negative progress toward the west wall. This distribution of randomly controlled events is typical of fifty/fifty chance (probability), and it is called the "normal" or "bell" shaped distribution. It is also called the Gaussian distribution after the 18th century mathematician Karl Friedrich Gauss, who discovered the importance of this distribution.

Examples of nature's respect for the Gaussian distribution abound. Consider the trivial example of popcorn popping. The pop, pop-pop, popity-popity-popity, pop-pop, pop so familiar to the connoisseur of popcorn is merely an audible expression of the Gaussian nature of the popcorn kernels' popping ability. Many of nature's randomly controlled events produce skewed but equally predictable

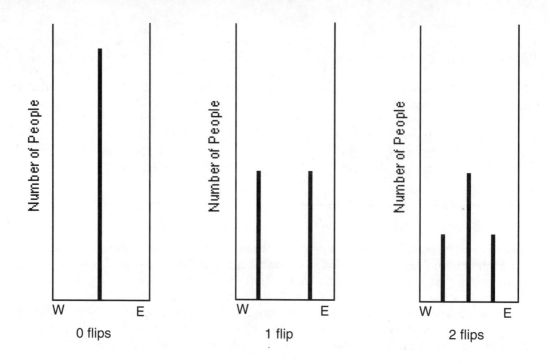

Figure 7.10 Distribution of Individuals Playing Coin Flip Game

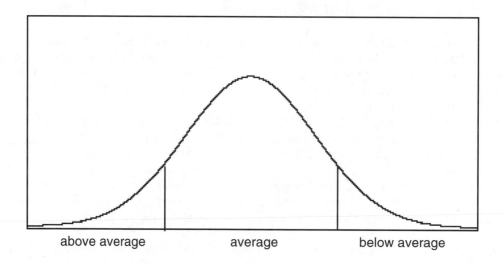

above average   average   below average

Figure 7.11 Gaussian Distribution

distributions. Since molecules moving in a random way are subjected to changes governed by the laws of probability, their motion can also be represented by a type of skewed normal distribution. In the quantitative applications of kinetic theory, these distributions give meaning to the term *average*.

7.14 The Second and Third Laws of Thermodynamics

A discussion of energy relationships in chemical reactions requires a brief mention of the second and third laws of thermodynamics.

Consider a closed box containing one hundred pennies all arranged heads up. If the box is shaken violently for one minute, what are the chances of opening the box and finding all the pennies still arranged heads up? The answer to this question is, of course, almost no chance at all. The pennies arranged in heads up positions are ordered, and this is a situation of low probability. The act of violent shaking quickly corrects this situation. And upon opening the box, a state of much higher probability is discovered—a state of disorder. If the number of pennies is large enough, probability will reign supreme with 50% heads and 50% tails.

In the world of molecules, there is constant violent shaking, and the number of molecules is unimaginably immense. Hence in the molecular world, probability does reign supreme—disorder rules (see Section 7.13). This is the essence of the second law of thermodynamics, a law that summarizes nature's rule of change.

The Second Law of Thermodynamics

Nature does not allow change to take place unless there is a
net increase in the probability status of matter/energy.

At the molecular level, an increase in the probability status means an increase in the disorder. The term "matter/energy" is used in the statement of the second law to indicate that nature keeps track of both quantities, and that it is the net probability status that must increase. To see how this works, consider two examples.

Imagine an insulated compressed gas cylinder with two chambers. One chamber contains a compressed gas and a temperature probe. The other chamber contains a vacuum (Fig. 7.12).

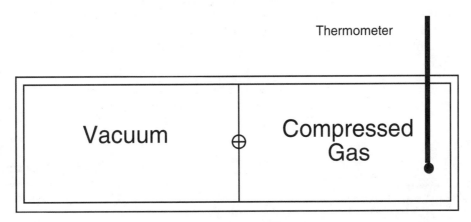

Thermometer

Vacuum   Compressed Gas

Insulated Double Chamber
Gas Cylinder

Figure 7.12   Expansion of a Compressed Gas

If a valve separating the two chambers is opened, the gas will expand into the vacuum chamber. At the same time, the temperature probe will register a decrease in the temperature of the gas.[10] These facts are understandable in terms of the kinetic molecular theory and the first law of thermodynamics. As the gas expands, the gas molecules move further apart. Since there are weak forces of attraction between the molecules of the gas, this increase in the average intermolecular distance corresponds to an increase in the average molecular potential energy. The law of energy conservation demands that this increase in potential energy be accounted for. Since the gas cylinder is insulated, the gas molecules draw on their own kinetic energy in order to provide increased potential energy after expansion. A decrease in the average kinetic energy of the gas molecules, however, corresponds to a decrease in temperature. Hence the gas cools itself. All of this is perfectly

---

10.   Actually, this temperature decrease is only observed when the gas is initially below a characteristic temperature called its "inversion temperature." In this discussion, it is assumed that the gas is below this inversion temperature.

understandable. With the proper placement of a compressor and an expansion chamber, a little engineering wizardry results in one kind of refrigerator.

There is one small problem in all of this. Why does the gas expand at all? After all, a book does not spontaneously leap from the surface of a table to increase its potential energy at the expense of its internal kinetic energy. Consider this question from the standpoint of energy flow during the gas expansion. The expansion results in a conversion of kinetic energy into potential energy. Kinetic energy is more chaotic than potential energy. Hence the conversion of kinetic energy to potential energy represents a movement of energy to higher order (low probability). This is the reason that nature does not allow a book to spontaneously lift itself off of a table. Well then, why does the gas expand? The answer to this question is that nature is a scrupulous matter/energy bookkeeper, and in the case of gas expansion below a characteristic temperature called the inversion temperature, the increased material disorder—the spread-out molecules are more disordered than the confined molecules—more than compensates for the energetic ordering. The gas expands because it meets nature's requirement for change—the second law of thermodynamics.

Keeping track of net matter/energy disorder in complicated systems presents a serious challenge. But using principles developed in the nineteenth century, the concept of matter/energy disorder can be quantified. Discussion of this aspect of the second law of thermodynamics is beyond the scope of this text, but because the scientific measure of matter/energy disorder has found its way into popular writing, this quantitative measure will be mentioned. An increase in the probability of matter/energy is said to be an increase in **entropy**.

Now consider a second example of the second law of thermodynamics. Although the book mentioned above will not spontaneously lift itself off a table, it can be lifted off the table. This lifting requires some external agent, which will be called an engine. In the context of this discussion, a **heat engine** is a device that burns a fuel to produce heat so that the heat can be converted to work (Figs. 7.13 and 7.14).

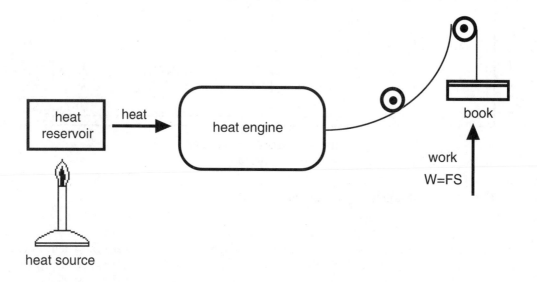

Figure 7.13 An Impossible Heat Engine

Figure 7.13 above represents the schematic of a heat engine that is impossible. Consider the following entropy bookkeeping:

HEAT RESERVOIR: The heat reservoir loses heat; its molecules become less chaotic and more ordered.

ENGINE: The engine goes round and round and round *ad nauseam*. Engines work in cycles; they do not become more or less ordered.

BOOK: The book has been moved to an improbable position. (Cut the string
          and see what happens.) The storage of energy in the lifted book
          represents more order.

Order! Order! Order! Where is the disorder? There is none, and nature will have none of it. Yet somehow books get lifted. How? The answer to this question is that nature does, indeed, allow engines to bring order into the universe, but only at a price. One small change in Figure 7.13 brings it back into accord with the second law of thermodynamics (Fig. 7.14).

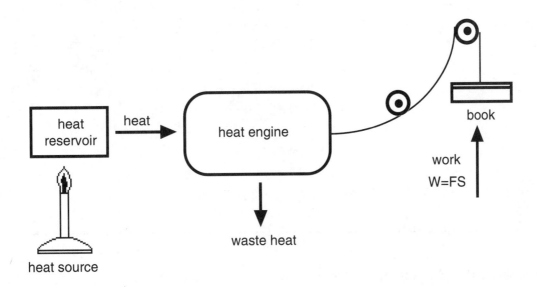

Figure 7.14   A Possible Heat Engine

The waste heat is not due to any flaw in the engine. It is, rather, the price that nature demands for all of the ordering. The waste heat goes into the environment to create more disorder than the combined order. The price is not cheap. A modern fossil fuel plant diverts about 60% of its heat to waste. According to the second law of thermodynamics, a good portion of this heat is diverted to waste even if the engine is perfect. The second law of thermodynamics can be used to calculate the ideal efficiency (work output/heat input) of a heat engine. This efficiency is determined by the Kelvin temperature of the engine's heat source reservoir and the Kelvin temperature of the waste heat sink.

Expression 7.I
The Ideal Efficiency of a Heat Engine

$$\frac{W}{Q} = \frac{T_h - T_l}{T_h}$$

W = Engine Work Output (joules, J)
Q = Engine Heat Input (joules, J)
T_h = Temperature of Heat Source Reservoir (kelvin, °K)
T_l = Temperature of Heat Sink (kelvin, °K)

EXAMPLE XIII

Calculation of the Ideal Efficiency of a Heat Engine

Calculate the ideal efficiency of a steam engine that operates with a heat source reservoir at 120°C (superheated steam) and a waste heat sink at 20°C (room temperature).

SOLUTION: 25.4%

$$°K = °C + 273$$

Therefore: $T_h = 393°K$ and $T_l = 293°K$

$$\text{Efficiency} = \frac{T_h - T_l}{T_h}$$

$$\text{Efficiency} = \frac{393 - 293}{393}$$

$$\text{Efficiency} = 0.254 \text{ or } 25.4\%$$

This is a sobering thought. Nature provides us with a constant amount of energy (first law), and then nature severely restricts the way in which we use it (second law).

The third law of thermodynamics is rather esoteric, and it is only included in this discussion for the sake of completeness. As heat is taken away from matter, the atoms and molecules that make up the matter become more ordered. The second law of thermodynamics allows for such ordering if the proper price is paid. The third law of thermodynamics states that all pure substances have a residual chaos that cannot be removed. Although it is not obvious, this fact implies a theoretical reference point for the quantitative evaluation of entropy. It also implies that zero degrees kelvin is an unattainable temperature.

7.15 The Solution to an Old Problem

As indicated in Chapter 5, the determination of the atomic masses demanded some knowledge of molecular formulas. Early nineteenth century chemists had to guess at simple formulas in order to do atomic weight calculations. At the Karlsruhe Congress (Section 5.6) held in 1860, methods for making educated guesses about molecular compositions were discussed. One method that involved the elements that are gases at room temperature was suggested by the Italian Chemist Avogadro in 1812. Another important method involving the specific heats of heavy solid elements was suggested by the French chemists Dulong and Petit in 1814. After the Karlsruhe Congress, world chemists were ready to accept the logic of these methods. Since these two methods involve concepts discussed in this chapter, it is appropriate to examine their logic.

The method of Avogadro is based on a discovery that was made by the French chemist Gay Lussac in 1811. This discovery, known as the law of Gay Lussac, involves the decomposition of chemical compounds that are composed of gaseous elements. Whenever such compounds are quantitatively decomposed, the gaseous elements formed will have volumes that are related by simple whole number ratios—provided that the volumes are measured at the same temperature and pressure. For example, water can be decomposed into hydrogen and oxygen by passing an electric current

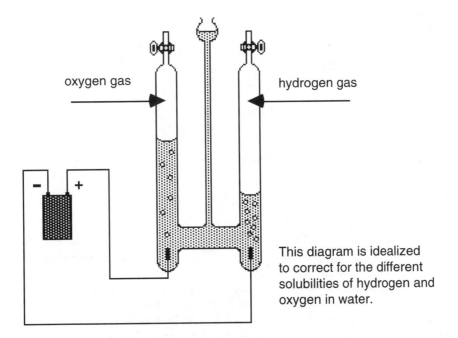

Figure 7.15 The Electrolysis of Water

oxygen gas

hydrogen gas

This diagram is idealized to correct for the different solubilities of hydrogen and oxygen in water.

through certain water solutions. This process, which is called electrolysis, is discussed in Chapter 9. When water is decomposed into hydrogen and oxygen, two volumes of hydrogen are formed for every one volume of oxygen (Fig. 7.15).

The fixed volume ratio also holds for the synthesis of water from hydrogen and oxygen. Could this experimental ratio be interpreted to mean that the formula of water is really H_2O? This would be the case if the number of molecules in two volumes of hydrogen is twice as great as the number of molecules in one volume of oxygen. Another way of stating this assumption is that equal volumes of gases contain the same number of molecules at the same conditions of temperature and pressure. This is essentially **Avogadro's hypothesis**.[11]

EXAMPLE XIV

Calculation of Atomic Mass: The Method of Avogadro

Using the following factual data, calculate the atomic mass of the unknown gaseous element, X. Assume that the Avogadro hypothesis is correct.

a. 125 mL of unknown gas, X, exactly reacts with 250 mL of oxygen. The gas volumes were measured at the same temperature and pressure.

b. The resulting binary compound is 69.6% oxygen.

SOLUTION: 14.0 amu

$$\frac{125}{250} = \frac{1}{2}$$

---

11. When the compound that is composed of gaseous elements is itself a gas, the equal volume—equal number of molecules hypothesis leads to some problems. These problems disappear if it is assumed that the gaseous elements involved are diatomic. The suggestion that some elements exist as diatomic molecules is an important part of the Avogadro hypothesis.

If the Avogadro hypothesis is correct, then the formula of the compound must be XO_2. By the method of Section 5.6 in Chapter 5, the atomic mass of X can now be calculated:

$$\frac{30.4 \text{ g X}}{69.6 \text{ g O}} = \frac{X \text{ amu X}}{2(16.0 \text{ amu O})}$$

FACT   THEORY

X = 14.0 amu

Since the example above assumes that the Avogadro hypothesis is correct, the result is still based on a guess. But this guess is an educated guess. Today there is considerable evidence for the Avogadro hypothesis, and chemists refer to it as a scientific law.

Although it was postulated prior to the kinetic molecular theory, the method of Dulong and Petit involved an atomic explanation of the specific heat differences of solid chemical elements. The specific heat of an element is the amount of heat required to increase the temperature of one gram of the element one degree centigrade. If the temperature increase of a solid element is linked by direct proportion to some atomic characteristic affected by the heat, then the amount of heat required to increase the temperature by one degree should depend on the number of atoms in the one gram sample of the solid. If one gram of solid element X contains twice as many atoms as one gram of solid element Y, then the amount of heat required to increase the temperature of one gram of element X by one degree should be twice the amount of heat required by one gram of element Y. According to this view, the specific heat ratios for two solid elements should equal the ratio of the number of atoms in one gram samples of these elements. Since one mole of all elements contain the same number of atoms, the same amount of heat should be required to increase the temperature of one mole of all of the solid elements. This is essentially the **Dulong and Petit law**.

By modern standards, the above atomic analysis of specific heat is very crude. According to the kinetic molecular theory, temperature is directly proportional to an atomic characteristic (*i.e.* average kinetic energy of atoms), but thermal energy that enters an element is almost always partitioned between kinetic and potential energy. Further, this partitioning of energy is not the same in solids, liquids, and gases. Recall that it is only atomic translational kinetic energy that is reflected in the measurable temperature change. Since the atoms of a solid element vibrate about fixed positions, they should store energy as both kinetic and potential in a very predictable way—fifty percent stored as potential and fifty percent stored as kinetic. Because of the complexities of the atomic world, this fifty/fifty distribution is only approximate, and in the case of the lighter elements, it is a very poor approximation. Nevertheless, since approximately half of the heat entering heavy solid elements is stored as translational kinetic energy, heavy solid elements require approximately the same quantity of heat to raise the temperature of one mole of solid one degree Celsius. This approximate relationship does allow for atomic ratio determination.

EXAMPLE XV

Determination of an Atomic Ratio: The Method of Dulong and Petit

When a chemical compound containing only zinc and an unknown element X is decomposed, approximately two grams of zinc is obtained for every gram of X. One calorie of heat increases the temperature of 1 gram of X and 2 grams of zinc approximately the same number of degrees. What is the formula for this compound?

SOLUTION: ZnX

The approximate atomic explanation of specific heat relates temperature increase to number of atoms. Since the temperature increase is the same for the 1 gram sample of X and the two gram sample of Zn, these samples must contain the same number of atoms. Hence, the formula is 1:1—ZnX.

7.16 A Final Note

In order to understand energy relationships in chemical reactions, the nineteenth century chemists brought their atomic/molecular theory into the realm of physics. At first, the physicists' response was a cool one, for the physics of thermodynamics was born in a pragmatic world of industrial machines, devoid of a molecular hypothesis. As late as 1882, there was still much resistance to the atomic/molecular view. Perhaps a quote by Max Planck, a physicist who ultimately contributed immensely to our understanding of atoms and molecules, best conveys the spirit of the times:

> "The second law of thermodynamics, logically developed, is incompatible with the assumption of finite atoms. Hence it is to be expected that in the course of the further development of the theory, there will be a battle between these two hypotheses, which will cost one of them its life. The result of this battle can be predicted with certainty, but it would be premature; meanwhile there seems to be present many kinds of indications that in spite of the great successes of atomic theory up to now, it will finally have to be given up and one will have to decide in favor of the assumption of a continuous matter."

Eventually, things fell into place—the kinetic molecular theory, the molecular interpretation of the laws of thermodynamics, and an understanding of the origin of chemical energy. Atoms and molecules were to become important fixtures of both chemistry and physics. Atomic/molecular theory was to flourish, and Max Planck was destined to play a major role in the development of atomic theory. Max Planck's original prediction was premature. But in his own case, his significant contribution to atomic theory proved him premature on at least one other occasion:

> "An important scientific innovation rarely makes its way by gradually winning over and converting its opponents. . . . What does happen is that its opponents gradually die out, and that the growing generation is familiarized with the ideas from the beginning." – Max Planck, German Physicist

Chapter Seven
Performance Objectives

P.O. 7.0

Review all of the boldfaced terminology in this chapter, and make certain that you understand the use of each term.

anaerobic glycolysis
bond dissociation
calorimeter
energy
fermentation
heat
heat of vaporization
kinetic energy
photosynthesis
respiration
work

Avogadro's hypothesis
calorie
Dulong/Petit law
exothermic
force
heat engine
Hess' law
melting point
potential energy
temperature

boiling point
Calorie
endothermic
entropy
Gaussian distribution
heat of melting
joule
phase change
power
thermochemical equation

P.O. 7.1

This chapter introduced seven mathematical equations that define and describe energy and energy related concepts. You must be able to solve simple problems involving the following seven equations:

$$F = ma \qquad\qquad W = FS$$

$$K.E. = \frac{1}{2}mv^2 \qquad\qquad P = \frac{W}{t}$$

$$Q = P_c m \qquad\qquad Q = cm\Delta T$$

$$\frac{W}{Q} = \frac{T_h - T_l}{T_h}$$

EXAMPLE:
A 25.0 pound weight has a metric weight of 111 newtons. How many joules of work are stored as potential energy when this weight is lifted 2.00 meters above the surface of the earth?

SOLUTION: 222 joules

Textbook Reference: Sections 7.3, 7.4, 7.5, 7.8, and 7.14

ADDITIONAL EXAMPLE:
For an average person, the metabolic rate during a vigorous walk is about 290 watts. How many kilocalories does this metabolic rate correspond to during a 15.0 minute walk (Note: 4.18 J/cal)?

ANSWER: 62.4 kcal

P.O. 7.2

You must be able to work with conversion factors involving the measurement units that are used in the seven mathematical equations that define and describe energy and energy related concepts.

EXAMPLE:
How many calories are equivalent to 425 joules (Note: 4.18 J/cal)

SOLUTION: 102 calories

Textbook Reference: Section 7.3

ADDITIONAL EXAMPLE:
How many joules are equivalent to 350 nutritional Calories. Recall that one nutritional Calorie is equivalent to one kilocalorie (Note: 4.18 J/cal)?

ANSWER: 1460000 J (rounded off to three significant figures)

P.O. 7.3

You must demonstrate an understanding of the energy flow that accompanies a phase change.

EXAMPLE:
Will the phase change described by the following schematic liberate energy or consume energy?

Liquid Substance → Solid Substance

SOLUTION: liberate

Textbook Reference: Sections 7.7 and 7.8

ADDITIONAL EXAMPLE:
When water freezes, is heat energy liberated or consumed by the water? Is the freezing of water an exothermic or an endothermic process?

ANSWER: liberated; exothermic

P.O. 7.4

You must demonstrate an understanding of the energy flow that accompanies a temperature change.

EXAMPLE:
A chemical reaction takes place inside of a metal container that is immersed in water. The chemical reaction causes the temperature of the water to decrease from 20°C to 15°C. Is the chemical reaction exothermic or endothermic?

SOLUTION: endothermic

Textbook Reference: Sections 7.7 and 7.8

ADDITIONAL EXAMPLE:
When the temperature of water changes from 25°C to 35°C, is the water experiencing an exothermic change or an endothermic change?

ANSWER: endothermic

P.O. 7.5

You must be able to calculate the quantity of heat liberated or consumed by a chemical reaction using calorimetric data.

EXAMPLE:
A chemical reaction liberates enough heat to raise the temperature of 700 grams of water from 20.00°C to 25.00°C. How much heat is liberated by the chemical reaction? The specific heat of water is 1.00 cal/g•°C.

SOLUTION: 3500 calories

Textbook Reference: Sections 7.8 and 7.9

ADDITIONAL EXAMPLE:
A chemical reaction liberates enough heat to raise the temperature of a 250 gram sample of water from 20.00°C to 40.00°C. How much heat is liberated by the chemical reaction. The specific heat of water is 1.00 cal/g•°C.

ANSWER: 5000 calories

P.O. 7.6

You must demonstrate the ability to do a simple thermochemical calculation.

EXAMPLE:
The combustion of 4.00 grams of octane (a component of gasoline) releases 46.0 kilocalories of heat energy. How many kilocalories will be released by the combustion of 32.0 grams of octane?

SOLUTION: 368 calories

Textbook Reference: Sections 7.9 and 7.10

ADDITIONAL EXAMPLE:
The combustion of 45.0 grams of glucose results in the formation of 66.0 grams of carbon dioxide and the release of 172 kilocalories of heat energy. If a larger quantity of glucose is burned, how much carbon dioxide would accompany the release of 300 kilocalories of heat energy?

ANSWER: 115 grams

P.O. 7.7

You must be able to perform a thermochemical equation calculation.

EXAMPLE:
Given the following thermochemical equation:

doubla

28   6

$$N_2(g) + 3 H_2(g) \rightarrow 2 NH_3(g) \qquad Q = -22.0 \text{ kcal}$$

Using this thermochemical equation, calculate the amount of heat that would be released by the formation of 68.0 grams of ammonia, NH_3. Atomic Masses: N = 14.0   H = 1.0 amu

SOLUTION: 44.0 kilocalories

Textbook Reference: Sections 7.9 and 7.10

1/1

ADDITIONAL EXAMPLE:
Consider the following incomplete thermochemical equation:

$$4 Fe(s) + 3 O_2(g) \rightarrow 2 Fe_2O_3(s) \qquad Q = -?$$

55.8

If the reaction of 4.00 grams of iron according to this equation results in the release of 7.17 kilocalories, what absolute value for Q is required to complete the thermochemical equation? Atomic Masses: Fe = 55.8   O = 16.0 amu

ANSWER: 400 kilocalories

P.O. 7.8

You must demonstrate an understanding of the kinetic molecular theory.

EXAMPLE:
The diagram below represents the expansion of a gas into a vacuum. The walls of the gas container are insulated (no heat transfer). Using the KMT as described in Chapter 7, answer questions A–C.

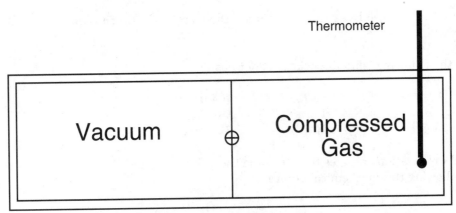

Thermometer

Vacuum   ⊕   Compressed Gas

Insulated Double Chamber
Gas Cylinder

a. Are the gas molecules moving faster before or after the expansion?
b. Is the gas temperature higher or lower after the expansion?
c. After the expansion is over, will the temperature be equal in both chambers (right and left)?

SOLUTION: a) before   b) lower   c) no

Textbook Reference: Sections 7.7, 7.12, 7.13, and 7.14

ADDITIONAL EXAMPLE:
Classify each of the following processes as exothermic or endothermic. In each case, identify the molecular origin or repository of the energy.

a. H_2O (s) → H_2O (l)
b. Liquid Substance → Solid Substance
c. Water (20°C) → Water (40°C)

ANSWER: a) endo   b) exo   c) endo

P.O. 7.9

You must demonstrate an understanding of the molecular origin of chemical energy.

EXAMPLE:
A water molecule is decomposed into individual atoms. Is this process exothermic or endothermic?

SOLUTION: endothermic

Textbook Reference: Sections 7.7 and 7.12

ADDITIONAL EXAMPLE:
Is the following process exothermic or endothermic?

$$H + H → H–H \qquad Q = ?$$

ANSWER: exothermic

P.O. 7.10

You must be able to use Hess' law to calculate the energy of a thermochemical equation.

EXAMPLE:
Consider the following bond dissociation energies:

H–H	104.0 kcal/mole
I–I	36.1 kcal/mole
H–I	71.4 kcal/mol

Use these bond dissociation energies to determine the value for Q in kilocalories that is required to complete the following thermochemical equation:

$$H_2 (g) + I_2 (g) → 2 HI (g) \quad Q = ?$$

SOLUTION: –2.7 kcal

Textbook Reference: Sections 7.7 and 7.12

ADDITIONAL EXAMPLE:
The thermochemical equations for the combustion of methane (CH_4) and methyl alcohol (CH_4O) are shown below:

$$CH_4 (g) + 2\ O_2 (g) \rightarrow CO_2 (g) + 2\ H_2O\ (l) \qquad Q = -223\ \text{kcal/mole}$$

$$CH_4O\ (l) + 3\ O_2 (g) \rightarrow 2\ CO_2 (g) + 4\ H_2O\ (l) \qquad Q = -436\ \text{kcal/mol}$$

Use the above two thermochemical equations to calculate Q for the following reaction:

$$CH_4 (g) + CO_2 (g) + 2\ H_2O\ (l) \rightarrow 2\ CH_4O\ (l) + O_2 (g) \qquad Q = ?$$

ANSWER: 213 kcal/mole

P.O. 7.11

You must demonstrate an understanding of the maximum work consequence of the second law of thermodynamics.

EXAMPLE:
Calculate the maximum efficiency of a heat engine that has a heat reservoir temperature of 200°C and a waste heat sink temperature of 20°C.

SOLUTION: 38.1%

Textbook Reference: Section 7.14

ADDITIONAL EXAMPLE:
The diagram below represents the schematic of a heat engine. What essential component of all heat engines is missing in this diagram?

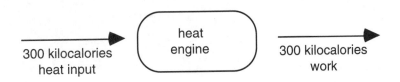

ANSWER: The engine must waste some heat.

P.O. 7.12

You must be able to calculate atom ratios based on the laws of Dulong/Petit and Avogadro.

EXAMPLE:
Solid element X has a specific heat of 0.1 cal/g•°C and solid element Y has a specific heat of 0.2 cal/g•°C. A binary compound of X and Y is 50% X and 50% Y. If these elements obey the law of Dulong/Petit, what is the formula of this binary compound?

SOLUTION: XY_2

Textbook Reference: Section 7.15

ADDITIONAL EXAMPLE:

How did Avogadro's hypothesis "prove" that a molecule of water contained two atoms of hydrogen and one atom of oxygen?

ANSWER:

The decomposition of water into the gaseous elements hydrogen and oxygen produces two volumes of hydrogen for every volume of oxygen at a given temperature and pressure.

Chapter Seven
Problems

Calculations Involving Basic Energy Relationships

1. The label on a frozen dinner states that the dinner contains 350 nutritional Calories (kilocalories). Convert 350 kilocalories to joules (Note: 4.18 J/cal).

   STUDENT SOLUTION:

   1460 round off to lowest
   1463

2. On the surface of the earth, a 25.5 kilogram mass has a weight of 250 newtons. How many joules of work are stored as potential energy when this weight is lifted 4.50 meters above the surface of the earth? How many kilocalories of work does this represent (Note: 4.18 J/cal)? W = FS 1 N = 1 kg

   STUDENT SOLUTION:

   4.50 × 250 ÷ 4.18 250 N

   1125 J 0.269 kilcal

3. An 80.0 watt light bulb consumes 4800 joules of electrical energy every minute. If 4800 joules of energy could be converted into work, how high in meters could an object weighing 250 newtons be lifted above the surface of the earth?

   STUDENT SOLUTION:

   $$\frac{4800 \text{ J}}{250 \text{ N}}$$
   19.2 m

   19.2 m

4. A soda contains 100 kilocalories of nutritional energy. If all of this energy could be converted into work, how high in meters could an object weighing 600 newtons be lifted above the surface of the earth (Note: 4.18 J/cal)?

   STUDENT SOLUTION:

   4.18 × 100 \ 600

   0.6966 round to nearist
   .697

5. What is the mass of an object that weighs 750 newtons? The acceleration due to gravity is 9.80 m/s².

   STUDENT SOLUTION:

   $$\frac{750}{9.80}$$

   76.5

6. How much kinetic energy will a 76.5 kg mass have if it is traveling at a velocity of 25.0 m/s?

   STUDENT SOLUTION:

7. How many calories are required to raise the temperature of 40.0 grams of iron from 20.00°C to 80.00°C? The specific heat of iron is 0.106 cal/g•°C.

   STUDENT SOLUTION:

8. A 35.0 gram sample of the element nickel is heated by the addition of 555 calories of heat energy. The heat addition does not produce a phase change in the solid sample of nickel. What temperature change in degrees Celsius will this amount of heat produce in the sample of nickel. The specific heat of solid nickel is 0.100 cal/g•°C.

   STUDENT SOLUTION:

9. How much heat is required to convert 15.5 grams of liquid water at 100°C into water vapor at 100°C? The P_c for the liquid to gas phase change for water is 540 cal/g.

   STUDENT SOLUTION:

10. How many calories are required to convert 30.0 grams of ice (solid) at 0.00°C to steam (gas) at 100.00°C? The P_c for the solid to liquid and the liquid to gas phase changes for water are 80.0 cal/g and 540 cal/g, respectively.

    STUDENT SOLUTION:

11. A 75.0 kg mass is lifted 5.00 m above the surface of the earth. How much potential energy is stored by this lifting process? The acceleration due to gravity is 9.80 m/s².

    STUDENT SOLUTION:

12. What will the velocity of the mass in the previous question be after the mass falls 5.00 meters (*i.e.* at the instant before impact with the earth)?

    STUDENT SOLUTION:

13. A 150 watt light bulb is turned on for 5.00 hours. Calculate the number of calories of energy that the light bulb consumes in this time period (4.18 J/cal).

    STUDENT SOLUTION:

14. If electrical energy cost $0.12 per kilowatt hour, how much does it cost to light the light bulb in the previous question?

    STUDENT SOLUTION:

15. The efficiency at which the human body converts kilocalories of nutritional energy to work varies with the type of work being done. One of the least efficient human activities is weight lifting, which humans execute with about 9% efficiency. Calculate the maximum theoretical efficiency of a heat engine that is operating with a heat reservoir temperature of 100°C and a waste heat sink temperature of 20°C (a classical steam engine). How does the maximum efficiency of this classical steam engine compare to a weight lifting human?

    STUDENT SOLUTION:

16. A shot-put with a mass of 6.00 kilograms has a weight of 58.8 newtons. How much work is required to lift this shot-put 5.00 meters above the ground? If the shot-put is dropped from the height of 5.00 meters, how much kinetic energy will it have at the instant before impact with the ground?

    STUDENT SOLUTION:

17. A diet soda contains 1000 calories of nutritional energy. This corresponds to 4180 joules of energy. If all of this energy could be converted into work, how high in meters could an object weighing 600 newtons be lifted above the surface of the earth?

STUDENT SOLUTION:

18. Which of the following will have more kinetic energy?

    a. a 1 pound weight after falling 4 feet (initial speed zero)
    b. a 4 pound weight after falling 2 feet (initial speed zero)

    Both objects were released twenty feet above the surface of the earth.

    STUDENT SOLUTION:

19. Which of the following will have more kinetic energy?

    a. a 2 pound weight after falling 6 feet (initial speed zero)
    b. a 4 pound weight after falling 3 feet (initial speed zero)

    Both objects were released twenty feet above the surface of the earth.

    STUDENT SOLUTION:

Energy Transfer during Physical and Chemical Change

20. Under proper conditions, solid water can be converted directly to gaseous water. This process is called sublimation. Is sublimation an exothermic process or an endothermic process?

    STUDENT SOLUTION:

21. If the following physical changes take place at constant temperature, will the changes liberate heat or consume heat?

   a. H_2O (g) → H_2O (l)
   b. CO_2 (s) → CO_2 (g)
   c. Pb (l) → Pb (s)
   d. liquid substance → gaseous substance

   STUDENT SOLUTION:

22. A chemical reaction takes place inside of a metal container that is immersed in water. The initial temperature of the water is 29.0°C. The chemical reaction causes the temperature of the water to change. After the chemical reaction, the final temperature is 25.0°C. Is the chemical reaction exothermic or endothermic?

   STUDENT SOLUTION:

23. A chemical reaction takes place inside of a metal container that is immersed in 1500 grams of a liquid chlorofluorocarbon. The initial temperature of the chlorofluorocarbon is 20.2°C. The chemical reaction causes the temperature of the chlorofluorocarbon to change. After the chemical reaction, the final temperature of the chlorofluorocarbon is 25.5°C. Is the chemical reaction exothermic or endothermic, and how many calories of heat are exchanged as a result of the chemical reaction? The specific heat of the liquid chlorofluorocarbon is 0.218 cal/g•°C.

   Note: Answer must reflect correct sign and magnitude.

   STUDENT SOLUTION:

24. It takes more energy to break the chemical bond between a hydrogen atom and a fluorine atom than to break the chemical bond between two hydrogen atoms. It also takes more energy to break the chemical bond between a hydrogen atom and a fluorine atom than to break the chemical bond between two fluorine atoms. Based on this relative bond strength information, is the chemical reaction represented by the following equation exothermic or endothermic?

$$H_2 \text{ (g)} + F_2 \text{ (g)} \rightarrow 2 \text{ HF (g)} \qquad Q = \text{? kcal}$$

   STUDENT SOLUTION:

25. Indicate if each of the following thermochemical equations are endothermic or exothermic?

a. $2\ ClBr\ (g) \rightarrow Cl_2\ (g) + Br_2\ (l)$ $Q = -6.27$ kcal
b. $2\ H_2O\ (g) \rightarrow 2\ H_2(g) + O_2(g)$ $Q = +116$ kcal

STUDENT SOLUTION:

Thermochemical Calculations

26. The combustion of 2.25 grams of ethyl alcohol releases 15.1 kilocalories of heat energy. How many kilocalories will be released by the combustion of 12.5 grams of ethyl alcohol?

STUDENT SOLUTION:

27. The combustion of 1.54 grams of a hydrocarbon results in the formation of 4.57 grams of carbon dioxide and the release of 18.2 kilocalories of heat energy. If 42.2 grams of the hydrocarbon were burned, how many kilocalories of heat energy would be released?

STUDENT SOLUTION:

28. The combustion of 8.00 grams of a liquid fuel releases 50.0 kilocalories of energy. How many grams of the fuel would be required to release 967 kilocalories of energy?

STUDENT SOLUTION:

Calculations Involving Thermochemical Equations

29. Given the following thermochemical equation, calculate the number of calories liberated when 1.00 gram of hydrogen is burned:

$$2\ H_2(g) + O_2(g) \rightarrow 2\ H_2O\ (l) \qquad\qquad Q = -136\ \text{kcal}$$

STUDENT SOLUTION:

30. The following thermochemical equation represents the combustion of carbon monoxide:

$$2\ CO\ (g) + O_2\ (g) \rightarrow 2\ CO_2\ (g) \qquad Q = -135\ kcal$$

Calculate the mass of carbon monoxide in grams that would need to be burned in order to release 338 kilocalories.

STUDENT SOLUTION:

31. The following thermochemical equation represents the combustion of butane:

$$2\ C_4H_{10}\ (g) + 13\ O_2\ (g) \rightarrow 8\ CO_2\ (g) + 10\ H_2O\ (l) \qquad Q = -1380\ kcal$$

Using this thermochemical equation, calculate the mass of carbon dioxide in grams that would be released when 251 kilocalories is produced by this combustion. Atomic Masses: C = 12.0 O = 16.0 H = 1.01 amu

STUDENT SOLUTION:

32. Consider the following incomplete thermochemical equation:

$$2\ C_2H_4O\ (l) + 5\ O_2\ (g) \rightarrow 4\ CO_2\ (g) + 4\ H_2O\ (l) \qquad Q = -?\ kcals$$

If the combustion of 3.21 grams of acetaldehyde (C_2H_4O) according to this equation results in the release of 20.4 kilocalories, what absolute value for Q in kilocalories is required to complete the thermochemical equation? Atomic Masses: C = 12.0 H = 1.01 O = 16.0 amu

STUDENT SOLUTION:

33. Consider the following thermochemical equation:

$$CO\ (g) + 2\ H_2\ (g) \rightarrow CH_3OH\ (l) \qquad Q = -?\ kcal$$

If the formation of 55.5 grams of methanol, CH_3OH, according to this equation results in the release of 53.1 kilocalories, what absolute value of Q in kilocalories is required to complete the thermochemical equation?

STUDENT SOLUTION:

34. In order to gain additional practice with thermochemical equations, calculate the mass of glucose that would have to be "burned" in each of the following reactions in order to lift a 444 newton weight (100 pounds) 2.00 meters above the surface of the earth with 9% efficiency:

a. Anaerobic Glycolysis (Chemical Equation 7.H)
b. Respiration (Chemical Equation 7.E)

STUDENT SOLUTION:

Kinetic Molecular Theory

35. According to the kinetic molecular theory, an increase in the temperature of a substance corresponds to an increase in the average potential energy of the molecules of the substance.

a) true   b) false

STUDENT SOLUTION:

36. If the following reactions take place at 25°C, which will be endothermic? Which will be exothermic?

a. O_2 (g) → 2 O (g)
b. H (g) + Cl (g) → HCl (g) Exo
c. H_2O (g) → 2 H (g) + O (g) End

STUDENT SOLUTION:

37. Will the smell of fresh coffee spread faster on a hot day or a cold day?

STUDENT SOLUTION:

38. The thermochemical equation in problem number 29 indicates that the combustion of 4.00 grams of hydrogen will liberate 136 kcal. Will the combustion of the same amount of hydrogen according to the following thermochemical equation liberate more or less than 136 kcal?

$$2 H_2 (g) + O_2 (g) → 2 H_2O (g) \qquad Q = ? \text{ kcal}$$

STUDENT SOLUTION:

Hess' Law of Constant Heat Summation

39. Given the following bond dissociation energies:

I–Br	41.9 kcal/mole
I–I	36.1 kcal/mole
Br–Br	46.2 kcal/mole

    subtract

    Calculate the value for Q in kilocalories for the following thermochemical equation:

    Double

    $$2 \text{ IBr (g)} \rightarrow I_2\text{(g)} + Br_2\text{(g)} \qquad Q = ? \quad .150$$

    STUDENT SOLUTION:

40. Given the following bond dissociation energies:

H–H	104.0 kcal/mole
F–F	38.0 kcal/mole
H–F	135.2 kcal/mole

    Calculate the value for Q in kilocalories for the following thermochemical equation:

    $$H_2\text{(g)} + F_2\text{(g)} \rightarrow 2 \text{ HF (g)} \qquad Q = ? \quad 13.4$$

    STUDENT SOLUTION:

41. Given the following bond dissociation energies:

O_2	118.2 kcal/mole
N_2	225.1 kcal/mole
NO	145.2 kcal/mole

    Calculate the value for Q in kilocalories for the following thermochemical equation:

    $$O_2\text{(g)} + N_2\text{(g)} \rightarrow 2 \text{ NO (g)} \qquad Q = ?$$

    STUDENT SOLUTION:

42. Coal gasification involves the conversion of carbon in coal to methane, CH_4, which is the major component of natural gas. This reaction is illustrated below:

$$C + 2\ H_2\ (g) \rightarrow CH_4\ (g) \qquad Q = ?$$

Calculate the value for Q in kilocalories for the above reaction from the following heats of combustion:

$$C + O_2\ (g) \rightarrow CO_2\ (g) \qquad\qquad\qquad Q = -94.1\ kcal$$
$$2\ H_2 + O_2\ (g) \rightarrow 2\ H_2O\ (g) \qquad\qquad Q = -136.8\ kcal$$
$$CH_4(g) + 2\ O_2\ (g) \rightarrow CO_2\ (g) + 2\ H_2O\ (g) \qquad Q = -223.0\ kcal$$

STUDENT SOLUTION:

43. Given the following reactions and their values for Q:

$$CaO\ (s) + H_2O\ (l) \rightarrow Ca(OH)_2\ (s) \qquad\qquad Q = -15.3\ kcal$$
$$2\ H_2\ (g) + O_2\ (g) \rightarrow 2\ H_2O\ (l) \qquad\qquad Q = -136.6\ kcal$$
$$2\ Ca(s) + O_2\ (g) \rightarrow 2\ CaO\ (s) \qquad\qquad Q = -303.6\ kcal$$

Use Hess' law to calculate the value of Q in kilocalories for the following reaction:

$$Ca\ (s) + O_2\ (g) + H_2\ (g) \rightarrow Ca(OH)_2\ (s) \qquad Q = ?$$

STUDENT SOLUTION:

Laws of Thermodynamics

44. Which of the following heat engines is in violation of the first law of thermodynamics? Which is in violation of the second law of thermodynamics?

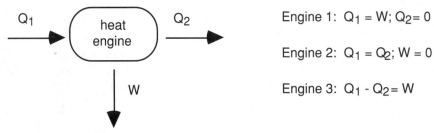

Engine 1: $Q_1 = W;\ Q_2 = 0$

Engine 2: $Q_1 = Q_2;\ W = 0$

Engine 3: $Q_1 - Q_2 = W$

Q_1 = heat from high temperature reservoir
Q_2 = waste heat
W = work

STUDENT SOLUTION:

45. In theory, can all of the energy from reactions in problem number 32 be converted to work?

    STUDENT SOLUTION:

Methods of Avogadro and Dulong/Petit

46. It is possible to make a very explosive solid compound with two simple household chemicals. An analysis of this explosive compound indicates that it is composed of the elements nitrogen, hydrogen, and iodine. If the combining volumes of these elements are measured as gases at the same temperature and pressure, then the nitrogen, hydrogen, iodine volume ratio is 1:1.5:1.5. What is the empirical formula of this compound?

    STUDENT SOLUTION:

47. The addition of ten calories of heat changes the temperature of one gram of metal element X from 20.0°C to 21.0°C. The addition of ten calories of heat changes the temperature of one gram of metal element Y from 20.0°C to 22.4 °C. Which one gram sample contains the greater number of atoms?

    STUDENT SOLUTION:

48. The addition of ten calories of heat changes the temperature of one gram of metal element X from 20.0°C to 21.0°C. The addition of ten calories of heat changes the temperature of one gram of metal element Y from 20.0°C to 22.4°C. Which element has the higher atomic mass?

    STUDENT SOLUTION:

Problems Involving Household Chemistry and Science

49. A Styrofoam cup can serve as a reasonably good calorimeter. Try measuring the heat of melting for ice using a Styrofoam cup calorimeter according to the following procedure.

    Form a 15 gram ice cube by freezing 2 tablespoons (15 mL = 15 g) of water in a small plastic container. After the water is frozen, add ½ cup (120 mL = 120 g) of hot tap water to a Styrofoam cup (8–12 oz). Record the temperature of the water. If a Fahrenheit thermometer is being used, convert the temperature to Celsius using the relationship given in Figure 7.4. Remove the ice cube from the freezer, and quickly place it into the water in the Styrofoam cup. Rapidly stir the mixture. When all of the ice has melted, measure the final temperature of the water in the

Styrofoam cup. Use the following equation to calculate the heat of melting for ice. This equation assumes that no heat is transferred to the Styrofoam cup or the air surrounding the cup:

$$(cm\Delta T)_{water} = P_c m + (cm\Delta T)_{ice}$$

Heat Lost by Water   Heat Gained by Ice

This equation is not as complicated as it seems. It is simply a combination of Expressions 7.F and 7.G that describes the law of energy conservation for this experiment. In the experiment, the 120 g of water in the Styrofoam cup lost heat to the 15 g of water in the ice cube. The 15 g of water in the ice cube gained heat in two ways. First, it gained heat to melt at 0°C—$P_c m$. Second, it gained heat to increase the liquid temperature from 0°C to the final temperature—$(cm\Delta T)_{ice}$. If the initial and final temperatures of the water in the cup are represented as T_i and T_f in °C, then the equation that describes this experiment can be expressed as shown below:

$$(1.0 \text{ cal/g} \bullet °C)(120 \text{ g})(T_i - T_f) = P_c(15 \text{ g}) + (1.0 \text{ cal/g} \bullet °C)(15 \text{ g})(T_f - 0°C)$$

In this equation, the temperature changes have been expressed as positive values. The equation, therefore, states that the magnitude of the energy gained must equal the magnitude of the energy lost. After using this equation to calculate P_c for ice from the measured temperatures, compare your value to the value given in the text. What assumptions and/or errors can account for a low or high experimental value?

STUDENT SOLUTION:

50. In Section 7.11 of this chapter, the process of respiration was discussed. During the process of respiration, animal and plant organisms "burn" carbohydrates or fats to produce biological energy. Chemical Equation 7.E represented the combustion of glucose, $C_6H_{12}O_6$.

Chemical Equation 7.E
Biological "Burning" or Respiration

$$C_6H_{12}O_6 (aq) + 6 \text{ } O_2 (g) \rightarrow 6 \text{ } CO_2 (g) + 6H_2O (l) \qquad Q = -686 \text{ kcal}$$

The first library problem in the next section explores the power limitations associated with the human body using respiration to do actual work. To get a feel for the power rates in this article, it might be interesting to determine your power during vigorous exercise. The experiment is really quite easy. All that is required is a watch with a second hand, a metric ruler, and several flights of steps.

Start the experiment at the top of several flights of steps. Use a metric ruler to measure the height of a single step, and then count the total number of steps as you walk to the bottom of all of the flights. At the bottom of the steps, note the second hand position, and then run up the steps at a rate commensurate with your level of health and training. Note the second hand position at the top of the steps. Calculate the power you developed by doing the following calculations:

a. Convert your weight from pounds to newtons (4.44 N/lb)
b. Calculate the height in meters that you lifted your weight by multiplying the height of one step times the number of steps you ran.

Chapter Seven Notes

Chapter Seven Notes

Chapter Eight

Chemical Bonding and the Geometry of Molecules _____

8.1 Introduction

The final triumph of nineteenth century chemical atomic/molecular theory was the emergence, toward the end of the century, of a three dimensional model of molecular structures. This three dimensional or stereo viewpoint was a giant conceptual leap. To the early nineteenth century chemist, the molecule was an abstraction that "explained" mass changes that occurred during the chemical reaction process. To this extent, the formula CH_4 was a mere bookkeeping statement encoding the definite mass composition of a chemical compound called methane. With the advent of the kinetic molecular theory came the belief that perhaps these abstractions were real after all, but this physical theory viewed molecules as simple points in space. Finally, certain facts began to emerge that demanded an extension of the atomic/molecular theory—an extension that pictured the molecule as a three dimensional entity.

The purpose of this chapter is to investigate the emergence of this three dimensional model of molecular structure and also to investigate the utility of this new point of view to the modern chemist.

8.2 Philosophic Origins: The Regular Polyhedra

The term "polygon" refers to any closed two dimensional figure bounded by three or more line segments. If all of the angles and all of the sides of a polygon are equal, then it is called a regular polygon. For example, consider the following familiar regular polygons (Fig. 8.1):

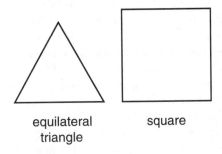

equilateral
triangle   square

Figure 8.1   Familiar Regular Polygons

Now consider a question about the series of regular polygons shown in Figure 8.2.

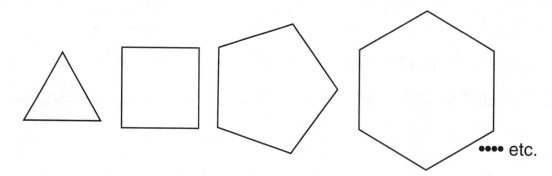

Figure 8.2 A Series of Regular Polygons

How long could this series be continued? That is, how many regular polygons are there? Most people can answer this question intuitively. There are an infinite number of regular polygons.

Now let us move into another dimension—the third dimension. The three dimensional counterpart of a regular polygon is called a **regular polyhedron**. Consider three examples (Fig. 8.3):

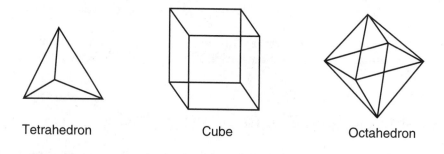

Tetrahedron   Cube   Octahedron

Figure 8.3 Three Regular Polyhedrons

In each of these regular polyhedra, all corner angles are equivalent, all edges are equivalent, and all faces are equivalent. How many different regular polyhedra are there? The answer to this question is not as intuitively obvious as the answer to the previous question about polygons. In fact, only five regular polyhedra can be constructed. In addition to the three regular polyhedra pictured above, there is also the icosahedron (twenty equilateral triangular faces, thirty edges, and twelve corners) and the dodecahedron (twelve regular pentagonal faces, thirty edges, and twenty corners).

That the three dimensional world only allows five regular polyhedra is, indeed, a very curious fact. Evidently, there is something limiting freedom of symmetrical construction in the three dimensional world.[1] The study of the restrictions that nature places on symmetrical construction involves some fairly sophisticated mathematics. Although this area of mathematics is beyond the scope of this text, it is important for you to appreciate that the construction limitation exists. Further, it should not be surprising if some of the regular polyhedra are encountered in our study of the three dimensionality of molecules.

---

1. The word symmetrical is used in this chapter without rigorous definition. Most people have an intuitive understanding of this word, and the following definition is given to reinforce this intuitive understanding:

   Symmetrical objects possess an exact correspondence of form on opposite sides of a dividing plane, or line, or point.

Although it is far from obvious that there are only five regular polyhedra, the ancient Greeks were aware of this fact as early as 500 B.C. The followers of the Greek philosopher Pythagoras incorporated the five regular polyhedra into their chemical philosophy. Each polyhedra was assigned as a symbol for the basic "elements" of nature: the tetrahedron for *fire*, the cube for *earth*, the octahedron for *air*, the icosahedron for *water*, and the dodecahedron for *ether*. Fire, earth, air, water, and ether were the "elements" of ancient Greek philosophy. Although the Pythagoreans were not doing chemistry by modern standards, their philosophy had brought them to an important conclusion. The regular polyhedra must be important to the very fabric of nature.

Modern natural science has rediscovered this latter conclusion and established it as a scientific "truth." How this was accomplished by the application of the scientific method is one of the subjects of the present chapter. This is not the first time that we have encountered the scientific rediscovery of a philosophic thought. Recall that the atom itself is the product of pure philosophic thought. Before proceeding, it may be wise to read again the quotation at the end of Chapter 5.

8.3 Crystals: A Window to the Atomic/Molecular World

The previous topic dealing with symmetrical construction may seem to be unrelated to the discipline of chemistry. What do symmetrical three dimensional objects have to do with elements and compounds? The answer to this question is that the solidification of many substances results in the formation of three dimensional solid forms called **crystals**. For example, if you look very closely at crystals of table salt, NaCl, you will observe that each crystal is a perfect little cube. Although the cube is one of the regular polyhedra, an investigation of other crystalline substances will reveal that nature forms a much larger variety of three dimensional crystals than the five that would correspond to the regular polyhedra. In other words, the crystals of nature are *often* symmetrical polyhedra, but they are not necessarily regular polyhedra with corners, edges, and faces equivalent (Fig. 8.4).

Well then, is there any limit to the number of crystalline shapes that can exist in nature? The answer to this question is yes. The great variety of crystalline shapes that exist in nature fit into one of seven crystalline systems. A detailed study of these crystalline systems is quite complicated. For example, each crystalline system (family) contains many different polyhedra, and in some cases the recognition of a family relationship between two crystals requires knowledge of a sophisticated branch of mathematics called group theory.[2] The important thing for you to understand here is not detail but essence. The essential point is that something in nature limits the construction of the crystals of chemical substances, and it restricts their shape to one of only seven crystalline forms. What is this limiting factor?

As early as 1784, the French scientist Rene Just Hauy recognized that the answer to this question involved the invisible structural units that make up crystals. Today we recognize these invisible structural units as atoms. To see how this limitation of construction works, consider a two dimensional analogy. There are an infinite number of two dimensional figures called polygons—regular and irregular (Fig. 8.5).

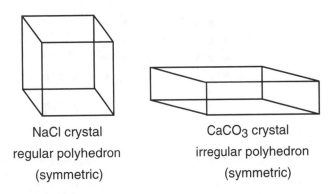

NaCl crystal
regular polyhedron
(symmetric)

CaCO₃ crystal
irregular polyhedron
(symmetric)

Figure 8.4 Two Crystalline Shapes

---

2. Each crystalline system (family) is characterized by a simple symmetrical polyhedron. For example, there is a cubic crystalline system. Some crystals that belong to the cubic system, however, are not shaped like perfect little cubes. That such crystals belong to the cubic family is not obvious to the untrained observer.

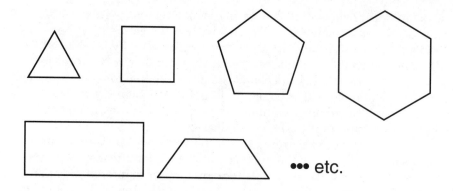

Figure 8.5 A Series of Polygons

Now suppose that we place a restriction on the construction of polygons. The restriction is that polygons must be constructed so that they exactly contain a repeating arrangement of points in which each point is identical (each point has identical surroundings). This may seem to be a strange restriction, but this is an analogy, and its chemical logic will be clear shortly. Well, how many ways can repeating points be arranged on a piece of paper so that each point in the arrangement is identical? Although, once again, the answer is far from obvious, there are only five such arrangements of points. These five possible arrangements of points are shown below (Fig. 8.6):

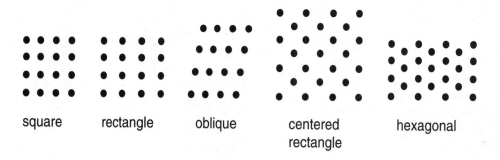

Figure 8.6 Five Unique Arrangements of Points in Space

Notice that each arrangement defines a building block polygon (square, rectangle, oblique diamond, centered rectangle, or hexagon shape) that is repeated over and over again in the array. Under the restriction, therefore, the only allowed polygons would be the ones that could be constructed from these building block polygons. For example, any attempt to construct the following polygon from a set of one of the allowed building block polygons is doomed from the start (Fig. 8.7):

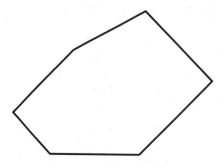

Figure 8.7 A Polygon that Cannot be Constructed
from Building Block Polygons

274

In 1848, the French physicist Bravais extended the concept of repeating arrangements of identical points to include three dimensional arrangements. He showed that only fourteen different arrangements are possible. Again, this is not obvious, but it is, nevertheless, true. But here is the really interesting thing. If we treat each of the Bravais arrangements like the two dimensional arrangements above, then each of the Bravais arrangements will define a building block polyhedron. Further, the fourteen Bravais arrangements define only seven building block polyhedra, and these correspond exactly to the polyhedra of nature's seven crystalline systems. (Notice that the five two dimensional arrangements define five building polygons.)

We are ready to draw an elegant conclusion, but first, a review of the *facts* is in order:

FACT A:	There are only five regular polyhedra (corners, edges, and faces equivalent).
FACT B:	There are an infinite number of irregular polyhedra.
FACT C:	If the construction of polyhedra is restricted so that they must be capable of exactly containing a repeating three dimensional arrangement of identical points (each point has identical surroundings), then only polyhedra corresponding to seven basic polyhedral building blocks can be constructed.[3]
FACT D:	The crystals of nature's substances are polyhedra that correspond to the seven basic polyhedral building blocks of fact C.
CONCLUSION:	Facts C and D suggest that crystals on nature's substances contain repeating arrangements of identical points.

The conclusion is a theory because we cannot "see" the points. The points are, of course, the atoms that make up the solid substance, and the restriction of their positioning (fact C) makes chemical sense. For example, a crystal of sodium chloride must contain exactly as many sodium atoms as chlorine atoms (*i.e.* NaCl). Further, a repeating three dimensional arrangement of identical sodium atoms simply means that all of the sodium atoms in a sample of sodium chloride are identical. This is chemically reasonable, and it is implied by the formula NaCl. The same reasoning applies to the chlorine atoms in a crystal of sodium chloride.[4] This is an elegant conclusion, indeed. The very existence of crystalline solids is a "proof" of the atomic/molecular theory.

8.4 "Seeing" Atoms

The previous section referred to crystals as a window to the atomic/molecular world. To get a better feel for how this window works, let us use the analogy of the previous section. Assume that you live in a fictitious two dimensional world. If you have trouble imagining this, please reread the quotes at the end of Chapter 4! The crystals of the chemical substances in your two dimensional world are, of course, polygons subject to the restrictions discussed in the previous section. As a chemist (indeed!), you are intrigued by the crystal of the elemental substance shown in Figure 8.8.

What do you "see" when you look at this crystal? Well, if you look closely, you can "see" atoms. Consider how this is accomplished.

---

3. Only one of these building block polyhedra is a regular polyhedron. It is the cube. The other six building block polyhedra are symmetrical irregular polyhedra.
4. In the crystal of an elemental substance, the atoms are arranged according to one of the fourteen Bravais arrangements (*i.e.*, seven basic polyhedral building block types), and each atom is equivalent. For a compound substance like NaCl, the atoms of each element form their own Bravais arrangements. These arrangements then overlap to form the crystal. Obviously, for larger molecules, these overlapped arrangements can become quite complex. Hence, there are really a large variety of individual crystals in nature, and atomic arrangements can only be deduced by visual inspection of very simple crystals. The modern chemists, however, have another trick up their sleeves—read on.

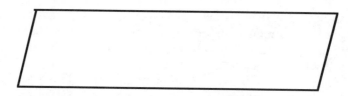

Figure 8.8 Crystal of an Elemental Substance from Fictitious Two Dimensional World

In your two dimensional world, there are five "Bravais" arrangements (Fig. 8.6) that define five building block polygons. Your crystal must correspond to one of the building block polygons, with the atoms contained within the crystal according to the "Bravais" arrangement (Fig. 8.9).

Figure 8.9 Crystal of an Elemental Substance Showing "Bravais" Points

The appropriate "Bravais" arrangement in this case is the oblique diamond shown in Figure 8.6. The dots in this representation of the crystal represent atoms. Hence, you can "see" atoms if you're looking through the eyes of a chemist.

There is one small problem in the foregoing discussion. There are an infinite number of similar diamonds that can be made to fit into the crystal (Fig. 8.10).

These "Bravais" arrangements differ only in the atom spacing, and each can be made to fit into the crystal over and over again. Hence, although we can determine the relative arrangement of the atoms in the crystal, we cannot determine their exact position unless we can determine the length of the side of the repeating diamond (*i.e.* the distance between the atoms in the crystal).[5] Using real three dimensional crystals, the modern chemist can actually accomplish this, *even for complex crystals.* These atomic distance measurements are made by reflecting X-rays off of crystal surfaces. The X-ray reflection patterns are captured on photographic film, and while they are not pictures of atoms, these reflection patterns can be used to

Figure 8.10 Two Diamond "Bravais" Arrangements for Filling Crystal

calculate the distances between atoms in the crystal.[6] Thus the chemist "sees" the exact positions of the atoms in a crystal. For compound substances, this also means it is possible to "see" the exact positions of atoms in a molecule.

---

5.   The word "exact" is being used very loosely in this discussion.
6.   There is an analogy that might be useful in attempting to understand this X-ray technique. Imagine a miniature golf course obstacle that contains five posts placed to interfere with the passage of a golf ball. To make the obstacle more challenging, a roof is constructed so that the posts are hidden from view. It is not hard to imagine that given a large bucket of golf balls, a putter, and a lot of patience the posts under the roof could be located. All one would have to do is observe the golf balls entering the obstacle and bouncing back. Eventually, the golf ball reflections would allow one to "see" the hidden posts in the same way that X-ray reflections allow the chemist to "see" hidden atoms.

8.5 Louis Pasteur and Molecular Shape

The X-ray technique mentioned in the previous section is truly a modern technique. Since the technique involves immense amounts of data and thousands of calculations, chemists could not really apply this method of "seeing" atoms in molecules until the advent of the computer age. The nineteenth century chemist, therefore, had to be content with the detailed visual inspection of crystals. As indicated in footnote four, this technique revealed atomic arrangements only in the case of very simple crystals. Actually, the Bravais arrangements were unknown until 1848, and prior to this date, chemists scrutinized crystals with a kind of blind faith that nature would eventually reveal the secret hidden within these structures. In 1847, the young chemist Louis Pasteur, while studying crystalline shapes, discovered a part of this secret. His discovery did not involve the Bravais arrangements, but it did allow him to describe for the first time the shape of a molecule. To understand the logic of this discovery, it is necessary to consider some historical background.

During the fermentation of grape juice (wine making), a sediment is formed that can be processed to produce a pure chemical compound called tartaric acid. This substance has a number of industrial uses, and its manufacture predates atomic/molecular theory. With the advent of atomic/molecular theory, chemists began to view compounds in terms of their molecular composition (*i.e.* chemical formula). For the sake of clarity in this discussion, the correct modern formula will be used. Hence, the nineteenth century chemist came to view tartaric acid as being composed of molecules with the formula $C_4H_6O_6$ (Chapter 5).

Sometime around 1819, the French chemical manufacturer Paul Kestner produced an anomalous batch of tartaric acid. The material appeared to be a totally different chemical compound. Ordinarily, this would have been no great cause for concern. A small change in the manufacturing procedure could have resulted in the formation of a different compound, and this fact would not have been very upsetting. This point of view was eventually adopted, and Kestner's new compound was named racemic acid. There was, however, one fly in the ointment, at least in the mind of early nineteenth century chemists. Tartaric acid and racemic acid had the same chemical formula, $C_4H_6O_6$.

In order to understand why this presented the early nineteenth century chemist with a problem, you must remember that originally chemical formulas were bookkeeping statements that kept track of and "explained" the law of constant composition (Chapters 4 and 5). Since the law of constant composition suggested that each compound had a unique constant composition, each compound should also have had a unique formula. At this point, a solution to this dilemma may have already occurred to you, but you must understand that the early nineteenth century chemist's mind had not been primed to think the thoughts that you may be thinking. The solution to this dilemma is that two different compounds can have the same formula, and hence the same constant composition, provided that the atoms in their respective molecules are arranged differently. According to this view, the properties of a compound are determined by the arrangements of the atoms in its molecule. But early nineteenth century chemists were not thinking very much about the arrangement of atoms in molecules. The discovery made by Louis Pasteur changed all of this. Consider now the details of this discovery.

In 1847, Louis Pasteur was involved in the process of gaining practice in research methods. In order to accomplish this goal, he decided to repeat some of the experiments of earlier investigators pertaining to the problem of racemic and tartaric acids. Pasteur, while examining the crystals of racemic acid and tartaric acid salts, discovered something that earlier investigators had overlooked— the crystals of the tartaric acid salt differed in a subtle and interesting way from the crystals of the racemic acid salt. Specifically, he observed that some of the faces of the tartaric acid salt crystals were distorted in such a way that each crystal was dissymmetrical. The three dimensional shape of each crystal possessed a kind of "handedness." This latter statement requires a brief explanation.

If you hold a symmetrical object like a cube up to a mirror, the mirror image of the object (cube) is identical to the object itself. If, however, the same experiment is performed with a dissymmetrical object, then the mirror image of the object is not identical to the object itself. This mirror test can be used to distinguish between symmetrical and dissymmetrical objects. One of the easiest ways to illustrate this mirror characteristic of a dissymmetrical object is to consider the human hand—an example of a dissymmetrical shape. If you hold your left hand and your right hand in a "praying" position, it is clear that your left hand is the mirror image of your right hand. But a left hand and a right hand are different. If you attempt to superimpose your left hand with your right hand, you will prove this obvious fact for yourself. Because of this familiar example, all dissymmetrical objects are said to possess "handedness." The technical word to express three dimensional "handedness" is **chiral**. This word is derived from the Latin word for "hand."

Expressed in modern terms, therefore, Pasteur discovered that the crystals of the tartaric acid salt were chiral. For the sake of convenience, these crystals will be referred to as right "handed" crystals. When Pasteur turned his attention to the crystals of the racemic acid salt, he discovered that the racemic acid salt was composed of two types of crystals. One type was identical to the right "handed" crystals of the tartaric acid salt. The other type was almost identical to the crystals of the tartaric acid salt, except that it was of the left "handed" form. The racemic acid salt, and hence racemic acid itself, was a mixture. Using tweezers, Pasteur separated the right "handed" and left "handed" crystals into two piles. The two piles of crystals, indeed, were different compounds with the same chemical formula. One of the piles was identical to the tartaric acid salt. The other was a different compound, but because of the difference in crystalline shape, the reason for the difference became clear. Since the crystals of the two compounds were different, the arrangements of the atoms were also different.

Actually, Pasteur went a bit further in his analysis of the cause of the differences between the *two* chemical compounds that make up the racemic acid salt mixture. Since both of these compounds retained certain unique optical characteristics inherent in their crystals even when dissolved in water (crystal state destroyed), Pasteur concluded that the arrangements of the atoms in the crystals were different *because the arrangements of the atoms in the molecules were different*. Further, since each type of crystal was chiral, Pasteur concluded that the individual molecules were also chiral. This represents a tremendous intellectual achievement. The shape of a molecule had been related to factual observation. Although there was still much to be learned about the detailed three dimensional arrangement of the atoms in molecules, Pasteur had given the world of chemistry its first "look" at a molecular shape.

Different chemical compounds that have the same chemical formula are really quite common in nature. In all cases, these compounds differ from one another because of differences in the arrangement of atoms within their respective molecules. Chemical compounds that have the same molecular formula, but different atomic arrangements are called **isomers**. The two components of racemic acid are isomers of one another. Some isomers, like the racemic acid isomers, have chiral molecules. The definition of isomer, however, does not require that the molecules be chiral. There are many examples of symmetrical molecules of identical formula with differing atomic arrangements. The term isomer was suggested by the Swedish chemist Berzelius (Chapter 4). It is derived from the Greek word meaning "composed of equal parts." Several examples of isomeric chemical compounds are discussed later in this chapter.

8.6 Directed Valence

One way to account for the regular geometric shape of molecules is to imagine that an atom is a sphere that directs from its surface potential chemical bonds. Using the language of Chapter 6, we could say that each atom directs potential chemical bonds from its surface in specific spatial directions—*i.e.* each atom directs its valences. This idea of **directed valence** became quite popular after the work of Louis Pasteur. Today, using modern techniques, we have learned much about the directed valences of the various atoms. The purpose of this section is to consider the first example of directed valence that was discovered in the latter part of the nineteenth century. In the last part of this chapter, we will see how the modern chemist uses the concept of directed valence to understand the chemical reaction process.

Consider the case of the element carbon. According to the method of Edward Frankland (Chapter 6), the valence of carbon is four. For convenience, this could be expressed as follows (Fig. 8.11):

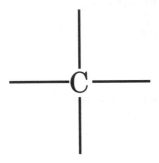

Figure 8.11 Carbon with Valence of Four

This is exactly the way the German chemist Kekule (Chapter 6) represented a carbon atom following the discovery of valence by Frankland. But now suppose that this representation is taken literally. Suppose that it is interpreted to mean that a carbon atom actually directs its valences toward the corners of a square. *If this is actually the case*, then certain chemical predictions can be made to test the square hypothesis (Fig. 8.12).

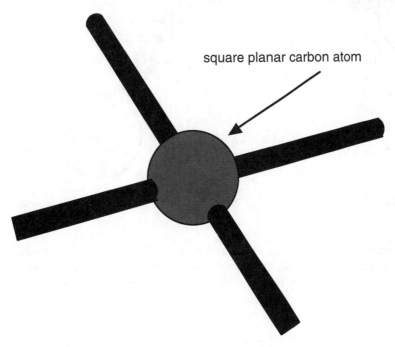

square planar carbon atom

Figure 8.12 Carbon with a Directed Valence of Four Square Planar Assumption

An example of one of these chemical predictions follows from the chemical reactions shown below:

Reactions 8.A and 8.B

$$CH_4 + Cl_2 \rightarrow CH_3Cl + HCl \qquad 8.A$$
$$CH_3Cl + Cl_2 \rightarrow CH_2Cl_2 + HCl \qquad 8.B$$

This reaction sequence could be continued, and eventually, all of the monovalent hydrogen atoms that were originally in the molecule of methane, CH_4, could be replaced by monovalent chlorine atoms. Consider now the structure that square directed valence hypotheses predicts for CH_4, CH_3Cl, and CH_2Cl_2.

The structure for the molecule of methane, CH_4, according to the square hypothesis is really very straightforward (Fig. 8.13).

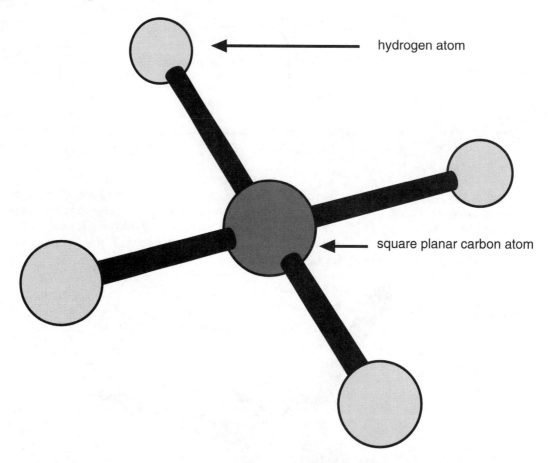

hydrogen atom

square planar carbon atom

Figure 8.13   Three Dimensional View of Methane Square Planar Assumption

This **structural formula** can be written more simply by making use of the symbols for the various atoms in the molecule and by eliminating attempts to show three dimensional perspective. Using this approach the structural formulas of CH_4 and CH_3Cl are equally straightforward (Fig. 8.14):

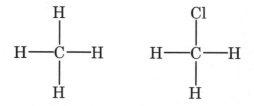

Figure 8.14   Structural Formulas
Square, Planar Assumption

The structural formula for the compound CH_2Cl_2 is a bit more complicated. In this case, it is possible to arrange the five atoms (2 H, 2 Cl, and 1 C) according to the square hypothesis in two entirely different ways (Fig. 8.15):

Cl   Cl
| |
H—C—Cl   H—C—H
| |
H   Cl

Figure 8.15 Two Square
Arrangements of CH_2Cl_2, Square
Planar Assumption

Since the square directed valence hypothesis for carbon allows for two different arrangements of the atoms in the molecule, two compounds with this formula should be formed by Reaction 8.B. In other words, the square directed valence hypothesis for an atom of carbon predicts that there should be two isomers (Section 8.5) of the compound CH_2Cl_2.

There is only one chemical compound with the formula CH_2Cl_2. This means that the square hypothesis is not correct. This bit of negative information does not represent a failure in the attempt to determine the direction of the carbon valence—*i.e.* the direction in space of a carbon atoms four potential bonds. Rather it represents the first step in a process of elimination. Using the method of isomer number described above and extending this method to include an analysis of many more carbon-containing compounds, one geometric shape after another can be tested as a possible directed valence model of the carbon atom.

In 1874, the Dutch chemist Jacobus Hendricus van't Hoff and the French chemist Joseph Achille Le Bel independently reported a directed valence shape for the carbon atom that was consistent with all of the known isomer numbers for carbon-containing compounds.[7] They proposed that a carbon atom directs its valence (four chemical bonds) toward the corners of a regular tetrahedron. Modern techniques have indicated that this view is indeed correct (Fig. 8.16).

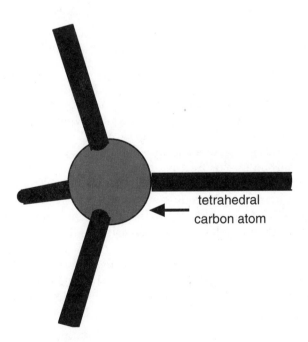

← tetrahedral carbon atom

Figure 8.16   Carbon with a Directed Valence of
Four, Tetrahedral Assumption

---

7.   In 1901, van't Hoff became the first recipient of the Nobel Prize for chemistry. He was only 22 years old in 1874, and his Nobel Prize was awarded for work he did in later years concerning the theory of solution chemistry.

The three dimensional nature of carbon's directed valence creates a spatial problem for the chemist. For example, the structural formula of methane, CH_4, should be drawn as shown in Figure 8.17A. Perspective drawing, however, is a nuisance. Consequently, modern chemists draw a planar structural formula for methane. In the mind of a chemist, this planar structural formula is equated to the tetrahedral three dimensional structure (Fig. 8.17B). In this sense, the planar structural formula is a kind of molecular blueprint.

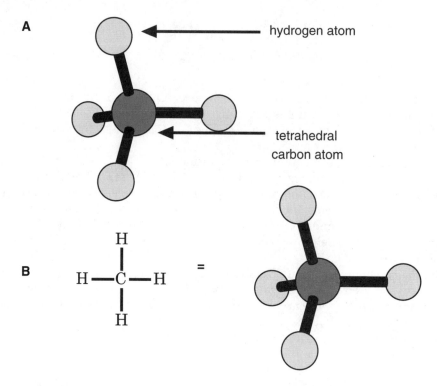

A — hydrogen atom

tetrahedral carbon atom

B $H\!-\!\overset{\displaystyle H}{\underset{\displaystyle H}{C}}\!-\!H$ =

Figure 8.17   Molecular Blueprint Structural Formula for Methane, Tetrahedral Assumption

Since the writing of planar structural formulas to represent three dimensional molecules is a common practice in chemical communication, chemists must develop an ability to "think" in three dimensions.

Incidentally, the tetrahedral shape of a carbon atom's valence can account for the existence of chiral molecules containing carbon atoms (Section 8.5). For example, if a carbon atom is bonded to four different atoms, then the five atoms can be arranged into two subtly different structures. The molecules that these structures represent are non-superimposable mirror images. Each is, therefore, chiral. Figure 8.18 and Figure 8.19 show the two structural arrangements for the formula CHFClBr. Isomers that differ because their molecules are non-superimposable mirror images are called enantiomers.

The above situation may not be easy for the nonchemist to visualize. It does, however, further illustrate the problem of "three dimensional thinking" that the modern chemist must face. Most modern chemists rely on wooden, plastic, or computer-generated three dimensional models in dealing with this problem.

8.7 A Guide to Chemical Bonding

A tetrahedral theory of carbon represents the first success of assigning to an atom directed valence. Since the original success of van't Hoff and Le Bel, chemists have learned much about the directed valence of other atoms. Figure 8.20 is a schematic of the periodic table that shows directed

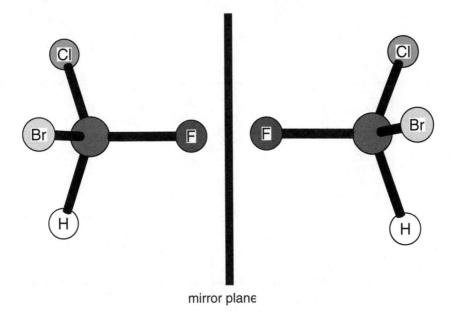

mirror plane

Figure 8.18 Two Chiral Structural Arrangements for CHFClBr
Mirror View

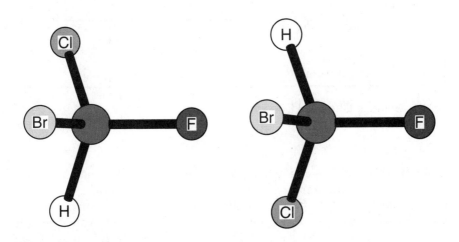

Figure 8.19 Two Chiral Structural Arrangements for CHFClBr
After Rotation

valence information for selected elements on the periodic table. Specifically, the geometric shape of the directed valence is indicated for each element in the first two periods of the table. Some of the more common directed valence shapes are also indicated in the region of the transition elements. The purpose of this section is to demonstrate how the concept of directed valence, when extended to elements other than carbon, allows the chemist to draw planar structural formulas for molecules. These planar structural formulas, called **Lewis structures**, serve as blueprints for the construction of three dimensional models of molecules.[8] The discussion will be limited to the elements of the first two periods on the periodic table.

---

8.   Named for the American chemist Gilbert Newton Lewis who contributed much to our understanding of these molecular pictures.

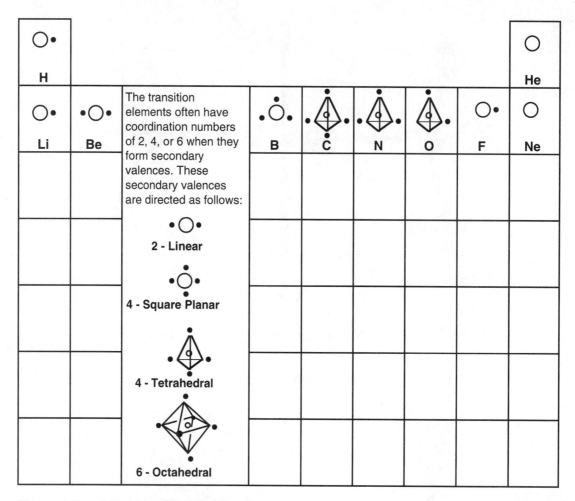

Figure 8.20 A Guide to Chemical Bonding

Each elemental block in periods one and two of this periodic table schematic (Fig. 8.20) contains a circle that represents a spherical atom. In each case, dots indicate valence directed from the surface of the atom. These dots are placed so that they indicate actual geometric shape. Notice that the directed valences of carbon, nitrogen, and oxygen are all related to the tetrahedron, although the actual bond angles for nitrogen and oxygen vary from the pure tetrahedral angle of 109.5°. Each of the directed valence figures allows for the construction of a directed valence model of the atom it represents. The model is made by drilling holes in wooden spheres so that sticks placed in the holes project in the directions indicated by the dots. This principle is illustrated for each of the selected elements in the Figure 8.21.

Using the models described above, the chemist can use tinker toy logic to construct three dimensional representations of molecules. This chemical tinker toy set, however, has an important building restriction. When it is possible, all of the attachment holes on a given atom model must be used (the rule of maximum chemical bond formation). The Lewis structure fragments listed in Figure 8.21 can also be used to construct a planar blueprint of a three dimensional molecular model. This construction of Lewis structures (Lewis structure drawing) is illustrated in the following examples.

Symbol	Model	Lewis Structure Fragment

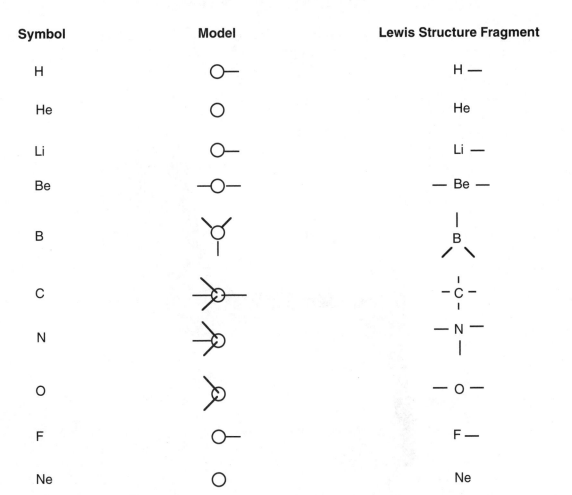

Figure 8.21   Directed Valence of Selected Atoms

EXAMPLE I

Lewis Structure of Hydrogen Peroxide

Draw the Lewis structure of hydrogen peroxide, H_2O_2.

SOLUTION:

I. Consider the Lewis structure fragments:

$$H— \qquad H— \qquad —O \qquad —O$$

II. Connect these fragments according to the rule of maximum chemical bond formation:

$$H\diagdown O\diagup O\diagdown H$$

With an appropriate set of models, this Lewis structure serves as a blueprint for the construction of a three dimensional molecular model for hydrogen peroxide (Fig. 8.22):

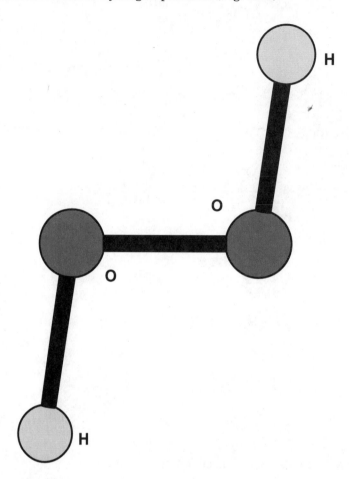

Figure 8.22 Three Dimensional Model of Hydrogen Peroxide

EXAMPLE II

Lewis Structure of Formaldehyde

Draw the Lewis structure of formaldehyde, CH_2O.

SOLUTION:

I. Consider the Lewis structure fragments:

II. Connect these fragments according to the rule of maximum chemical bond formation:

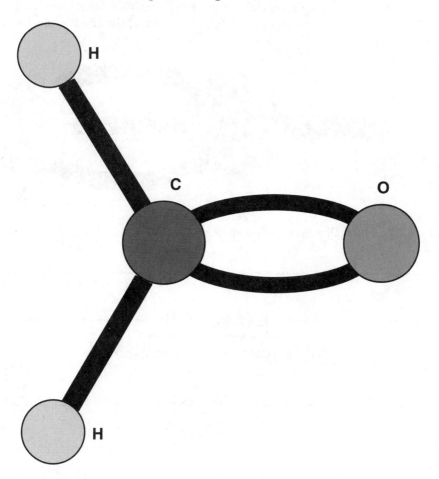

In this case, the only way to use all of the potential chemical bonds is to link the carbon atom and the oxygen atom by two chemical bonds. Such a linkage between atoms is called a double bond. Since the carbon atom (valence four) is only connected to three other atoms, formaldehyde is said to be an unsaturated molecule. Notice, however, that the carbon atom in formaldehyde still forms four chemical bonds (*i.e.* valence four). With an appropriate set of models, this Lewis structure also serves as a blueprint for the construction of a three dimensional molecular model. In this case, the double bond necessitates the use of flexible connecting sticks (Fig. 8.23).

Figure 8.23 Three Dimensional Model of Formaldehyde

EXAMPLE III

Lewis Structure of Hydrogen Cyanide

Draw the Lewis structure of hydrogen cyanide, HCN.

SOLUTION:

$$H\!-\!C\!\equiv\!N$$

The only way to use all of the potential chemical bonds is to link the carbon atom and the nitrogen atom by three chemical bonds. Such a linkage between atoms is called a triple bond. As in the case of molecules containing the double bond, molecules that contain triple bonds are also said to be unsaturated. With an appropriate set of models, this Lewis structure also serves as a blueprint for the construction of a three dimensional molecular model. In this case, the triple bond necessitates the use of flexible connecting sticks (Fig. 8.24).

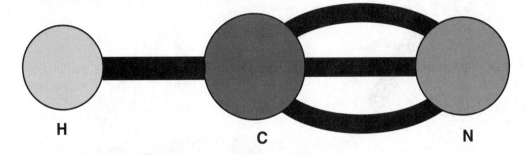

Figure 8.24 Three Dimensional Model of Hydrogen Cyanide

EXAMPLE IV

Lewis Structure Involving Isomers

There are two different chemical compounds (isomers) with the formula C_3H_7F. Draw the Lewis structure for each of these compounds.

SOLUTION:

structure A   structure B

These are the only possible arrangements for the formula C_3H_7F. Because the Lewis structure is a planar figure that represents a three dimensional structure, this fact is not obvious to the beginner. For example, if the following Lewis structure is used as a blueprint to construct a three dimensional model, this model will be identical to the three dimensional model corresponding to structure B. Hence, structure B and structure C are equivalent:

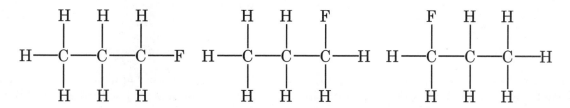

structure C

The remaining planar arrangements of the formula C_3H_7F are all equivalent to structure A:

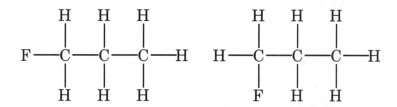

Demonstrating the equivalency of the above structures with three dimensional models may require the rotation of equivalent models or the rotation of carbon-to-carbon single bonds within a given model. Problem 32 at the end of this chapter describes the construction of a simple model kit that can be used to visualize the equivalency of the above structures.

The two isomers of C_3H_7F in the above example differ in their atom-to-atom arrangement within the molecule. Isomers of this type are called structural isomers. Notice that the isomers called enantiomers that were introduced in Section 8.6 differ in a much more subtle way than structural isomers. The molecules of enantiomers have exactly the same atom-to-atom arrangement within their molecules. They differ only because they are chiral—they differ as a left hand differs from a right.

8.8 Lewis Structure Rules

In the previous section, a tinker toy logic approach was used to construct the Lewis structure of several molecules. Since the ability to draw Lewis structures is an important skill that is used in the remainder of this chapter, it is useful to reduce this task to a series of rules. In this section, these Lewis structure rules will be illustrated using the formula C_3H_6 as an example.

Lewis Structure Drawing Rules

1. Draw the molecular framework using the non-univalent atoms that are contained in the molecule.

   In this case, a framework consisting of three carbon atoms is constructed.

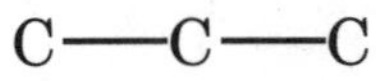

$$C—C—C$$

2. Draw bond fragments for each of the non-univalent framework atoms so that each atom reflects the correct valence.

   In this case, each framework carbon atom must reflect a valence of four.

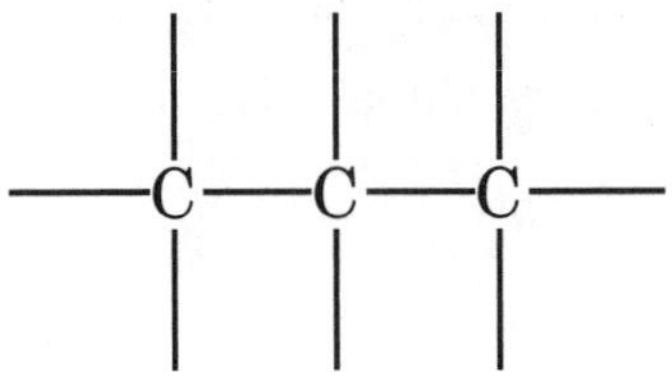

3. If the number of bond fragments added in step two exceeds the number of univalent atoms, use bond fragments to form rings and/or multiple bonds until the number of bond fragments equals the number of univalent atoms. Note that each new bond formed requires two bond fragments.

   In this case, two structures are possible. The three member ring structure may look strange, but it is possible to form this ring using the flexible bonds in a model kit.

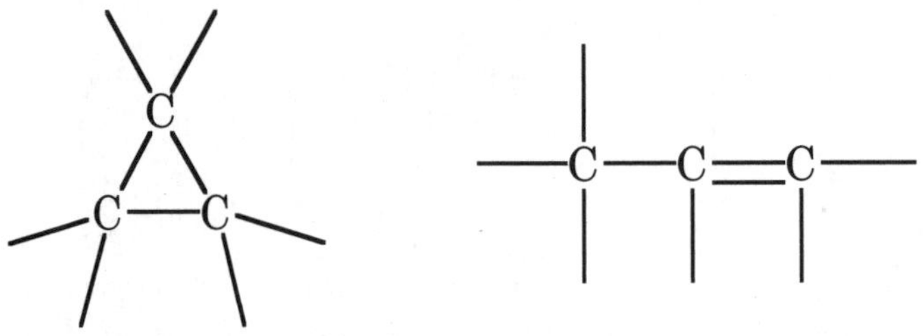

4. Finish drawing the Lewis structure, by adding the univalent atoms.

   In this case, the six hydrogen atoms are added to each structure.

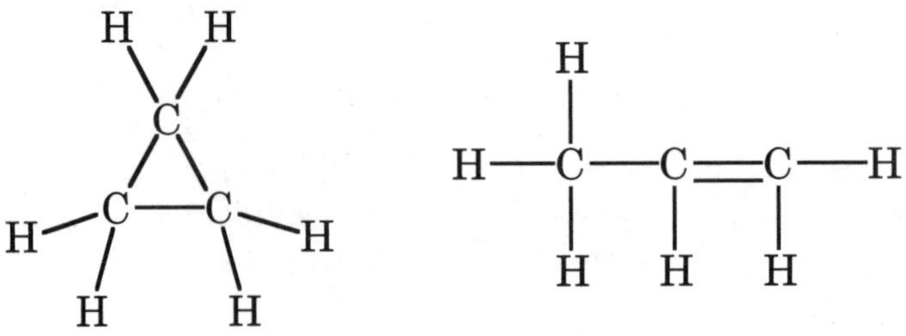

   The Lewis structure rules indicate that there are two isomers of C_3H_6, and two compounds with this formula are known. At this point, it is convenient to extend the definition of the term unsaturation to include multiple bonds and rings, hence both isomers of C_3H_6 are said to be **unsaturated**. Molecules that do not contain multiple bonds or rings are said to be **saturated**.

   The rules presented above allow the Lewis structures of simple compounds to be drawn quickly, and they offer an efficient alternative to the approach that was used in the previous section. In Chapter 9, Lewis structures will be placed on a more solid theoretical foundation. In the meantime, the very basic concepts associated with Lewis structures as put forth in this section allow chemistry students to explore the utility of structural formulas without any loss of essence.

8.9 Carbon: A Very Versatile Atom

In order to draw the two Lewis structures in the previous example, it was necessary to bond three carbon atoms into a chain. Although the tinker toy strategy suggested in Section 8.7 would seem to allow such chaining for any atom with a valence greater than one, carbon is the only atom to form a large number of molecules that involve extensive chaining. Carbon seems to be able to form chains without limit, and hence the number of potential carbon-containing compounds is theoretically infinite. This means that the carbon atom is a very versatile molecular building block. It is not surprising, therefore, that an entire area of the discipline of chemistry is devoted to the study of carbon (organic chemistry).[9] Carbon-containing compounds are central to the chemistry of life. Carbon-containing compounds also play a major role in the material and energy economic sectors of human society.

Representing structural formulas (Lewis structures) of carbon-containing molecules (organic molecules) presents the chemist with some interesting challenges. As the number of carbon atoms in a formula increases, the number of possible isomers increases at a greater rate. For example, although there are only two allowed arrangements for the atoms in the molecular formula C_4H_{10}, the molecular formula $C_{10}H_{22}$ corresponds to 75 atomic arrangements:

Two Molecular Arrangements for C_4H_{10}

Seventy-Five Molecular Arrangements for $C_{10}H_{22}$

74 other arrangements

---

9.   The term "organic chemistry" is somewhat of a misnomer. Because carbon forms so many chemical compounds, it is an essential element to life on this planet. The study of organic chemistry, however, goes beyond the study of life chemistry. The term organic chemistry has come to mean the study of the element carbon. The other major subdisciplines of chemistry are listed below:

Inorganic Chemistry—the study of elements other than carbon
Biochemistry—the study of the chemistry of life processes
Analytical Chemistry—the study of the composition of matter
Physical Chemistry—the application of the principles of physics and mechanics to the study of chemistry

Notice that carbon atoms can form both "straight" chains and branched chains. This effect tends to increase the number of possible atomic arrangements, particularly for molecular formulas that contain a large number of carbon atoms. The molecular formula $C_{20}H_{44}$ corresponds to 366,319 different arrangements.

Another factor that tends to increase the number of possible isomers is unsaturation—the tendency of carbon atoms to form multiple bonds and rings. This point was illustrated in the previous section. The presence of double bonds and/or rings in a molecule also provides the condition for another type of isomerism called geometric isomerism. Geometric isomerism is caused by the lack of rotation around a carbon-to-carbon double bond. This lack of rotation also exists for bonds that make up a ring. The ability of a carbon-to-carbon single bond to rotate and the inability of a carbon-to-carbon double bond to rotate can be illustrated with tetrahedral models of the carbon atom. The models indicate that rotation around a double bond requires that one bond in the double bond break and then reform after rotation. Since the breaking and reforming of bonds represents a chemical reaction, each of the rotated structures represents a stable isomer. Geometric isomerism is illustrated in Example V.

EXAMPLE V

Geometric Isomers

There are three different chemical compounds with the formula $C_2H_2Cl_2$, and there are two chemical compounds with the formula $C_2H_4Cl_2$. Draw the Lewis structure for each of these compounds.

SOLUTION:

I

II
cis geometric isomer

III
trans geometric isomer

IV

V

Structures I, II, and III represent isomers of $C_2H_2Cl_2$. Structures IV and V represent isomers of $C_2H_4Cl_2$. Structures II and III are geometric isomers of $C_2H_2Cl_2$. These isomers have the same atom-to-atom arrangement, but they cannot be interconverted without a chemical reaction because the carbon-to-carbon double bond cannot rotate like the carbon-to-carbon single bond. Notice the use of the words "cis" (same side) and "trans" (across) to describe geometric isomers. Because the carbon-to-carbon single bond can rotate, the following two structures for $C_2H_4Cl_2$ represent the same compound:

```
     H    Cl                    Cl   Cl
     |    |                     |    |
H —— C —— C —— H          H —— C —— C —— H
     |    |                     |    |
     Cl   H                     H    H
```

Before continuing, it might be useful to summarize the three types of isomers introduced in this chapter. **Structural isomers** are isomers with different atom-to-atom arrangements within their molecules. Enantiomers and geometric isomers are isomers with the same atom-to-atom arrangements within their molecules. **Enantiomers** differ because their molecules are non-superimposable mirror images. **Geometric isomers** differ because an interconverting molecular rotation is prevented by a double bond or a ring.

The molecules discussed in Examples IV and V are rather small when compared to the majority of organic molecules. In spite of this fact, the Lewis structures of these comparatively small molecules are somewhat messy to write. This written communication problem becomes quite acute in the case of larger molecules. In order to deal with this communication problem, chemists have developed a Lewis structure shorthand for organic molecules. Since the next section of this chapter deals with the importance of the Lewis structure to the modern chemist, an understanding of this Lewis structure shorthand for organic molecules is essential.

The Lewis structure shorthand for organic compounds is based on two characteristics of the carbon atom. First, a carbon atom almost always forms four chemical bonds. Second, organic molecules usually contain a large number of hydrogen atoms bonded to carbon atoms. In order to see how these two characteristics lead to a simple Lewis structure shorthand for carbon-containing molecules, consider drawing the Lewis structure of the "straight" chain isomer of $C_{10}H_{22}$. You will recall that the molecular formula $C_{10}H_{22}$ corresponds to 75 different arrangements.

Lewis Structure Shorthand Rules for Organic Molecules

1. Carbon atom chains ("straight," branched, and rings) are represented by jagged lines where angles and endpoints represent carbon atoms.

    In this case, the "straight" chain isomer of $C_{10}H_{22}$ would simply be written as follows:

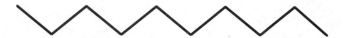

    In the figure below (Fig. 8.25), dots are placed on the jagged line to emphasize the positions of the ten carbon atoms. These dots are not used in the actual shorthand structures.

Carbon atoms indicated with dots.

Figure 8.25   Shorthand Lewis Structure

2. Double or triple bonds between carbon atoms are indicated by double or triple connecting lines (Example VI below).

    In this case, the molecule does not contain double or triple bonds.

3. Hydrogen atoms attached to carbon atoms are not actually written in the structure; rather, their presence is inferred from the valence of each carbon atom (C = valence 4).

The application of this rule is illustrated in Figure 8.26:

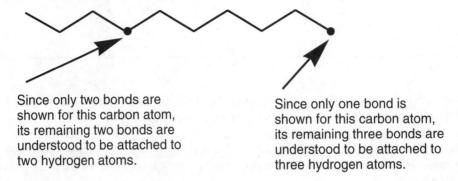

Since only two bonds are shown for this carbon atom, its remaining two bonds are understood to be attached to two hydrogen atoms.

Since only one bond is shown for this carbon atom, its remaining three bonds are understood to be attached to three hydrogen atoms.

Figure 8.26   Counting Hydrogen Atoms in a Lewis Shorthand Structure

In a similar manner, the number of hydrogen atoms attached to each carbon atom is inferred rather than written.

4. The previous rule necessitates that all atoms other than carbon atoms and hydrogen atoms attached to carbon atoms be written in standard Lewis structure notation (Example VI below).

In this case, there are no other atoms. The final shorthand Lewis structure is, therefore, simply a jagged line.

5. Occasionally, for the sake of communication ease, a carbon atom will actually be written in the shorthand Lewis structure. This is permissible, but all of the hydrogen atoms attached to this carbon atom must also be written.

In this case, if ease of communication demanded that the chain end carbon atoms actually be written, then it would be permissible to write:

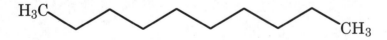

These five rules and the Lewis structure shorthand for organic molecules are further illustrated in Examples VI and VII.

EXAMPLE VI

Writing Shorthand Lewis Structures

Draw the shorthand Lewis structure for each of the following molecules:

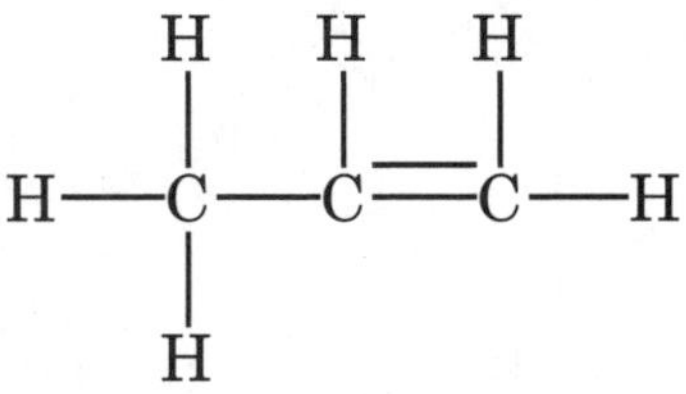

A

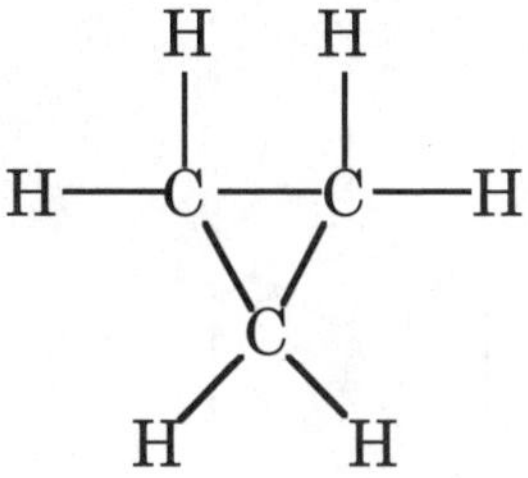

B

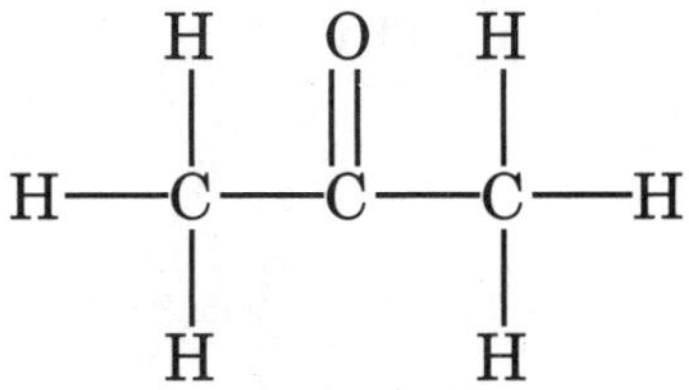

C

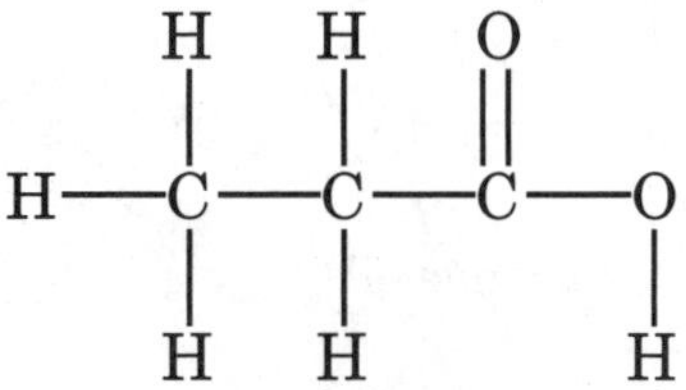

D

SOLUTION:

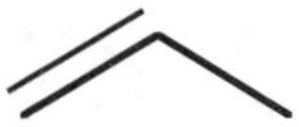

A    B 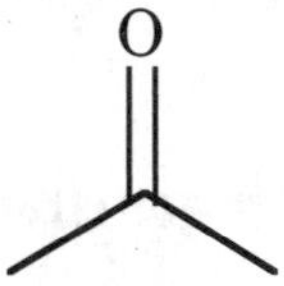   C 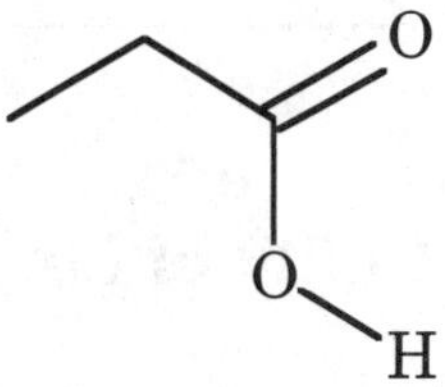   D

EXAMPLE VII

Reading Shorthand Lewis Structures

Write the molecular formula for each of the following molecules:

orange oil   aspirin

SOLUTION:

Orange oil: $C_{10}H_{16}$ Aspirin: $C_9H_8O_4$

8.10 The Utility of Lewis Structures

A Lewis structure represents a blueprint for the construction of a three dimensional model of a molecule. It also represents a final triumph of nineteenth century chemical atomic/molecular theory. Although the use of structural formulas was a tool developed in the latter part of the nineteenth century, their use remains an important tool of the modern chemist. The purpose of this section is to investigate this modern utility of Lewis structures.

The key to understanding the utility of the Lewis structure to the modern chemist is contained in a single statement: The chemical and physical properties of a chemical compound can be correlated to the Lewis structure of its molecule. The Lewis structure is not just an esoteric secretarial shorthand; rather, it is a description of the compound it represents. The easiest way to explain the correlation of compound function to molecular structure (Lewis structure) is by means of example.

Consider the following Lewis structures listed with the boiling points of the compounds they represent (Fig. 8.27).

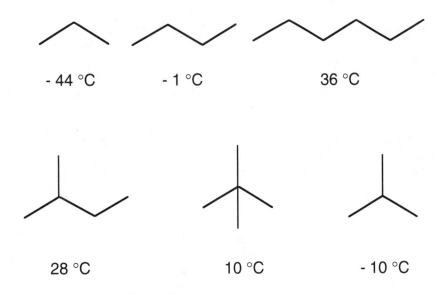

- 44 °C   - 1 °C   36 °C

28 °C   10 °C   - 10 °C

Figure 8.27 Lewis Structures and Boiling Points of Six Compounds

Although a casual inspection of this information is not particularly revealing, an elegant pattern does emerge if these compounds are arranged according to their Lewis structures. Specifically, if the compounds are listed in order of increasing number of carbon atoms in their molecular formulas, then the compounds fall into three groups—C_3H_8 (one compound or one isomer), C_4H_{10} (two compounds or two isomers), and C_5H_{12} (three compounds or three isomers). Further, if within each group the compounds are listed in order of decreasing branching represented by their Lewis structures, then the following ordered list results (Fig. 8.28).

Lewis Structure	Formula	Boiling Point
	C_3H_8	- 44 °C
	C_4H_{10}	- 10 °C
	C_4H_{10}	- 1 °C
	C_5H_{12}	10 °C
	C_5H_{12}	28 °C
	C_5H_{12}	36 °C

Figure 8.28 Ordered List of Lewis Structures for Six Compounds

Notice that listing the compounds according to molecular structure automatically arranges the compounds according to the physical property of boiling point (i.e. from low boiling point to high boiling point). The boiling point trends in Figure 8.28 do not apply only to the six compounds listed; the trends apply to all analogous chemical compounds. Analogous compounds in this case means compounds containing no unsaturation (multiple bonds or rings), and only carbon and hydrogen— saturated hydrocarbons. These trends can be summarized as follows:

1. In a series of saturated hydrocarbons, the hydrocarbon with the higher molecular weight will have the higher boiling point.
2. In a series of saturated hydrocarbon isomers, the boiling point decreases as the compound's Lewis structure becomes more branched.

EXAMPLE VIII

Using Lewis Structures to Predict Physical Characteristics

Arrange the following compounds in order of increasing boiling point:

SOLUTION:

This type of property to structure correlation is of immense use to the chemist, and it is by no means limited to physical property correlation.

Consider the following chemical equations written in terms of the Lewis structures of the compounds involved (Reactions 8.C and 8.D):

Reaction 8.C

Reaction 8.D

Notice that a molecule of water is being lost by the reacting molecules in both of these reactions. The dotted lines indicate that the water molecules are being lost in a similar way in both reactions. Further, in both of the reactions, the resulting organic product is formed by a similar joining of the resulting molecular fragments (indicated by the arrow). Although Reaction 8.E listed below is different from the ones listed above, it involves similar reactant molecules, and the organic product of the reaction can be predicted by analogy to Reactions 8.C and 8.D.

Reaction 8.E

EXAMPLE IX

Using Lewis Structures to Predict Chemical Reactions

Consider the following chemical reaction:

By analogy to the above reaction, predict the organic product of the following chemical reaction:

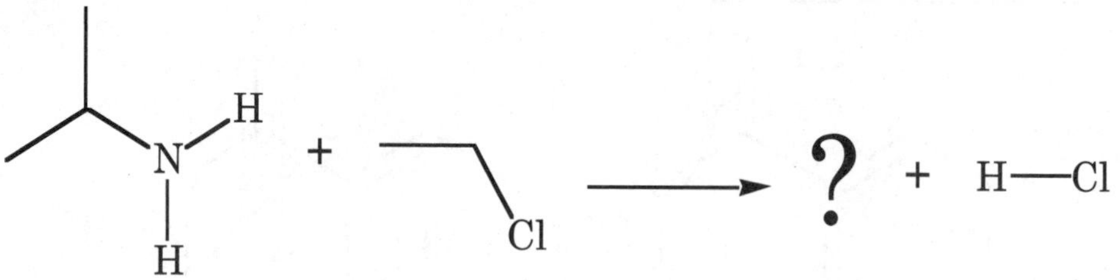

[chemical reaction structures]

SOLUTION:

[chemical structure of product]

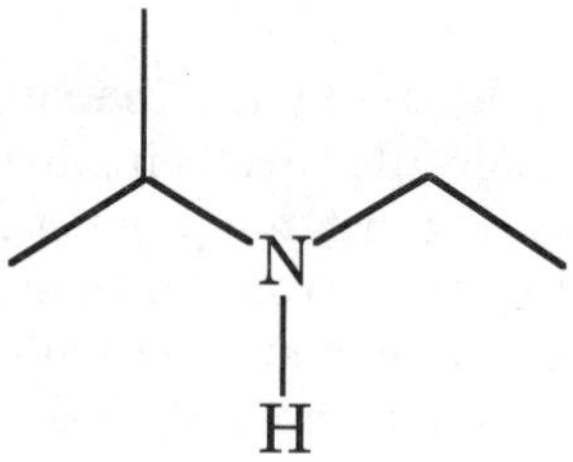

Over the years, chemists have correlated many chemical and physical characteristics to Lewis structural features. Knowledge of these correlations allows the modern chemist to predict the outcome of chemical reactions based on the Lewis structures of the reactants. The chemist becomes the architect of new chemical compounds by literally becoming a molecular architect.

8.11 Functional Groups

As chemists began to correlate chemical and physical characteristics of chemical compounds with molecular structure, they found it convenient to organize chemical compounds according to molecular structural features. A specific group of bonded atoms within a molecule is called a **functional group**. It was quickly discovered that the presence of a particular functional group in a complex molecular structure imparted functional group characteristics to the compound represented by the formula.

Modern chemists must be familiar with the characteristics of several dozen common functional groups. Twelve of these functional groups are useful in discussions involving basic organic chemistry—four hydrocarbon functional groups and eight functional groups containing carbon, hydrogen, oxygen and nitrogen atoms. These twelve functional groups are shown in Figure 8.29. In this figure, three of the hydrocarbon functional groups are listed with their homologous formulas. These homologous formulas indicate that each hydrocarbon functional group forms a series of compounds with predictable molecular formulas. For example, the alkane homologous formula C_nH_{2n+2} indicates the series of compounds CH_4 (n = 1), C_2H_6 (n = 2), etc. The hydrocarbon functional groups simply classify the hydrocarbons according to degree and type of unsaturation—alkanes (saturated), alkenes (double bond), alkynes (triple bond), and aromatic (benzene structure).

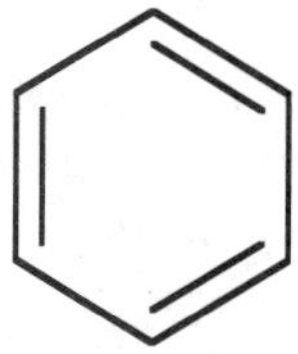

benzene
an aromatic hydrocarbon

Although benzene looks like a complex alkene with three double bonds, this specific arrangement of three double bonds in a six member ring has special functional group characteristics very much unlike alkene chemistry. The specific structures of eight oxygen and nitrogen functional groups are shown in Figure 8.29. In these functional group structures, the letter R is used to represent an organic molecular fragment. This is a convenient way of indicating that the functional group can be part of a more complex molecular structure.

Hydrocarbon Functional Groups

Alkane	C_nH_{2n+2}	saturated – single bonds
Alkene	C_nH_{2n}	unsaturated – double bond
Alkyne	C_nH_{2n-2}	unsaturated – triple bond
Aromatic	benzene type structure	

Nitrogen and Oxygen Functional Groups

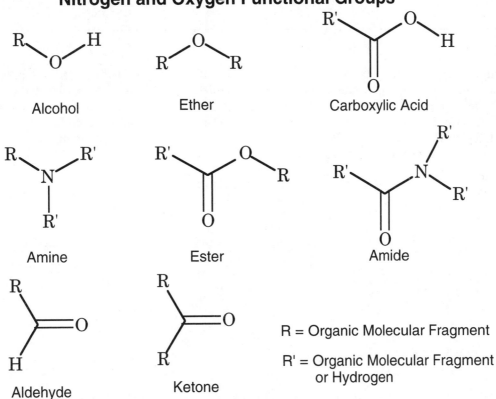

Figure 8.29 Functional Group Names and Structures

EXAMPLE X

Functional Group Identification

The chemical compound alanine can be prepared by the chemical decomposition of silk. The structural formula of alanine is shown below. What functional groups are contained in the alanine molecule?

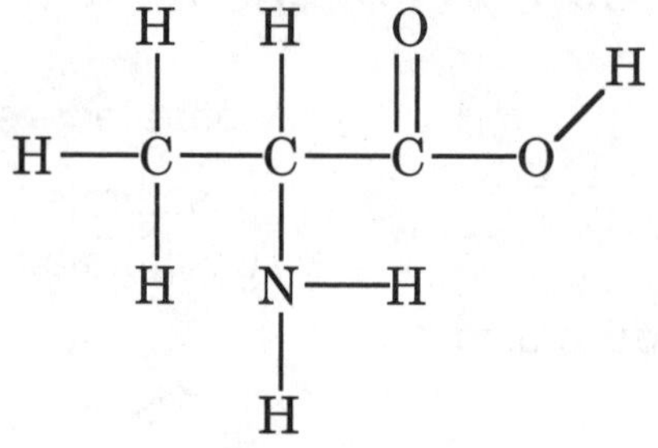

alanine

SOLUTION: carboxylic acid, amine, alkane

Alanine, the molecule in Example X, can be used to illustrate the utility of the functional group concept in understanding basic chemistry. Alanine is called an amino acid because it contains the amine functional group and the carboxylic acid functional group. Under the proper set of conditions, the amine functional group and the carboxylic acid functional group can react to form the amide functional group. Using the letters R and R' to represent any organic molecular fragments, the equation for this reaction is presented as a prototype in Reaction 8.F.

Reaction 8.F

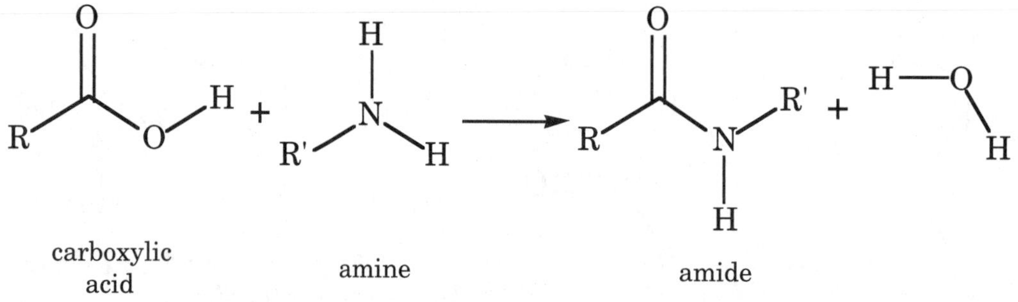

carboxylic
acid

amine

amide

Because alanine contains both the amine functional group and the carboxylic acid functional group, alanine molecules can bond together in long chains by forming amide functional group links between the amine function of one molecule and the carboxylic acid function in another molecule. Large molecules made up of smaller molecules that are chemically linked together by some type of functional group reaction are called **polymers**. The small molecules that make up a polymer are called **monomers**. The term polymer usually infers that only one type of monomer is contained in the large molecule. If large molecules are formed from a variety of monomers, these large molecules are usually called **co-polymers** or **macromolecules**.

Alanine is one of about twenty natural amino acids that nature uses as monomers to form the biological macromolecules known as proteins. Silk is a fibrous protein containing mainly the amino acids alanine, glycine, and serine. The structure of these amino acids and the structure of the macromolecule silk is shown in Figure 8.30.

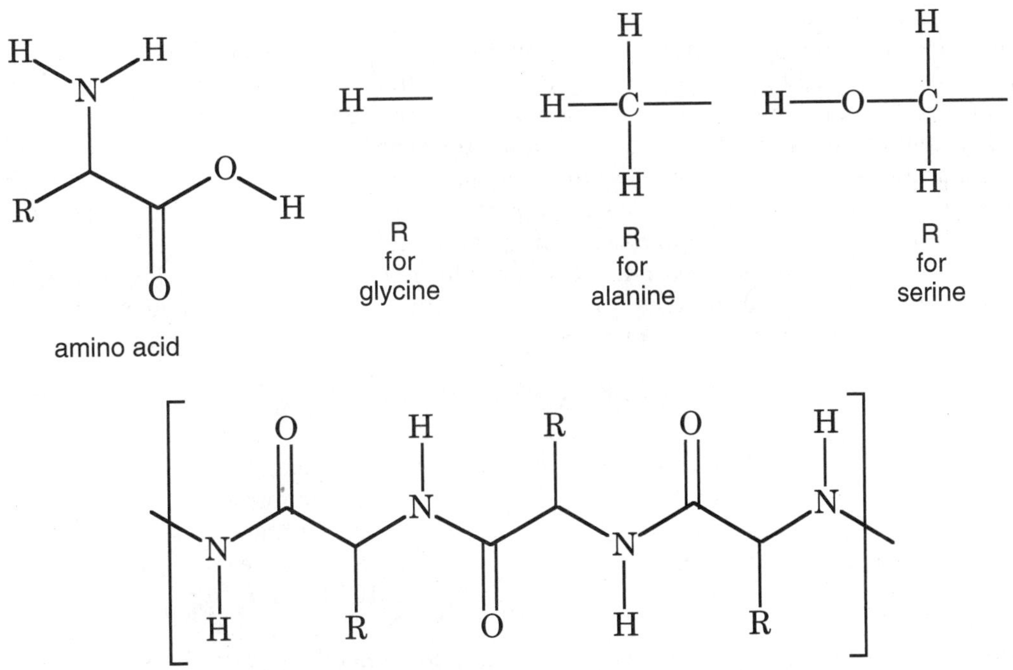

amino acid

R for glycine

R for alanine

R for serine

three amino acid units in protein macromolecule

The protein silk is composed mainly of the amino acids glycine, alanine, and serine.

Figure 8.30 The Structure of Silk

With the exception of glycine, the molecules of all of the natural amino acids are chiral. Nature only uses one of the two enantiomers to form proteins. This natural preference for one of two possible mirror image isomers is a common theme in many natural biological molecules. In the case of proteins, the single "handedness" of the amino acid molecules leads to a single "handedness" of protein molecules. The "handedness" of these molecules plays an important role in their function.

Natural and synthetic polymers can be formed by the reaction of a wide range of monomer functional groups. The resulting polymeric compounds include such familiar materials as starch, cellulose, polyethylene, polypropylene, styrene, nylon, Teflon, and polyesters. The library research problem in question 35 at the end of this chapter proposes further study of natural and synthetic polymers.

8.12 The Hydrocarbon Components of Petroleum

Since carbon-containing materials play a central role in the chemistry of life, it is not surprising that society is interested in the industrial development of carbon-containing materials. Garments can be made from the carbon-containing silk produced by a silkworm, and headaches can be cured by the carbon-containing drug contained in willow bark. As the theory of structural chemistry developed, chemists realized that the theory provided the tools required to imitate the production of nature's carbon-containing materials. In the late nineteenth century, the primary source of simple carbon compounds for the production of useful carbon-containing materials was coal tar. Coal tar is a byproduct that is formed when coal is converted to an impure form of elemental carbon called coke. Coke is used in the conversion of iron ore to iron. (See Sections 4.2 and 4.3 in Chapter 4.) Society's demand for useful carbon-containing materials has grown considerably since that late nineteenth century, and the modern chemist must turn to a more economic source of simple building block carbon compounds. Today the primary source of carbon compounds for the production of useful carbon-containing materials is petroleum.

Petroleum is one of the most important raw materials managed by the chemical profession. Petroleum can be processed to provide not only the wide variety of fuels required by the energy sector of society, but also the basic molecular building blocks required by the materials sector of society. Petroleum is a complex mixture of hydrocarbons and other chemical compounds. In question 34 at the end of this chapter, the composition, processing, and use of petroleum will be explored through a library research question. Since the alkanes are a major hydrocarbon component of petroleum, the library research problem requires an understanding of alkane nomenclature.

As mentioned in the previous section, the alkanes form a homologous series with the molecular formula C_nH_{2n+2}. In order to understand alkane nomenclature, it is necessary to learn the names of the first ten continuous chain (unbranched) alkanes.

Unbranched Alkanes
(n = 1 to 10)

1. Methane	6. Hexane
2. Ethane	7. Heptane
3. Propane	8. Octane
4. Butane	9. Nonane
5. Pentane	10. Decane

Using these names, more complex branched alkanes can be named by applying a simple set of rules. The following branched isomer of $C_{11}H_{24}$ will be used to illustrate these rules:

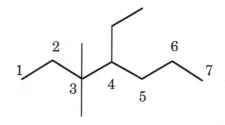

Alkane Nomenclature Rules

1. Identify the longest continuous chain in the molecule and use the alkane name of this continuous chain as the root name of the branched alkane. The branched alkane will eventually be named as a substituted version of this root alkane.

   In this case, the longest continuous chain contains seven carbon atoms. The root name of this alkane is, therefore, heptane. The branched alkane will be named as a substituted heptane.

2. Number the carbon atoms in the longest continuous chain. Assign numbers in the direction that assigns the lowest possible numbers to carbon atoms that contain a branch.

Numbering from left to right places branches at position 3 and 4. Numbering from right to left would have placed branches at positions 4 and 5.

3. Name each branch that connects to the longest continuous chain after the alkane with the same number of carbons as the branch. In naming the branch, drop the "ane" ending from the alkane name and add "yl." Alkane branches so named are referred to collectively as alkyl groups.

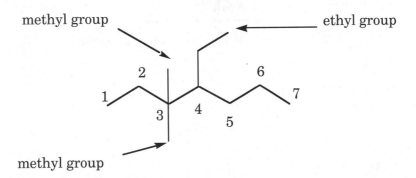

Note that the two one-carbon branches at position three are named methyl after methane. The single branch at position four is named ethyl after ethane.

4. Name the branched alkane by using the branch position numbers and names as prefixes to the name of the longest continuous chain.

In this case, the final name of the branched alkane is 3, 3-dimethyl-4-ethylheptane. Note that the final name is one continuous word with commas separating numbers and dashes separating numbers and letters. Note also that identical alkyl groups are designated with the prefixes di (2), tri (3), tetra (4), penta (5), etc.

The four structures shown below further illustrate the application of these rules to the naming of alkanes:

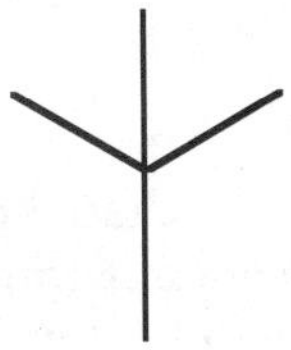

2,2-dimethylpropane

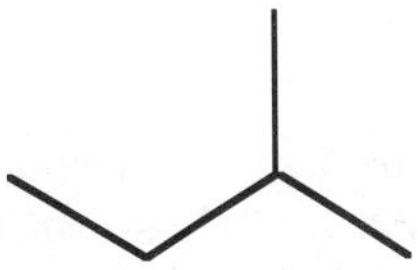

2-methylbutane

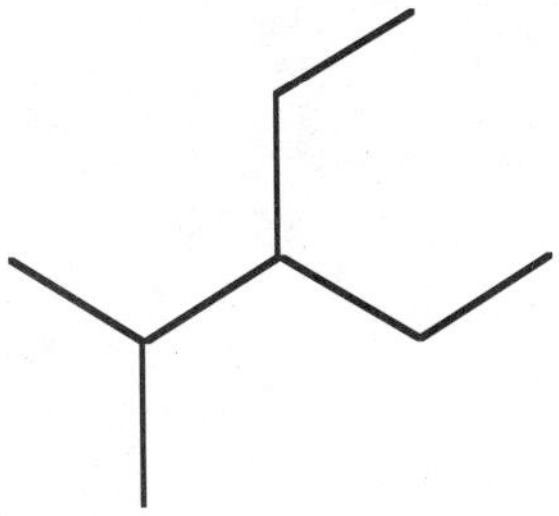

2-methyl-3-ethylpentane

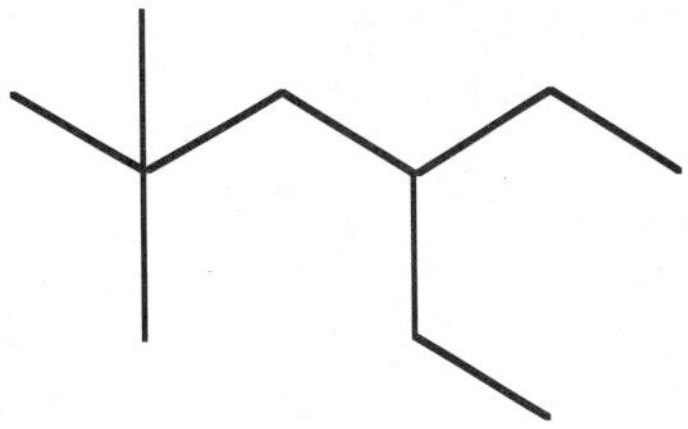

2,2-dimethyl-4-ethylhexane

Alkanes are the most fundamental organic compounds. Since they can be built up into more complex organic compounds containing other functional groups, the nomenclature of other functional groups is based on alkane nomenclature. The nomenclature of all organic compounds is too complex to be detailed in a text of this scope. It is instructive, however, to see at least one example of the extension of alkane nomenclature to other functional groups. For example, simple compounds containing the alcohol functional group are named by replacing the "ane" ending of the corresponding alkane to "ol." Hence the alcohol derived from ethane is called ethanol.

EXAMPLE XI

Structures of Substituted Alkanes

Name the following alkane:

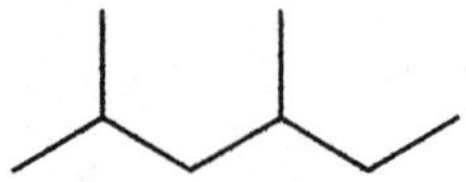

SOLUTION: 2,4-dimethylhexane

ADDITIONAL EXAMPLE:
Draw the structure of the following substituted alkane: 2,3,4-trimethyl-4-ethylheptane

ANSWER:

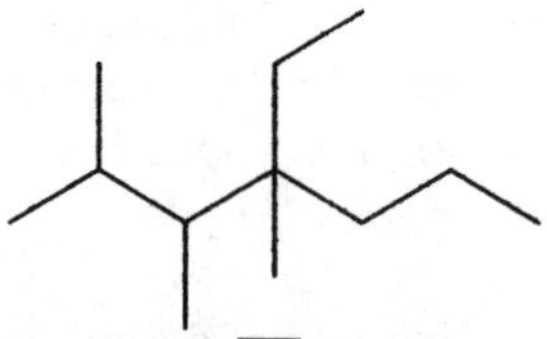

Ethanol can be prepared by the yeast fermentation of sugars. Ethanol can also be prepared from the ethane in petroleum. Both of these routes to ethanol are of economic importance, and both routes are managed by the chemical profession. Figure 8.31 shows the chemical reactions that are involved in these processes. Although the hydrocarbons in petroleum are used as building blocks for many complex organic molecules, the reactions in Figure 8.31 are a simple way to show the material value of a barrel of petroleum. The library research problem in question 34 at the end of this chapter suggests a more thorough investigation of petroleum chemistry.

Production of Ethanol by Yeast Fermentation:

$$C_6H_{12}O_6 \xrightarrow{\text{yeast}} 2 \text{ } CH_3CH_2OH + 2 \text{ } CO_2$$

glucose   ethanol

Production of Ethanol by Petroleum Processing:

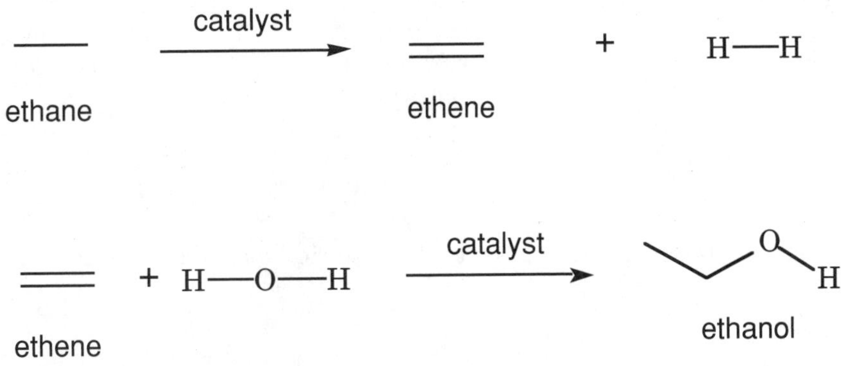

ethane $\xrightarrow{\text{catalyst}}$ ethene $+$ H—H

ethene $+$ H—O—H $\xrightarrow{\text{catalyst}}$ ethanol

Figure 8.31   Two Methods of Producing Ethanol

8.13 A Final Note

The first eight chapters of this textbook have attempted to detail the evolution of a human idea—the chemical atomic/molecular theory. At the same time, they have also attempted to convey the essence of what chemistry is and what chemists do. Although the teaching vehicle has been the nineteenth century, the message has been contemporary. Using atomic/molecular theory, the modern chemists continue to pursue their ancient goal:

$$\text{MATTER I} \xrightarrow{?} \text{MATTER II}$$

They also continue to serve the society in which they live by using atomic/molecular theory to manage material and energy resources.

CHEMISTS HAVE SOLUTIONS – American Chemical Society, T-shirt Message

Chapter Eight
Performance Objectives

P.O. 8.0

Review all of the boldfaced terminology in this chapter, and make certain that you understand the use of each term.

chiral	crystals	co-polymer
directed valence	enantiomers	functional group
geometric isomer	isomer	Lewis structure
macromolecule	monomer	polymer
regular polyhedron	saturated	structural formula
structural isomer	unsaturated	

P.O. 8.1

You must demonstrate an understanding of the historic evolution of the concept of molecular geometry.

EXAMPLE:
Although the ancient Greeks discussed the possibility of a three dimensional "atomic" world, a three dimensional theoretical world of atoms and molecules based on scientific evidence did not reach maturity until the nineteenth century. Which of the following experimental observations was most useful to Louis Pasteur in the early development of this theory of molecular geometry?

a. Measurement of density
b. Thermal properties of compounds
c. Optical properties of crystals
d. Measurement of color
e. Microscopic examination of molecules

SOLUTION: c

Textbook Reference: Section 8.1–8.5

ADDITIONAL EXAMPLE:
Assume that the directed valence of a carbon atom is square planar. How many isomers would such a square planar model predict for each of the following formulas:

CH_3Cl CH_2Cl_2 $CHCl_3$ $CHClBrI$

ANSWER: 1 2 1 2

P.O. 8.2

You must be able to identify a correct or incorrect Lewis structure.

EXAMPLE:
Which of the following Lewis structures is not correct?

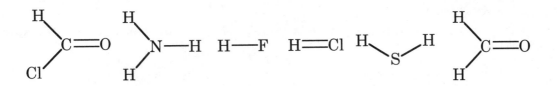

SOLUTION: $H=Cl$

Textbook Reference: Section 8.7

ADDITIONAL EXAMPLE:
According to the directed valence theory presented in the textbook, which of the following Lewis structures is not correct?

a) H–S–H   b) H–Be–H   c) H–S–H–F   d) all correct   e) all incorrect

ANSWER: c

P.O. 8.3

You must be able to identify a double or a triple bond in a simple molecule after drawing the Lewis structure of the molecule.

EXAMPLE:
Which of the following molecules contains a double bond? Which contains a triple bond?

$$C_2H_6 \quad C_3H_8 \quad C_2H_2 \quad CCl_2O \quad H_2O_3$$

SOLUTION: CCl_2O double C_2H_2 triple

Textbook Reference: Section 8.7

ADDITIONAL EXAMPLE:
After drawing Lewis structures, decide which of the following molecules contains a double bond?

a) CH_4 b) C_2H_6 c) CH_3N d) all   e) one

ANSWER: c

P.O. 8.4

You must be able to identify a structural feature in a simple molecule after drawing the Lewis structure of the molecule.

EXAMPLE:
Which of the following molecules could contain the -O-H atom group?

$$C_2H_6O \qquad CHClO$$

SOLUTION: C_2H_6O

Textbook Reference: Section 8.7

ADDITIONAL EXAMPLE:
How many atoms of hydrogen are bonded to the nitrogen atom in the molecule CH_5N? (Answer requires Lewis structure.)

a) 1   b) 2   c) 3   d) 4   e) 5

ANSWER: b

P.O. 8.5

You must be able to identify the number of possible isomers from a simple molecular formula.

EXAMPLE:
How many isomers are represented by the molecular formula CHClO?

SOLUTION: one

Textbook Reference: Section 8.7

ADDITIONAL EXAMPLE:
How many isomers are represented by the molecular formula C_3H_5Cl?

ANSWER: five

P.O. 8.6

You must be able to read a Lewis "line" structure.

EXAMPLE:
The figure below represents a molecule of the chemical compound Tylenol. What is the molecular formula of this compound?

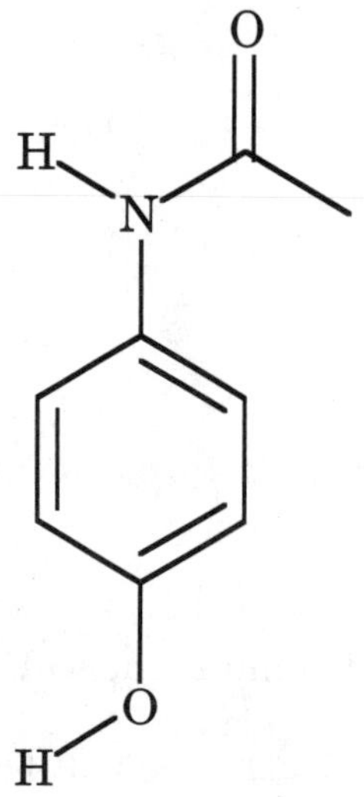

SOLUTION: $C_8H_9O_2N$

Textbook Reference: Section 8.9

ADDITIONAL EXAMPLE:
The figure below represents a molecule of the chemical compound eugenol (oil of cloves). What is the molecular formula of this compound?

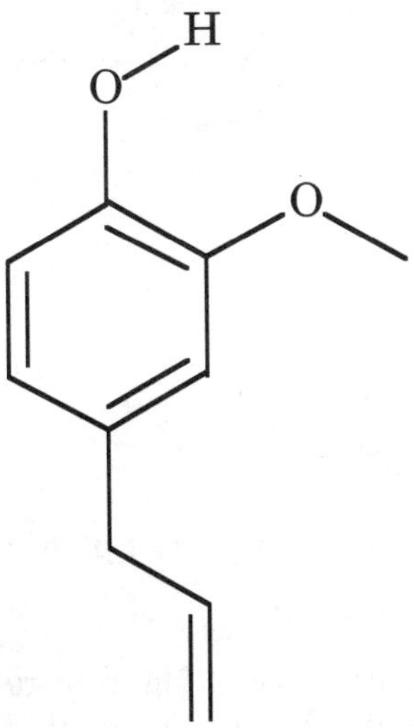

ANSWER: $C_{10}H_{12}O_2$

P.O. 8.7

You must be able to predict the product of a chemical reaction given the prototype of the reaction.

EXAMPLE:
The acid catalyzed chemical reaction of methyl alcohol with hydrogen chloride:

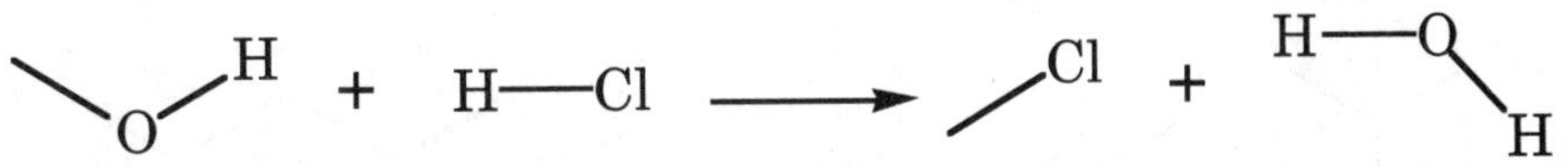

Ethyl alcohol undergoes the same acid catalyzed reaction. What is the structural formula of the organic product formed when ethyl alcohol undergoes this reaction?

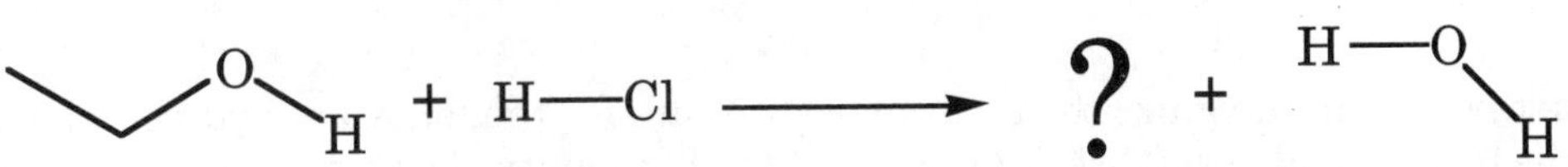

SOLUTION:

Textbook Reference: Section 8.10

ADDITIONAL EXAMPLE:
The chemical reaction between ethyl chloride and methyl amine is shown below:

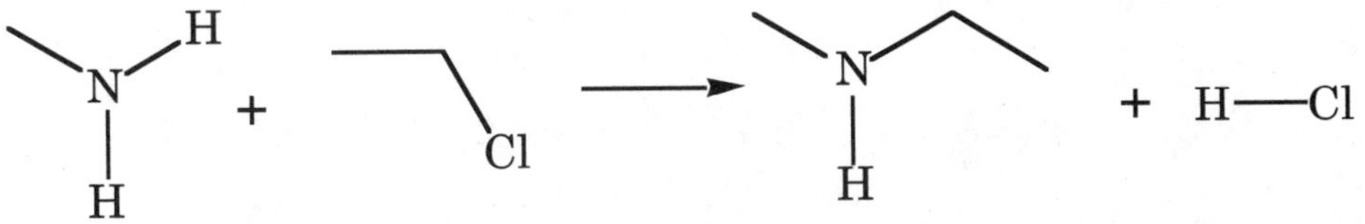

311

Ethyl amine undergoes the same chemical reaction. What is the structural formula of the organic product formed when ethyl amine undergoes this reaction?

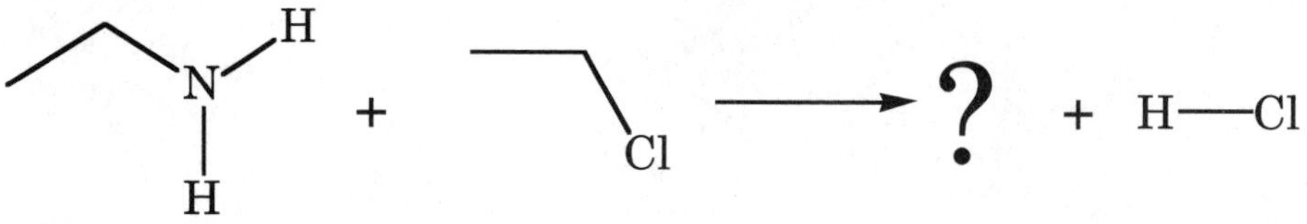

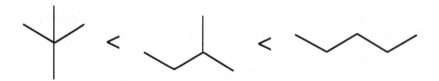

ANSWER:

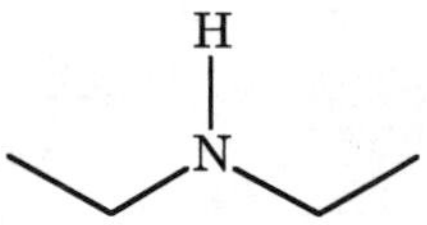

P.O. 8.8

You must be able to use Lewis structures to predict physical characteristics.

EXAMPLE:
Acetic acid (found in vinegar) and butyric acid (found in rancid butter) have the formulas, CH_3COOH and $CH_3CH_2CH_2COOH$, respectively. Which should have the higher boiling point?

SOLUTION: butyric acid

Textbook Reference: Section 8.10

ADDITIONAL EXAMPLE:
For a series of alkane isomers, the boiling point decreases as the number of branch chains increases. Use this empirical rule to organize the isomers of C_5H_{12} in order if increasing boiling point.

ANSWER:

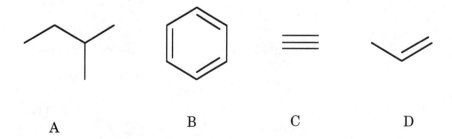

P.O. 8.9

You must be able to recognize the molecular structure for the following hydrocarbon functional groups: alkane (C_nH_{2n+2}), alkene (C_nH_{2n}), alkyne (C_nH_{2n-2}), aromatic (benzene type structure).

EXAMPLE:
Identify each of the following hydrocarbons as an alkane, alkene, alkyne, or aromatic:

A   B   C   D

SOLUTION: A = alkane; B = aromatic; C = alkyne; D = alkene

ADDITIONAL EXAMPLE:

Identify the two hydrocarbon functional groups contained in the following molecule:

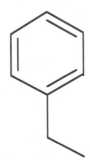

ANSWER: aromatic & alkane

P.O. 8.10

You must be able to recognize the structure of the following functional groups:

Hydrocarbon Functional Groups

Alkane	C_nH_{2n+2}	saturated – single bonds
Alkene	C_nH_{2n}	unsaturated – double bond
Alkyne	C_nH_{2n-2}	unsaturated – triple bond
Aromatic	benzene type structure	

Nitrogen and Oxygen Functional Groups

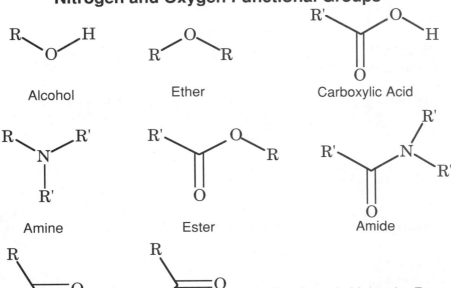

Alcohol   Ether   Carboxylic Acid

Amine   Ester   Amide

Aldehyde   Ketone

R = Organic Molecular Fragment

R' = Organic Molecular Fragment or Hydrogen

EXAMPLE:
Identify all of the functional groups in the following molecule:

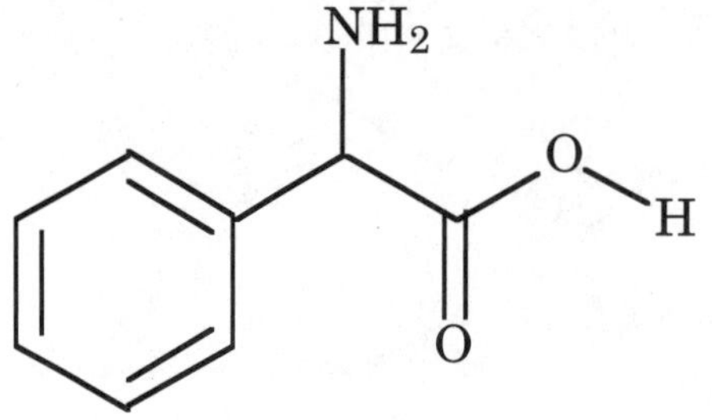

SOLUTION: alkane and aromatic hydrocarbon; amine; carboxylic acid

Textbook Reference: Section 8.11

ADDITIONAL EXAMPLE:
Draw the two isomers of C_2H_6O. What is the functional group of each isomer?

ANSWER: alcohol and ether

P.O. 8.11

You must be able to name the first ten normal alkanes (*i.e.* "straight" chain isomers for C_nH_{2n+2} with n=1 to n=10).

EXAMPLE:
What is the name of the following alkane shown below?

SOLUTION: butane

Textbook Reference: Section 8.12

ADDITIONAL EXAMPLE:
Name the first ten normal alkanes.

ANSWER: methane, ethane, propane, butane, pentane, hexane, heptane, octane, nonane, decane

P.O. 8.12

You must be able to name a substituted alkane given its structure.

EXAMPLE:
Name the following alkane:

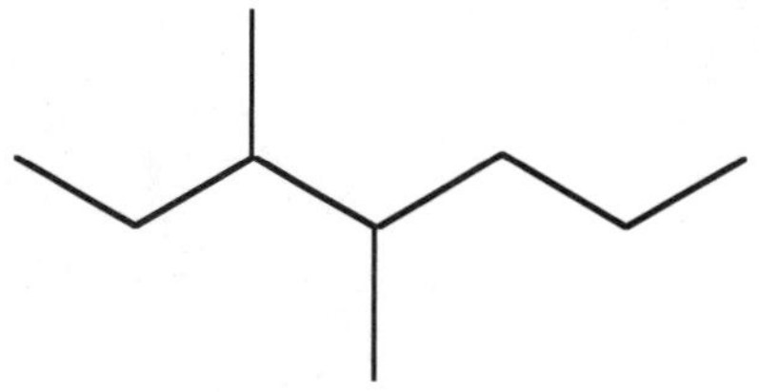

314

SOLUTION: 3,4-dimethylheptane

Textbook Reference: Section 8.12

ADDITIONAL EXAMPLE:
Name the following alkane:

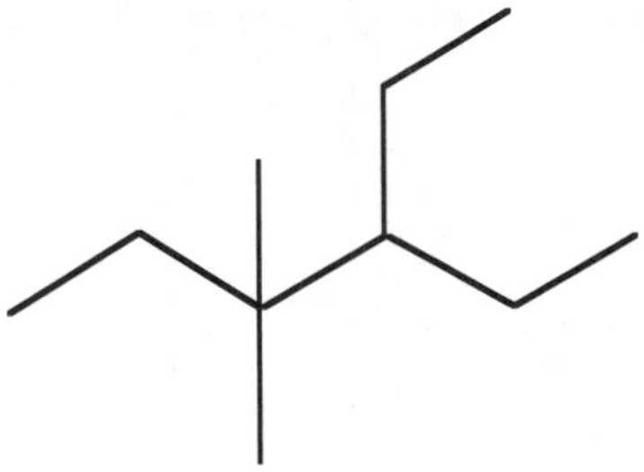

ANSWER: 3,3-dimethyl-4-ethylhexane

P.O. 8.13

You must be able to draw the structure of a substituted alkane given its name.

EXAMPLE:
Draw the structure of 2-methyl-4-propylnonane.

SOLUTION:

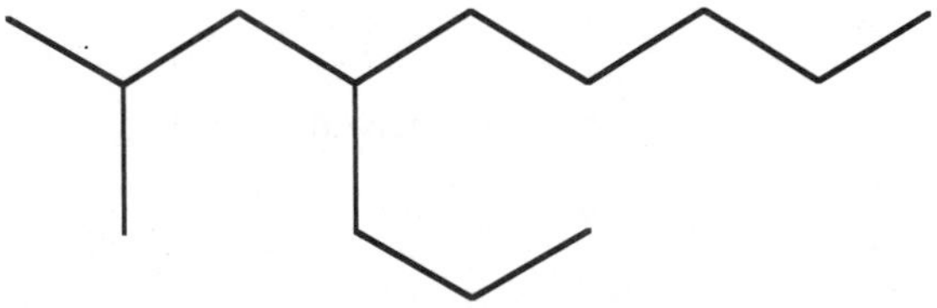

Textbook Reference: Section 8.12

ADDITIONAL EXAMPLE:
Draw the structure of 2,3-dimethyl-3-ethylnonane.

ANSWER:

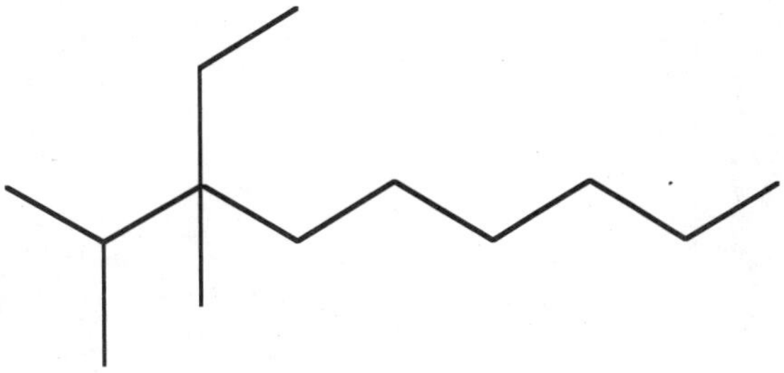

Chapter Eight
Problems

Problems on the Historical Evolution of Molecular Geometry

1. What facts suggest the theory that crystals of pure substances contain repeating arrangements of atoms?

    STUDENT SOLUTION:

2. Suppose you live in the fictitious two dimensional world described in Section 8.6 and you are interested in the crystal shown below. To what two dimensional "Bravais" arrangement (Fig. 8.9) do the atoms in this crystal belong?

    STUDENT SOLUTION:

3. Suppose that the valences of a boron atom are directed as indicated in the figure below:

    If the valences of boron are directed as indicated, how many isomers of BBr_2I could exist?

    STUDENT SOLUTION:

4. A transition metal, M, has a coordination number of six when this transition metal forms a complex ion with nonmetal elements A and B (Section 6.7). The formula of this complex ion is MA_4B_2. If the geometric shape of this complex ion is an octahedron (shown below), how many isomers are possible for this complex ion?

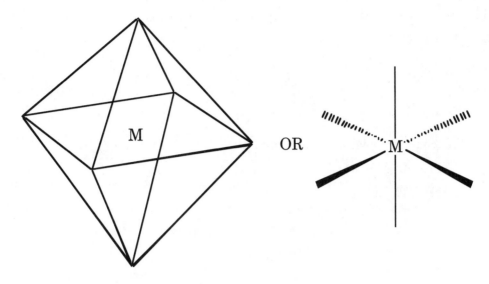

STUDENT SOLUTION:

Lewis Structures

5. Draw Lewis structures for each of the following molecules:

C_2H_6 H_2O_3 N_2 C_2H_2 CO_2 NH_3 BF_3 SiF_4 $OCCl_2$ BeH_2 H_2S

STUDENT SOLUTION:

6. Draw Lewis structure for each of the following molecules and indicate which have double and triple bonds:

CCl_2F_2 BCl_3 I_2 CS_2 C_2H_4 HCN N_2 NCl_3

STUDENT SOLUTION:

7. According to the directed valence theory presented in the textbook, which of the following Lewis structures is not correct?

a) H–N–Cl   b) H–N=C–H   c) O=C–O–H   d) O=C=C–H
    | | | |
    H   H   N–H   H

STUDENT SOLUTION:

8. Which of the following molecules could contain the C=O atom grouping?

a) C_2H_6O b) $C_2H_4O_2$ c) $C_4H_{10}O$ d) $C_2H_6O_2$

STUDENT SOLUTION:

Isomers

9. Draw the Lewis structures for the three isomers of C_3H_8O.

STUDENT SOLUTION:

10. Draw the Lewis structures for all the isomers of C_6H_{14}.

STUDENT SOLUTION:

11. Draw the Lewis structures for all the isomers of C_4H_9Cl.

STUDENT SOLUTION:

12. Draw Lewis structures for all the isomers of C_2H_4ClBr.

STUDENT SOLUTION:

13. Draw the Lewis structures for all the isomers of C_2H_2ClBr.

STUDENT SOLUTION:

14. Draw the Lewis structures for all the isomers of C_5H_{12}.

STUDENT SOLUTION:

15. Two of the proposed replacements for chlorofluorocarbons (e.g. Freon) have formulas of $C_2H_2F_4$ and $C_2H_3Cl_2F$. Draw the Lewis structures for all of the isomers of these compounds.

STUDENT SOLUTION

16. Recall that a carbon atom bonded to four different atoms or groups of atoms can result in a chiral molecule (Section 8.6). Do any of the structures in question fifteen have such a chiral carbon?

STUDENT SOLUTION:

Lewis Line Structures

17. Write the molecular formulas for each of the following chemical compounds:

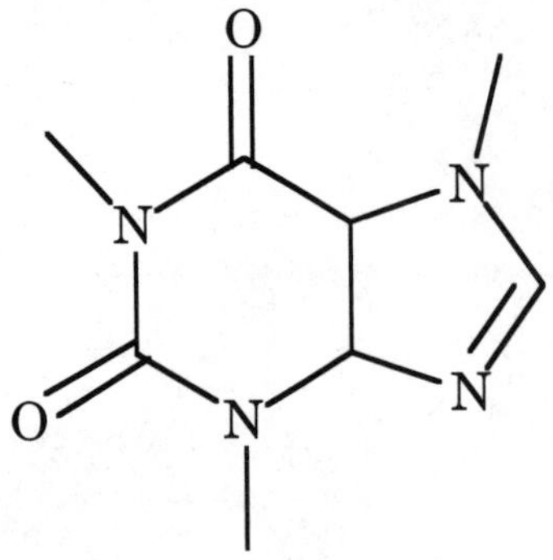

Caffeine

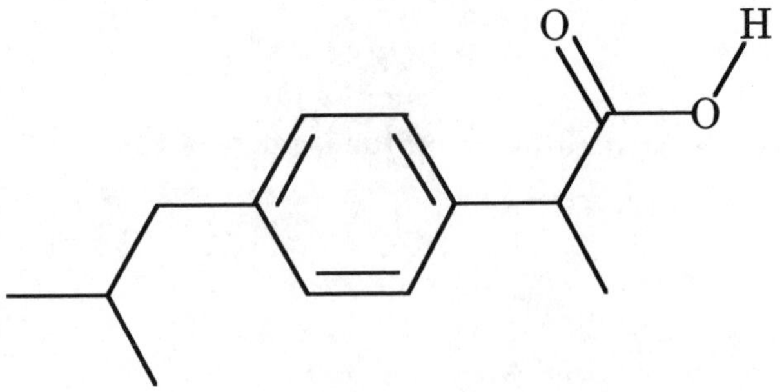

Ibuprofen

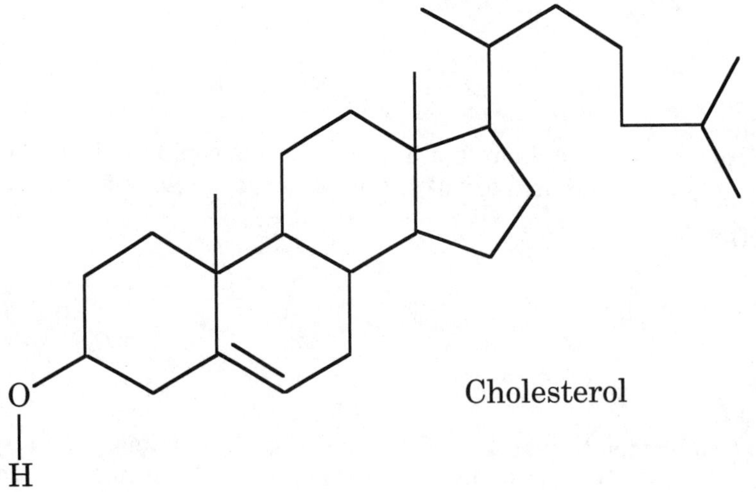

Cholesterol

STUDENT SOLUTION:

18. The Lewis structure of carvone (oil of spearmint) is shown below. What is the molecular formula of this compound?

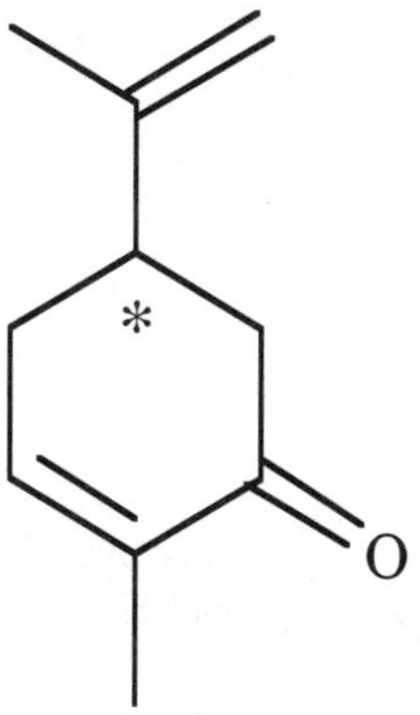

carvone

* This carbon is chiral (see problem sixteen). Because of this carbon atom, carvone exists as two different, non-superimposable mirror image isomers. One isomer is responsible for the odor of spearmint, and the other isomer is responsible for the odor of caraway.

STUDENT SOLUTION:

Alkane Nomenclature

19. What is the name for each of the alkanes shown below?

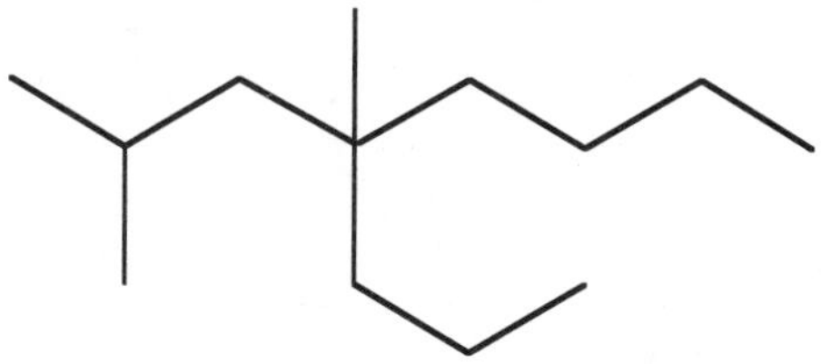

Structure A

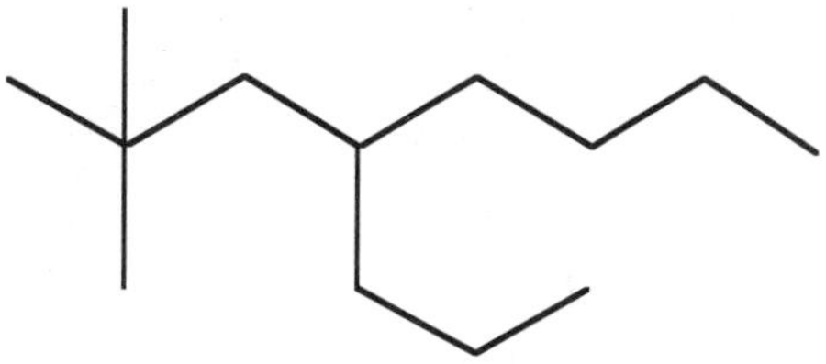

Structure B

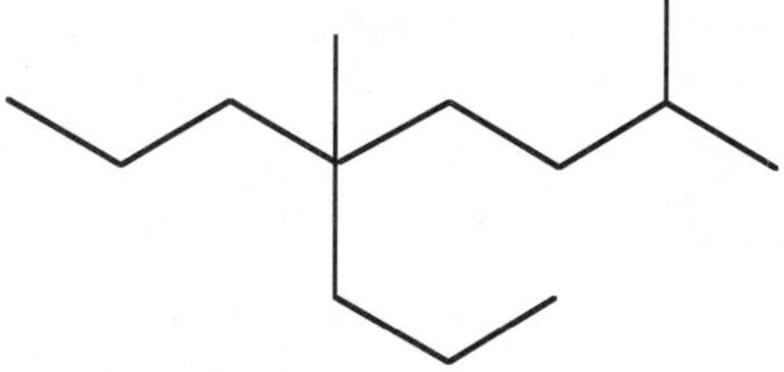

Structure C

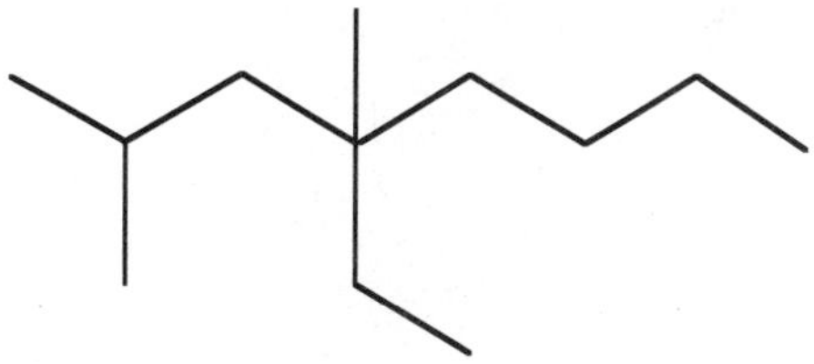

Structure D

STUDENT SOLUTION:

20. Draw the Lewis line structures for each of the following alkanes: 2,2,4-trimethylpentane, 2-methyl-4-ethylhexane, 2,3-dimethyl-4-propyloctane.

STUDENT SOLUTION:

Functional Groups

21. Draw the Lewis structures for all of the isomers of $C_4H_{10}O$. How many of these isomers are alcohols? How many of these isomers are ethers? How many of the isomeric structures contain a chiral carbon?

STUDENT SOLUTION:

22. Identify all of the functional groups in each of the following molecules.

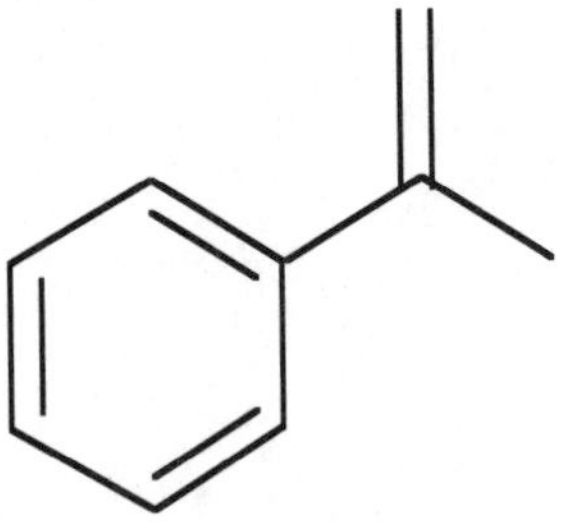

A 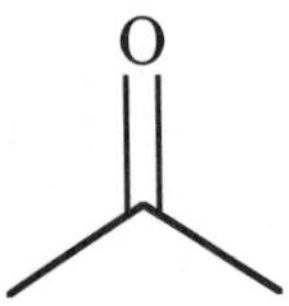   B 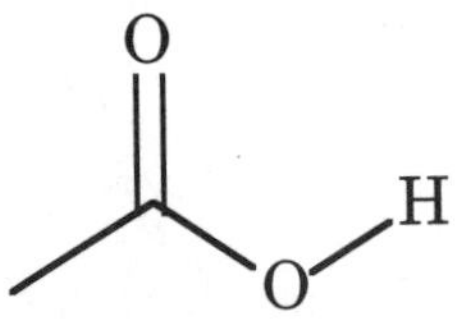   C 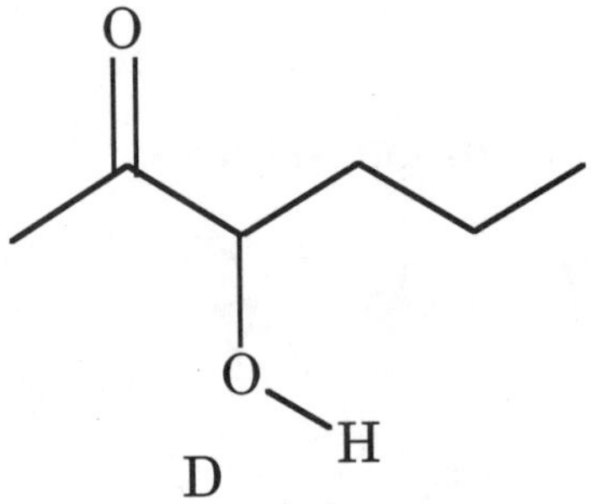   D

STUDENT SOLUTION:

23. What is the molecular formula of the compound represented by the following structural formula?

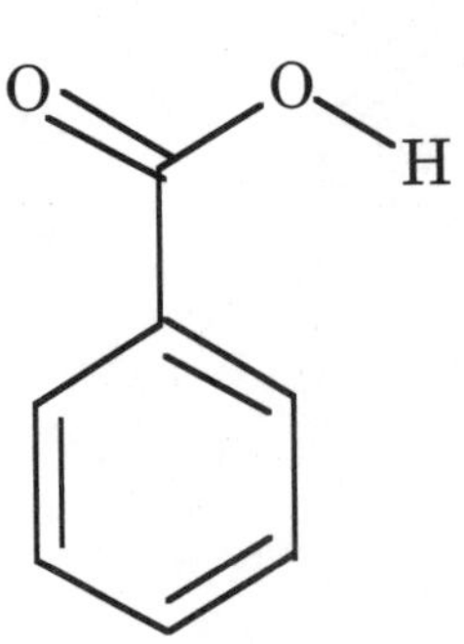

Which of the following functional group combinations best describes the molecule represented by this structural formula?

a) alkane/alcohol   b) aromatic/amine   c) aromatic/alcohol
d) aromatic/carboxylic acid   e) alkene/carboxylic acid

STUDENT SOLUTION:

Lewis Structures and Physical Characteristics

24. Which of the following compounds has the higher boiling point?

$$C_5H_{12} \qquad\qquad C_{10}H_{22}$$

STUDENT SOLUTION:

25. Arrange the isomers of C_6H_{14} (problem ten) in order of increasing boiling point.

STUDENT SOLUTION:

26. Simple alcohols are organic compounds where the -OH functional group replaces a hydrogen atom in an alkane. Although alcohols have higher boiling points than the corresponding alkanes, alcohols derived from alkanes follow the same boiling point trends as do the alkanes. Arrange the following alcohols in order of increasing boiling point?

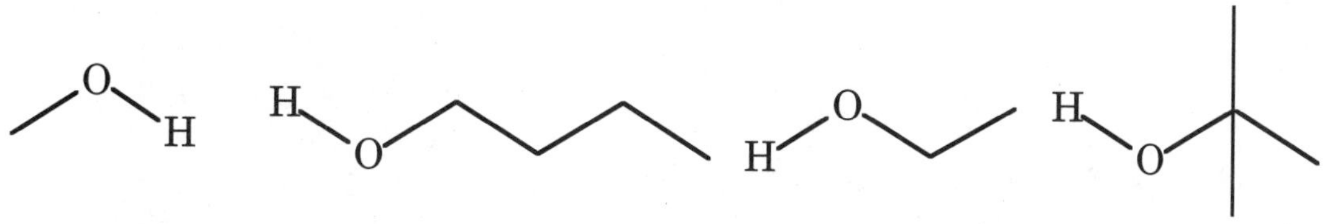

A   B   C   D

STUDENT SOLUTION:

27. A mixture containing ethanol (CH_3CH_2OH), methanol (CH_3OH), and propanol ($CH_3CH_2CH_2OH$) can be separated by distillation. Which component would have the lowest boiling point?

STUDENT SOLUTION:

Prototype Reactions

28. By analogy to the prototype reaction in Example IX (Section 8.10), predict the organic product of the following reaction:

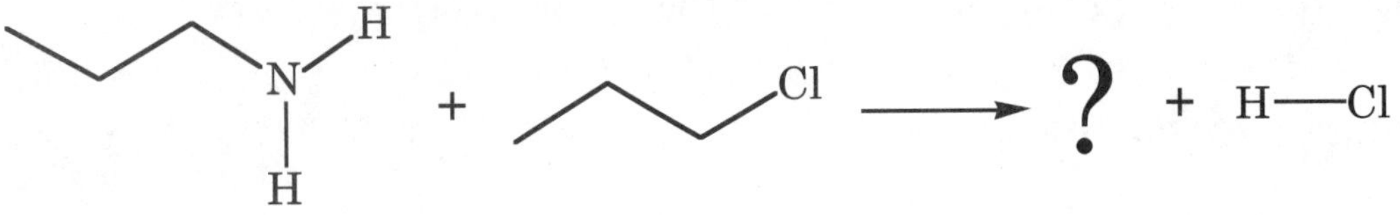

STUDENT SOLUTION:

29. Two analogous chemical reactions are represented by the equations shown below: In one of these equations, the structure of one of the products is missing. By analogy to the complete equation, which of the suggested structures represents the missing product?

TWO ANALOGOUS CHEMICAL REACTIONS

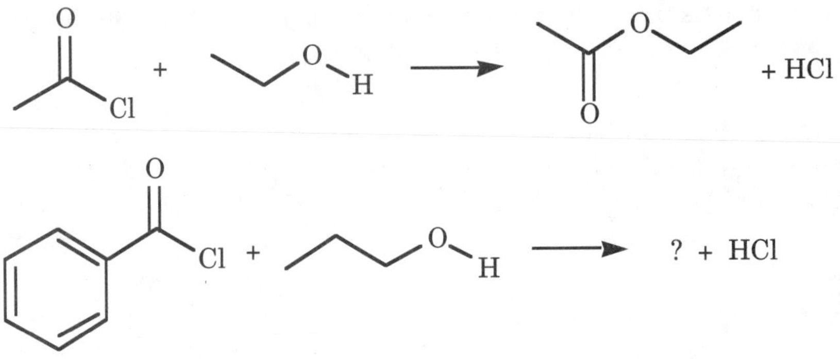

(continued)

324

SUGGESTED STRUCTURES FOR MISSING PRODUCT

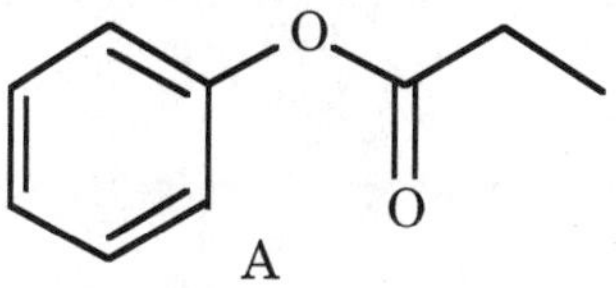

A — phenyl ester of propanoic acid (phenyl propanoate)

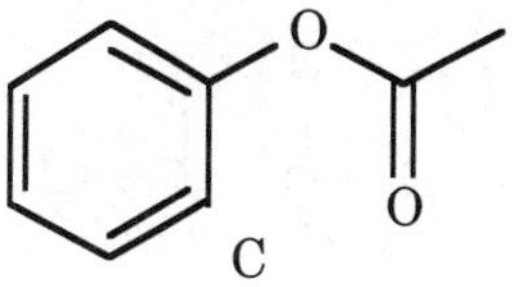

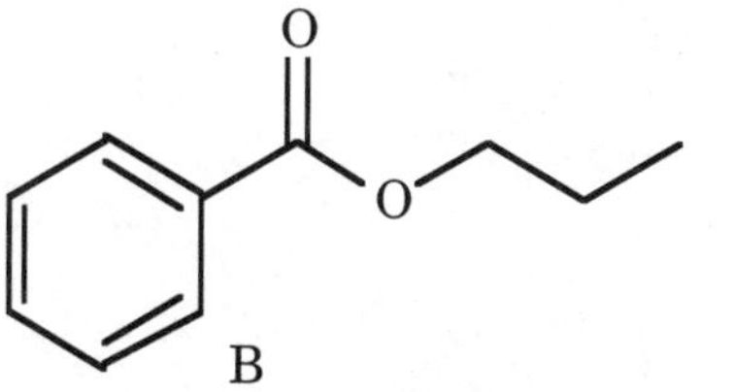

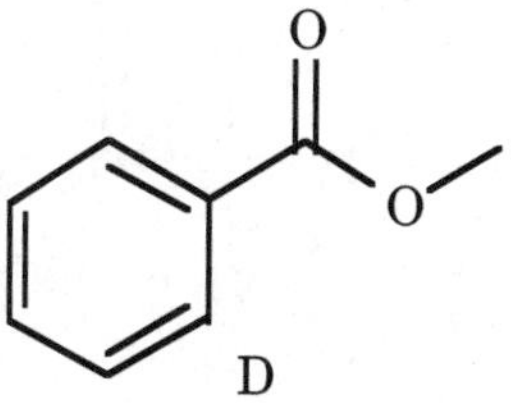

C — phenyl acetate

B — propyl benzoate

D — methyl benzoate

STUDENT SOLUTION:

30. In each of the following examples, a complete prototype reaction is followed by an incomplete reaction analogous to the prototype. Predict the organic product of the incomplete reaction in each example:

Example A:

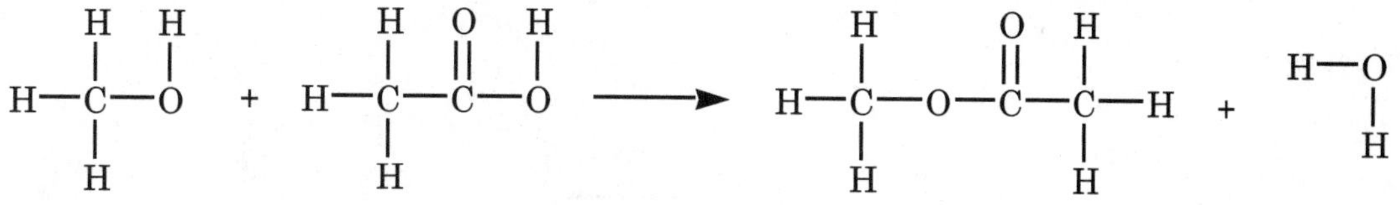

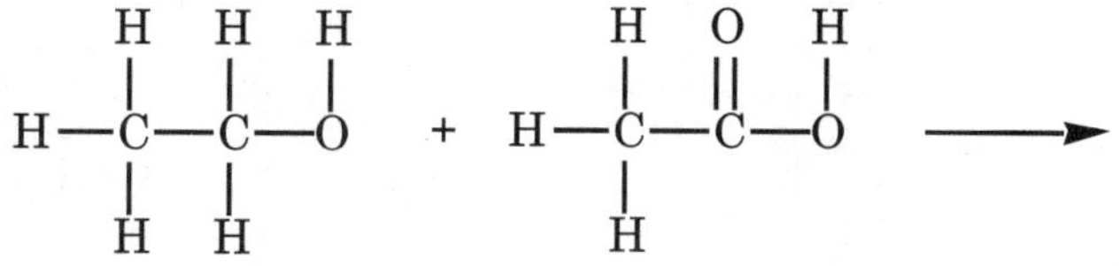

$$\text{CH}_3\text{OH} + \text{CH}_3\text{COOH} \longrightarrow \text{CH}_3\text{-O-CO-CH}_3 + \text{H}_2\text{O}$$

$$\text{CH}_3\text{CH}_2\text{OH} + \text{CH}_3\text{COOH} \longrightarrow$$

Example B:

$$\text{CH}_3\text{OH} + \text{CH}_3\text{OH} \longrightarrow \text{CH}_3\text{-O-CH}_3 + \text{H}_2\text{O}$$

$$\text{CH}_3\text{CH}_2\text{OH} + \text{CH}_3\text{CH}_2\text{OH} \longrightarrow$$

325

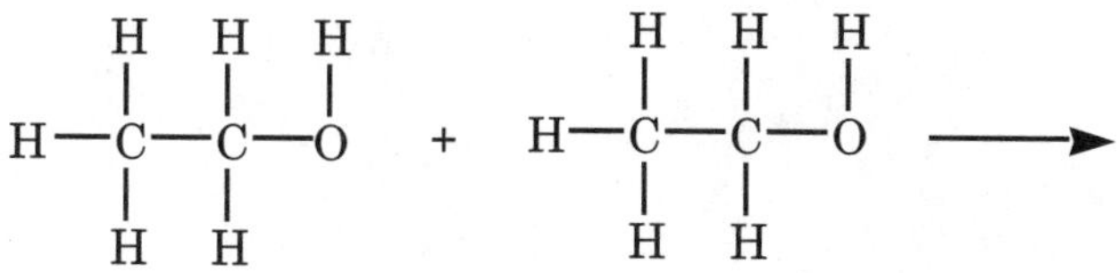

Example C:

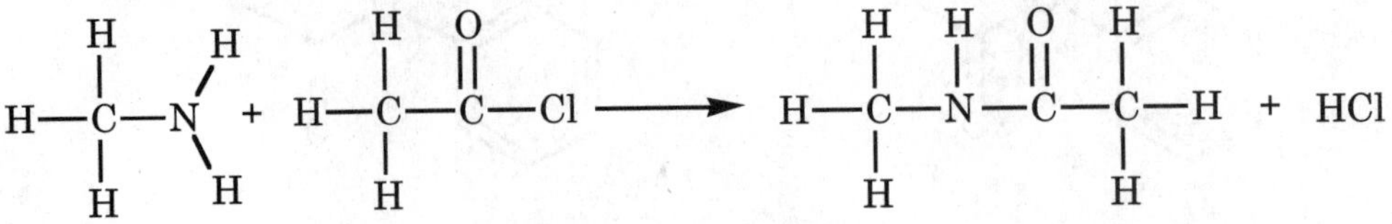

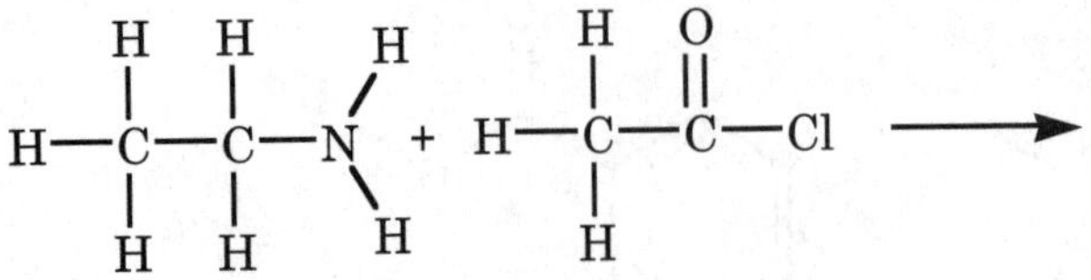

$$H-\underset{\underset{H}{|}}{\overset{\overset{H}{|}}{C}}-\underset{\underset{H}{|}}{\overset{\overset{H}{|}}{N}} \quad + \quad H-\underset{\underset{H}{|}}{\overset{\overset{H}{|}}{C}}-\overset{\overset{O}{||}}{C}-Cl \quad \longrightarrow \quad H-\underset{\underset{H}{|}}{\overset{\overset{H}{|}}{C}}-\overset{\overset{H}{|}}{N}-\overset{\overset{O}{||}}{C}-\underset{\underset{H}{|}}{\overset{\overset{H}{|}}{C}}-H \quad + \quad HCl$$

$$H-\underset{\underset{H}{|}}{\overset{\overset{H}{|}}{C}}-\underset{\underset{H}{|}}{\overset{\overset{H}{|}}{C}}-\underset{\underset{H}{|}}{\overset{\overset{H}{|}}{N}} \quad + \quad H-\underset{\underset{H}{|}}{\overset{\overset{H}{|}}{C}}-\overset{\overset{O}{||}}{C}-Cl \quad \longrightarrow$$

Example D:

(structure: ethylamine $CH_3CH_2-NH_2$) $+$ (structure: ethyl chloride CH_3CH_2-Cl) $\longrightarrow$ (structure: diethylamine) $+$ HCl

(structure: isopropylamine) $+$ (structure: ethyl chloride) $\longrightarrow$

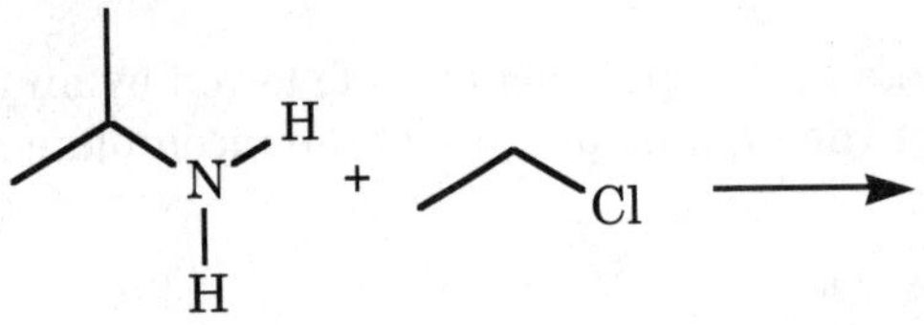

Example E:

(ethene) $+$ H_2O $\longrightarrow$ $-O-H$

(cyclohexene) $+$ H_2O $\longrightarrow$

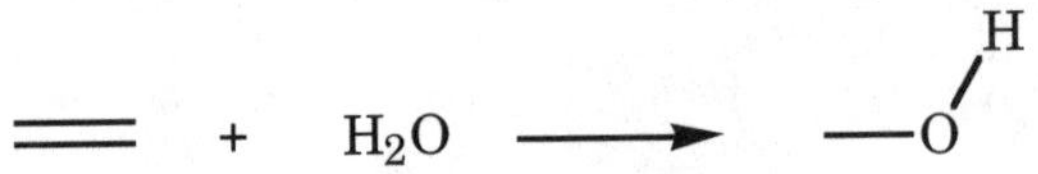

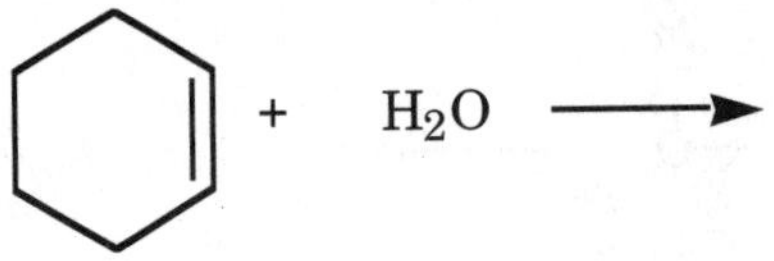

Example F:

STUDENT SOLUTION:

Problems Involving Household Chemistry and Science

31. There is only one chemical compound (isomer) with the formula CH_2Cl_2. How do chemists know that there is only one compound with this molecular formula? The answer to this question involves the analysis of the CH_2Cl_2 product(s) of the following chemical reaction:

$$CH_3Cl + Cl_2 \rightarrow CH_2Cl_2 + HCl$$

All attempts to separate different compounds from the CH_2Cl_2 component of the reaction product lead to failure. Chemists have always been interested in separating the components of mixtures. Distillation, for example, has been used even before the days of the alchemists, and, today it is an important technique in the production of petroleum products.

Another separation technique, developed early in the 20th century, is called chromatography. Its name stems from the Greek word, *chroma*, meaning "color" because it was first developed to separate the colored pigments of leaves. Chromatography is easy to demonstrate using paper coffee filters and water soluble inks.

Cut a coffee filter into strips of approximately 3 cm x 10 cm. Next, obtain a selection of water soluble felt-tip pens of different colors and brands. Start with a black pen and place a dot of ink (2–3 mm in diameter) about one cm from one end of a strip. Place ¼ cup of water into a small tumbler, then immerse the end of the strip containing the ink into the water. Do not allow the dot, itself, to contact the water. In a few seconds, you will note that the liquid begins to migrate up the strip and beyond the ink dot. Remove the strip from the water after the liquid has migrated for about six cm, then place the strip on a piece of paper towel. Can you determine the colors that the manufacturer used to formulate the black ink? Do different brands of black ink contain the same colors? Do other colors of ink, such as red or blue, consist of more than one component? Repeat the experiment with three tablespoons of rubbing alcohol added to the water. Do the colors appear to move faster or slower? Formulate an hypothesis relating the amount of alcohol in the water to the rate at which the colors move up the strip. Test your hypothesis by repeating the experiment with six tablespoons of alcohol added to ¼ cup of water. Chemists call the liquid portion in this experiment the "mobile phase" and the paper strip the "stationary phase." Can you predict the mobile phase (alcohol or water) that will be best for separating the components of water insoluble inks (permanent markers)? Test your hypothesis.

STUDENT SOLUTION:

32. Three dimensional molecular models are frequently used by chemists to apply structural atomic/molecular theory to the solution of real world problems. *The Double Helix: A Personal Account of the Discovery of the Structure of DNA* by James D. Watson [New York American Library, New York (1968)] represents an interesting case history involving the use of molecular models.

Molecular models are also useful to chemistry students. For example, some of the concepts in this chapter are easier to understand if they are illustrated with three dimensional models. The article "Molecular Models Constructed in an Easy Way" (He, Fu-cheng; Liu, Lu-bin; Li, Xiang-yuan. *J. Chem. Educ.* **1990**, 67, 556) describes the construction of five models from a strip of paper. One of these models is a tetrahedron that can be used to illustrate some of the concepts from this chapter. On page 558 of the same issue of the Journal, an article by Saqib All and M. Mazhar describes the construction of a tetrahedral model from cotton swabs. Using two of either of these tetrahedral models, examine the geometry of the molecules described in Figures 8.18, 8.22, 8.23, and 8.24. The univalent atoms in these structures can be represented by color coding the tetrahedron vertices (or cotton tips) with colored markers. With either tetrahedral model type, single, double, and triple bonds are formed according to the following rules.

1. Single Bonds—A vertex of one tetrahedron touches a vertex of a second tetrahedron.

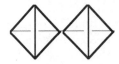

2. Double Bonds—An edge of one tetrahedron touches an edge of a second tetrahedron.

3. Triple Bonds—A face of one tetrahedron touches a face of a second tetrahedron.

After constructing the models shown in Figures 8.18, 8.22, 8.23, and 8.24, use three tetrahedrons to construct a model of the molecule allene:

$$\begin{array}{c}
H \\
\diagdown \\
\quad C = C = C \\
\diagup \\
H
\end{array}
\begin{array}{c}
H \\
\diagup \\
\\
\diagdown \\
H
\end{array}$$

allene

328

Describe the geometry of allene. In what way is the Lewis structure of allene misleading?

STUDENT SOLUTION:

Library Problems

33. The Preface of this book posed the question, "What services do chemists offer to society?" One of the services that chemists provide is the continual formulation and manufacture of products that provide a better quality of life. A quick check of the products in your medicine cabinet should illustrate this. Sometimes, however, a chemical product exhibits deleterious properties other than those that were intended. A pharmaceutical may have harmful side effects, or a pesticide may accumulate in the environment. It is then up to society to weigh the pros and cons of these materials and to decide whether they should continue to be produced. Also, it is the responsibility of chemists to formulate replacement materials with less harmful effects. For example, replacements for ozone depleting chlorofluorocarbons are now being proposed, and compostable disposable diapers are being developed. Chemists do have solutions!

Just as it is easy for scientists to be smug and criticize in hindsight earlier scientific theories such as phlogiston, it is also easy for society to criticize chemists for producing "harmful" materials. It is instructive, however, to review the history of the development of chemical products that are now either banned or have been reformulated. What properties did these chemicals originally have that made them so advantageous? What particular niche did they fill, and what problem did they solve? What is presently being done to reformulate these products or to develop replacement materials? Some examples and references are given below:

Product	Reference
DDT	Friedmen, Harold B. *J. Chem. Educ.* **1992**, 69, 362.
Freon	Midgley, Thomas and Henne, Albert L. *J. Chem. Educ.* **1962**, 39, 361.
	Zurer, Pamela S. *Chemical and Engineering News* **1993**, Nov. 15, p. 12.
Synthetic Detergents	Snell, Cornelia T. *J. Chem. Educ.* **1947**, 24, 505.
	Snell, Foster D. and Snell, Cornelia T. *J. Chem. Educ.* **1958**, 35, 271.
	Ainsworth, Susan J. *Chemical and Engineering News* **1994**, Jan. 24, p. 34.

STUDENT SOLUTION:

34. Chapter 7 explored an expansion of the atomic molecular theory that incorporated energy concepts into the theory. In Chapter 8, the concept of molecular structure was introduced. These two modifications of the atomic molecular theory provided chemists with the tools that were necessary to deal with society's energy sector and materials sector problems.

Because of its utility as a source of energy and material, petroleum is indeed one of the most important raw materials managed by the chemical profession. Petroleum can be processed to provide not only the wide variety of fuels required by the energy sector of society, but also the basic molecular building blocks required by the materials sector of society. Read the articles

"Hydrocarbons in Petroleum" (Rossini, Frederick D. *J. Chem. Educ.* **1960**, 37, 554), "Petroleum Chemistry" (Kolb, Doris; Kolb, Kenneth E. *J. Chem. Educ.* **1979**, 56, 465), and "The Chemistry of Modern Petroleum Product Additives" (Vartanian, Paul F. *J. Chem. Educ.* **1991**, 68, 1015). What are the three broad classes of hydrocarbons found in petroleum? What basic broad fractions of hydrocarbons are separated during distillation of petroleum? About how many compounds are typically found in the gasoline broad fraction? What is the purpose of petroleum refining? What processes are involved in the refining of petroleum? What is an industrial "feedstock"? Use some of the references given in the article by Vartanian to assemble a list of the top five industrial feedstock materials produced during the refining of petroleum.

STUDENT SOLUTION:

35. In order to gain a better understanding of natural and synthetic polymer chemistry, read the article "Polymer Structure—Organic Aspects (Definitions)" (Carraher, Charles E., Jr.; Seymour, Raymond B. *J. Chem. Educ.* **1988**, 65, 314). What is an addition polymer? What is a condensation polymer? Give an example of a common addition polymer and an example of a common condensation polymer. Polymers are often referred to by abbreviations. For example, recycling classification systems use polymer abbreviations. What polymer names correspond to the following abbreviations: PE, PMMA, PP, LDPE, HDPE, PVC?

STUDENT SOLUTION:

Formal Laboratory Exercise

The Molecular Structure of Substituted Allene
Working with Computer Generated Molecular Models

Introduction

As the concept of molecular geometry developed in the last half of the nineteenth century, chemists found that many of these concepts could be more easily understood with the help of molecular models. During the first half of the twentieth century, molecular models gradually evolved into a number of standard model types that were routinely used by chemists. These standard molecular model types are perhaps best described by simply listing the very descriptive names of the more popular model kits: ball and stick; wire frame; space filling. As chemists learned more about the atomic and molecular world, the model kits became more sophisticated. Standard color codes were established to represent various atoms; space filling models were scaled to represent relative atomic sizes; and wire frame models were scaled to represent relative bond lengths.

The computer age brings a new level of sophistication to modern molecular models. Although the visual computer interface retains the appearance of the popular ball and stick, wire frame, space filling models, the computer allows complex quantitative molecular dynamics calculations to be included with the computer "molecular model kit." These calculations allow chemists to explore the chemical and physical characteristics of hypothetical chemical compounds by a detailed examination of molecular dynamics. Since the communication of molecular modeling information is important to chemists, it is not surprising that the rapid growth of World

Wide Web (WWW) communication has included the development of tools that allow molecular modeling via the WWW. Many of these tools are in the public domain. Consequently, they can be freely used by chemistry students. In this formal laboratory exercise, one of these WWW molecular modeling tools will be used to explore a question associated with problem 32 in this current textbook chapter.

In problem 32 (Problems Involving Household Chemistry and Science), a crude model kit was used to analyze a number of issues associated with molecular structure. One of the questions presented in this molecular modeling problem involved the three dimensional geometry of the molecule of the compound allene:

$$A\ H \diagdown \atop B\ H \diagup \mathrm{C}\underset{1}{=\!=}\mathrm{C}\underset{2}{=\!=}\mathrm{C}\underset{3}{\diagup H\ C \atop \diagdown H\ D}$$

allene

The use of molecular models revealed that the Lewis structure of allene is misleading because the two dimensional Lewis structure does not reveal that the two hydrogen atoms on carbon atom 1 (A&B) are at right angles to the two hydrogen atoms on carbon atom 2 (C&D) as indicated in Figure 8.32.

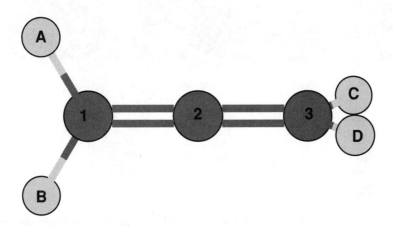

Figure 8.32 Model of Allene Molecule

Because of this peculiar geometry, some substituted allenes can exist as enatiomers. The purpose of this formal laboratory exercise is to use computer generated molecular models to determine the allene substitution pattern required to produce isomers that are non-superimposable mirror images.

Procedural Overview

The computer-generated molecular model kit that is used in this laboratory exercise is accessed via the Liberal Arts Chemistry Home Page at chemagic.com.

Figure 8.33 below is a picture of the Molecular Model Viewer page. Although the details of using this model viewer may change from time to time, the basic idea remains the same. Molecular models from the model selection menu can be loaded into the display windows. After loading, the models can be manipulated with computer mouse movements as if they were real three dimension models. The color of any atom in a model can also be changed to simulate atom substitution. The model selection menu contains all of the molecules discussed in chapter eight of the textbook. Note that allene is the last model in the menu. The specific instructions for using the molecular model viewer are somewhat platform and browser dependent. Before starting the procedure section of this formal laboratory exercise, the "Instructions" link on the Molecular Model Viewer page should be used to review the detailed instructions for the use of the model viewer. After reading the brief instructions, the model viewer should be very easy to use.

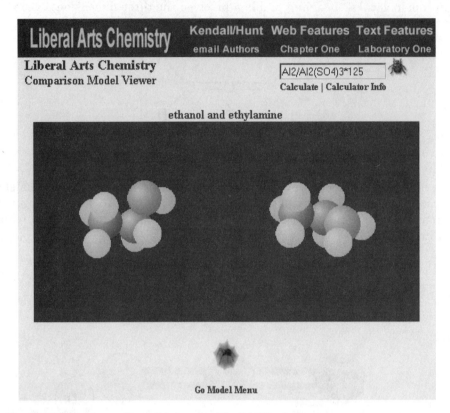

Figure 8.33 Home Page Molecular Model Window

Materials

computer set up to access WWW; appropriate WWW browser (Acceptable browsers list can be accessed via the "Instruction" link referenced in the Procedural Approach.)

Chemicals

Procedure

The following instructions assume a familiarity with computer mouse clicking and dragging.

Playing with the Model Viewer

1. Play! Load some of the models from the model selection menu. Click model manipulation links just to see the effects. Change the color of atoms in loaded models by selecting colors and clicking on atoms. Use the mouse to manipulate the models. The model display windows can always be refreshed by loading a model from the model selection menu.

2. After getting a feel for the use of the model viewer, load methane (CH_4) into both model display windows.

3. Change the color of the four hydrogen atoms in the top methane model to red, yellow, green, and cyan. The specific color arrangement is not important. When all four hydrogen atoms have a different color, the model can be viewed as a molecule of CHFClBr (Section 8.6).

4. Imagine that the divider between the model display windows is a mirror. Change the color of the four hydrogen atoms on in the bottom methane model so that each atom has the same color as its "mirror" image in the top model window. After this has been done, the two model windows can be viewed as a three dimensional representation of the two CHFClBr enatiomers in Figure 8.18.

5. Use the mouse click and drag movements to rotate the two models to verify that they are indeed non-superimposable mirror images.

Using the Model Viewer to Solve a Problem

1. Use the model selection menu to load allene (H_2CCCH_2) into both model display windows.

2. Use the hydrogen atom substitution approach described for methane above to find the minimum hydrogen atom substitution required to produce chirality (non-superimposable mirror images) on the allene framework. Start by replacing one hydrogen atom on each model. Make sure that each substitution results in mirror images.

3. HINT: The required substitution pattern will require more than one but less than four hydrogen atom substitutions.

4. Use this information to answer the questions in the conclusion and discussion section.

Data Table

Conclusion and Discussion

1. For methane, three of the four hydrogen atoms must be substituted with different atoms or groups in order to produce a chiral molecule. What is the minimum number of hydrogen atom substitutions required to produce a chiral molecule from allene by substitution?

2. In Figure 8.32, the four hydrogen atoms in allene are labeled A, B, C, and D. What specific hydrogen atoms must be substituted to produce a chiral molecule from allene by substitution?

3. To produce a chiral molecule starting with methane, all three hydrogen atom substitutions must be to *different* atoms or groups. Is this true of the hydrogen substitutions required to make allene a chiral molecule?

Chapter Eight Notes

Atomic Substructure _____

9.1 Introduction

The atoms of John Dalton were indestructible spheres. By the beginning of the twentieth century, it was clear to the world of natural science that this was an incomplete view of the atom. Experimental evidence was being collected by chemists and physicists that indicated that the atom indeed had a substructure. Could it be that by probing the nature of atomic substructure, the chemist would learn the secret of another 100-year-old mystery? That 100-year-old mystery was the nature and cause of chemical bonding, and its secret was indeed coded in the atomic substructure.

Almost all students have heard of electrons, protons, and neutrons. One of the purposes of this chapter is to acquaint the student with the experimental data that demanded the "existence" of the subatomic particles.

With the discovery of an atomic substructure came another important question. If a neutral atom is composed of electrons, protons, and neutrons, what is the arrangement of these particles in the neutral atom? This problem occupied the time and effort of the scientific community at the beginning of this century, and, in fact, it is a subject of continuing research even today.

The second purpose of this chapter is to examine the experimental basis for a model of subatomic structure known as the nuclear model of the atom. Understanding this model necessitates an exposure to several important contemporary concepts: Radioactivity, Nuclear Fission, and Nuclear Fusion.

9.2 Electrolysis

During the mid-nineteenth century, the English natural scientist Michael Faraday conducted a series of chemical experiments. These experiments revealed that an intimate relationship existed between the chemical reaction process and electricity. In these experiments, Faraday passed an electric current through the solution and/or liquid phase of certain chemical compounds. This process, called **electrolysis**, brought about the chemical decomposition of many of the compounds investigated.

Figure 9.1 shows the results of the electrolytic decomposition of three binary compounds—NaCl, KCl, and $CaCl_2$.

In each electrolysis vessel, the conducting rod (electrode) connected to the negative terminal of the battery will begin to accumulate a deposit of metal (Na, K, or Ca). If the temperature of the vessel is hot enough, the molten metal will collect on the bottom of the vessel. The electrodes connected to the positive terminal of the battery will all accumulate bubbles of chlorine gas. These qualitative observations indicate that a relationship exists between the chemical reaction process and electrical energy. If this electrolysis experiment is conducted quantitatively, this relationship becomes more evident.

If electric current is passed through the apparatus until one mole of sodium forms (23.0 g) in vessel I, then it will be observed that 1 mole of potassium (39.1 g) has also formed in vessel II. Electric charge appears to be moving in concert with mass in the chemical reaction process.[1] In vessel III, however, it will be observed that only ½ mole of calcium (20.0 g) forms. The electric charge appears to be only half as effective in bringing about the chemical reaction in vessel III. Why the difference?

---

1. Electric current is the movement of electric charge. The concept of electric charge is the subject of Section 9.3.

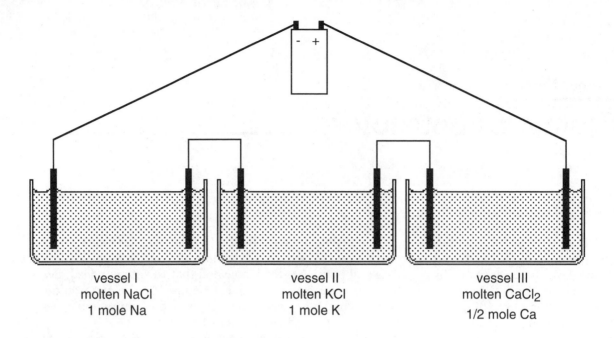

Figure 9.1 Electrolysis of Three Chemical Compounds

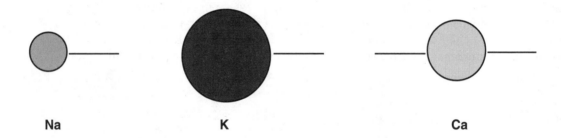

Na   **K**   **Ca**

Figure 9.2 Directed Valence "Pictures" of Na, K, and Ca

Consider the atomic "pictures" of the atoms of sodium, potassium, and calcium (Fig. 9.2).

One way of understanding why electric charge appears to be only half as effective in bringing about the chemical decomposition of calcium chloride is to attribute the difference to valence. There are twice as many chemical bonds to be broken for every calcium atom when compared to sodium and potassium. This relationship between electric charge and valence holds for the binary compounds of other elements. For example, in experiments similar to the one described in Figure 9.1, electric charge appears to be only a third as effective in bringing about the decomposition of aluminum (valence three) compounds when compared to similar sodium compounds.

When Faraday experimentally established the quantitative laws of electrolysis, valence was a very vague concept. He therefore missed the valence connection. Later in the century, the connection between atomic valence and the quantitative laws of electrolysis was not missed. And by the end of the nineteenth century, the time was right for a bold speculation. Was it possible that the directed valences of the atoms were really electrical structural features on the surface of atoms? This speculation was put forth by the English physicist G. Johnston Stoney in 1890. Stoney suggested that this substructural unit of the atom be called the **electron**.

9.3 The Laws of Electric Charge

The electrical interpretation of valence implies that atoms are not indestructible spheres. Atoms must possess an electrical substructure. Understanding how knowledge of this atomic substructure evolved necessitates a familiarity with the basic laws associated with electric charge—the laws of electric charge.

The ancient Greeks were aware that a static electric charge could be produced by friction. In fact, the modern word "electricity" is derived from the Greek word for amber, *elektros*. Amber is a resinous plant material that, when rubbed with fur, becomes electrically charged. A cursory study of electrically charged objects will reveal that they are capable of attracting other objects. For example, a plastic comb electrically charged by rubbing it with a piece of fur will attract small bits of paper. A more thorough study, however, reveals that electric charges come in two different forms designated as positive (+) and negative (−). These two different forms are revealed by the qualitative force laws that they obey (Fig. 9.3):

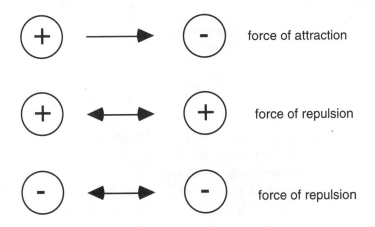

Figure 9.3 Qualitative Laws of Electric Charges

The type of charge that develops in a given situation depends on the nature of the material that is being rubbed to produce the charge. These qualitative laws, which were stated by Benjamin Franklin in 1780, can be summarized as follows:

Franklin's Laws, Qualitative Laws of Electric Charges

- Like electrically charged objects repel one another.
- Unlike electrically charged objects attract one another.

In 1790, the French physicist Charles Coulomb quantified the laws of electric charge. Coulomb proved that the magnitude of the force of attraction or repulsion between two charged objects obeyed a mathematical law. For this reason, the electric force is also called the Coulombic force. This mathematical expression is called **Coulomb's law**:

Expression 9.A
Coulomb's Law[2]
The Quantitative Law of Electric Charges

$$F_e = \frac{K_e Q_1 Q_2}{R^2}$$

---

2. In this text, Coulomb's law will not be used in any calculations. It is presented to emphasize the factors that influence electric force.

F_e = Electric Force of Attraction or Repulsion
Q_1 = Magnitude of Electric Charge on Object 1
Q_2 = Magnitude of Electric Charge on Object 2
R = Distance between Object 1 and Object 2
K_e = Universal Electric Force Constant

In Chapter 7, reference was made to nature's four basic forces—gravitational force, electromagnetic (electric or Coulombic) force, atomic nuclear strong force, atomic nuclear weak force. Early twentieth century natural scientists were only aware of the **gravitational force** and the **electrical force**, the gravitational force having been quantified by Sir Isaac Newton in the seventeenth century.

Qualitative Law Gravity

- All masses attract one another.

Expression 9.B
Newton's Law
The Quantitative Law of Gravity

$$F_g \;=\; \frac{K_g M_1 M_2}{R^2}$$

F_g = Gravitational Force of Attraction
M_1 = Magnitude of Mass 1
M_2 = Magnitude of Mass 2
R = Distance between Mass 1 and Mass 2
K_g = Universal Gravitational Force Constant

By the beginning of the twentieth century, it appeared that two forces were at work in nature. The gravitational force was a significant factor only in the world of large objects, and the electric force was dominant at the subatomic level. This view was very close to the "truth" as we know it today. The discovery of the two nuclear subatomic forces did not occur until later in the twentieth century (Section 9.9).

The similarity that exists between Newton's law of gravitational attraction and Coulomb's law of electric attraction and repulsion is striking. Expressions 9.A and 9.B indicate that both forces diminish rapidly with distance ($F \propto 1/R^2$), and that both forces increase with increasing quantity of attracting or repelling material ($F \propto M_1 M_2$ and $F \propto Q_1 Q_2$). The actual values of the universal constants in these two expressions indicate that the electric force is much stronger than the gravitational force. And, of course, the electric force can be both attractive and repulsive, but still there is a great similarity. Natural scientists view the similarity of Expressions 9.A and 9.B as both curious and pretty. For our immediate purpose, the similarity is also quite useful. In the case of electric particles that are attracting, we can use a more familiar gravitational situation as an analogy. Consider the case of energy being stored by electrically charged particles.

Whenever attracting or repelling objects are separated by a distance, they are also storing energy. The energy that they store is equal to the work required to separate them according to the equation W = FS (Chapter 7). Hence a 1 kilogram mass (particle one) suspended 1 meter above the surface of the earth (particle two?) is storing less gravitational energy than a 1 kilogram mass suspended 2 meters above the surface of the earth. For example, mass I in Figure 9.4 is storing twice as much energy as mass II. Most people who have ever dropped a heavy object on their toe have an intuitive feel for this gravitational situation. Because of the similarity of Expressions 9.A and 9.B, the energy stored in the electrical systems depicted in Figure 9.4 can be understood by analogy. Electrical system I is storing more energy.

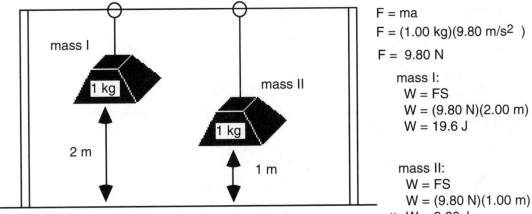

$$F = ma$$
$$F = (1.00 \text{ kg})(9.80 \text{ m/s}^2)$$
$$F = 9.80 \text{ N}$$

mass I:
$$W = FS$$
$$W = (9.80 \text{ N})(2.00 \text{ m})$$
$$W = 19.6 \text{ J}$$

mass II:
$$W = FS$$
$$W = (9.80 \text{ N})(1.00 \text{ m})$$

Mass I is storing twice as much gravitational energy as mass II. $W = 9.80$ J

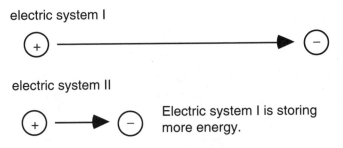

electric system I

electric system II

Electric system I is storing more energy.

Figure 9.4 Energy Stored by Gravitational and Electric Particles

---

EXAMPLE I

Energy Stored by Attracting Electric Charges

Which of the electrical systems depicted below is storing more energy? The charged particles in both systems are separated by the same distance, but the positive particle in system I has four charge units (Q), and the positive particle in system II has only two charge units (Q).

electric system I: (+ 4)   (− 1)

electric system II: (+ 2)   (− 1)

SOLUTION:
By analogy to gravitational reasoning, charge system I is storing more energy—A heavy book falling one meter hurts more than a light book falling one meter when it hits a person's toe.

339

EXAMPLE II

Energy Stored by Repelling Electric Charges

Which of the electrical systems depicted below is storing more energy? The charged particles in system IV are separated by a greater distance than the charged particles in system III, but the charge magnitudes are identical in both systems.

electric system III:

electric system IV:

SOLUTION:

Since the force in both of these systems is repulsive, the gravitational analogy breaks down. System I is storing the most energy. A useful "real world" analogy for understanding a repulsive charge system is a compressed spring—the more the spring is compressed (pushed) together, the more energy it stores.

9.4 The Cathode Ray Tube and the Electron

Toward the end of the nineteenth century, physicists and chemists became increasingly interested in the electric spark discharge phenomena. The atomic explanation of the laws of electrolysis and the subatomic particle speculations of G. Johnston Stoney suggested that a spark discharge might actually be a "beam" of charged subatomic particles. During the eighteenth century, Benjamin Franklin attempted to probe the spark discharge phenomenon of lightning with his famous (almost fatal) kite experiment. By the end of the nineteenth century, spark discharges were being investigated with a relatively safe piece of laboratory equipment called a cathode ray tube (electric discharge tube).

A schematic of a cathode ray tube is depicted in Figure 9.5. In constructing a cathode ray tube, electrodes from a high voltage source are embedded in a glass tube that is then evacuated. When the high voltage source is turned on, an electric discharge occurs in the tube that causes the entire low pressure gaseous contents of the tube to glow or fluoresce.[3] The color of this gaseous fluorescence depends on the nature of the residual gas in the tube. The glass tube also fluoresces. If the tube is appropriately constructed with a metal disk barrier, the fluorescence on the glass detection screen can be reduced to a spot of light that is located at the center of the detection screen. The light spot can also be enhanced by coating the detection screen with certain chemical substances that produce a brighter fluorescence than glass. Zinc sulfide is such a substance. If a magnet is brought close to the tube, the spot of light will move as indicated in Figure 9.5. A similar effect can be created by electrically charged plates; however, the magnetic deflection and the electric deflection of the spot of

---

3. Since this discharge appeared to come from the negative electrode, which was also called the cathode, early investigators dubbed the instrument "cathode ray tube." In 1895, Wilhelm Röntgen discovered that cathode ray tubes also emit a type of invisible high energy radiation or "light rays." Today these "light rays," called X-rays, are recognized as the result of subatomic movements within the atom.

340

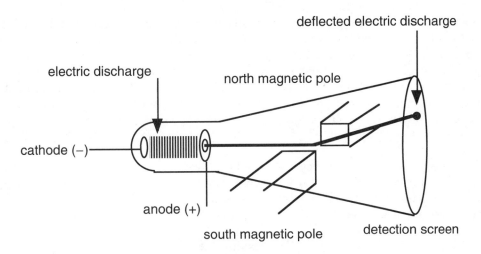

Figure 9.5 Cathode Ray Tube, Magnetic Deflection

light differ by 90 angular degrees. It is no coincidence that the cross sectional profile of the glass tube looks like a television picture tube or a computer display tube, since, indeed, both of these pieces of modern technology are sophisticated cathode ray tubes.

The effects described in the previous paragraph can be visualized with a television set and a relatively weak kitchen magnet. A television picture tube is a cathode ray tube where the spot of light moves so rapidly that the entire detection screen can be illuminated. The movement of the spot of light is created by electromagnetic effects inside the tube. These electromagnetic effects also code information into the beam so that the screen illumination appears as a picture. If a relatively weak kitchen magnet is placed next to an illuminated television screen, the picture will distort in the vicinity of the magnet. This is exactly the effect described in the previous paragraph.

The movement of a spot of light on the screen of a cathode ray tube represented fact, but what was the theoretical meaning of the spot? In 1897, the English physicist J. J. Thomson suggested an answer to this question. Using a cathode ray tube that was equipped with both magnetic and electric deflection plates, Thomson showed that the cathode rays could, indeed, be interpreted as a beam of charged subatomic particles moving from the negative electrode (cathode) toward and past the positive electrode (anode). The dual deflection system was required by Thomson's quantitative approach to the problem of analyzing the cathode ray beam.

A qualitative understanding of Thomson's approach is possible using the schematic of a single deflection cathode ray tube depicted in Figure 9.6.

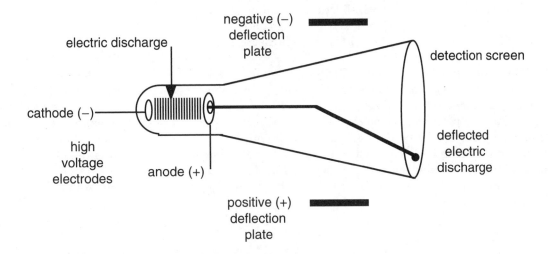

Figure 9.6 Cathode Ray Tube, Electric Deflection

When the electric deflection plates represented in this schematic are charged as indicated, the fact that the spot of light moves down can be interpreted to mean that the spot of light is caused by a negative beam of particles with sufficient kinetic energy to move past the positive electrode. The kinetic energy of these particles would depend on the voltage of the high voltage source. If this is the case, then the extent of this downward movement or deflection depends on the strength of the negative charge on each tiny particle in the beam. In the same sense that the compression of a spring can be used to measure the weight of the object causing the compression, the deflection of the light spot should allow the measurement of that which caused it—the charge on each tiny subatomic particle. Unfortunately, another characteristic of the subatomic particles can also affect the extent of particle deflection. The more mass each particle carries, the less the extent to which each particle will be deflected. This follows from simple logic. It is much easier to divert a 50 pound child moving at 5 miles per hour than a 250 pound nose tackle moving at the same speed. In terms of the mathematics of linear reasoning, the particle deflection could be expressed as follows:

Expression 9.C
Particle Deflection in Cathode Ray Tube

$$D \propto \frac{Q_e}{M_e}$$

D = Deflection of Spot of Light
Q_e = Electric Charge on Each Subatomic Particle
M_e = Mass of Each Subatomic Particle

Thomson's dual-deflection cathode ray tube was designed to measure the deflection of the light spot, not in terms of linear displacement, but in terms of massive, magnetic, and electric characteristics. In this way, Thomson converted Expression 9.C into an equation (Expression 9.D). He measured the charge to mass ratio of the subatomic particles in the cathode ray tube. For the first time in history, a subatomic particle had been characterized, albeit not completely. The cathode ray particles were ultimately named according to G. Johnston Stoney's suggestion. The particles were called **electrons**.

Expression 9.D
Thomson's Charge to Mass Ratio of the Electron

$$D = \text{Constant} \; \frac{Q_e}{M_e}$$

D = Deflection of Spot of Light
Q_e = Electric Charge on Each Electron
M_e = Mass of Each Electron

Thomson's equation (Expression 9.D) was an equation with two unknowns—the electron charge and the electron mass. To really characterize the electron, one of these unknowns had to be determined by an independent experiment. This was accomplished by the American physicist Robert Millikan in 1904. By placing electron charges on oil drops, Millikan designed an experiment where the rate of free fall of the charged oil drops under the influence of electrically charged plates depended only on the electron charge. Because of Expression 9.D, Millikan's experiment allowed for the calculation of the charge and the mass of the electron. The result of this calculation indicated that the electron was, indeed, a small subatomic particle. The mass of the electron was determined to be only 1/1837 atomic mass units!

9.5 The Electric Discharge Tube and the Proton

The least massive electrically neutral atom is hydrogen with an atomic mass of 1 atomic mass unit. The electron is a negatively charged subatomic particle with a mass of 1/1837 atomic mass units (0.0005 amu). These data would seem to indicate that atoms also contain some rather massive positively charged subatomic particles. Figure 9.7 shows a cathode ray tube that has been modified to characterize positive subatomic particles.

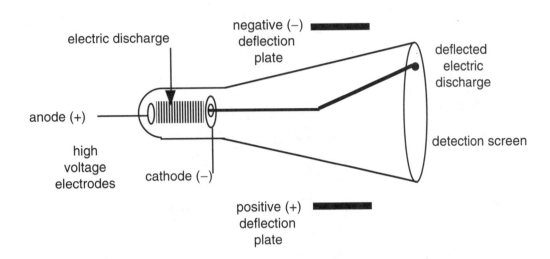

Figure 9.7   Modified Cathode Ray Tube, Positive Electric Discharge Tube

The only difference between Figures 9.6 and 9.7 is that the high voltage electrodes have been reversed. This change results in a change in the behavior of the detected spot of light. The spot is now deflected toward the top of the screen. It is deflected toward the negative deflection plate, and away from the positive deflection plate. This can be interpreted to mean that the spot of light is caused by a beam of positive subatomic particles. When an electric discharge tube is constructed according to the specifications in the previous section (Fig. 9.6), it always produces electrons (*i.e.* negative particles with a mass of 1/1837 amu). The characteristics of the positive subatomic particles produced by the modification suggested in Figure 9.7 depends on the nature of the residual gas in the tube.

Figure 9.8 represents a plausible theoretical explanation of these observations. This figure shows the high voltage electrode region of an electric discharge tube. The electric discharge tube is a primitive "atom smasher" and particle accelerator. According to this view, the high voltage electrode region of the cathode ray tube is actually an atom smashing region. In this region, electrons generated by the high voltage accelerate from the negative electrode toward the positive electrode. In transit, some of these electrons collide with neutral gaseous atoms and dislodge other electrons leaving behind an electrically charged atom (ion). These positively charged ions then accelerate toward the negative electrode. If the voltage of the high voltage source is sufficient, some these ions will reach the detection screen.

This modification of the cathode ray tube represented a most significant discovery for the discipline of chemistry. Since the problem of measuring charge had been solved by Millikan, this type of apparatus could be used to measure particle mass. But the particles in this case were atomic ions, and their mass was essentially equal to the atomic mass of the elemental atoms in the tube. This approach to atomic mass determination presented chemists with a major tool to help solve an old problem—the accurate determination of atomic masses. (See the discussion of the nineteenth century atomic mass crisis in Section 5.6.) Chemists could now determine atomic masses by chemical techniques (Section 5.6), by physical techniques (Section 7.15), and by the highly accurate method described above.

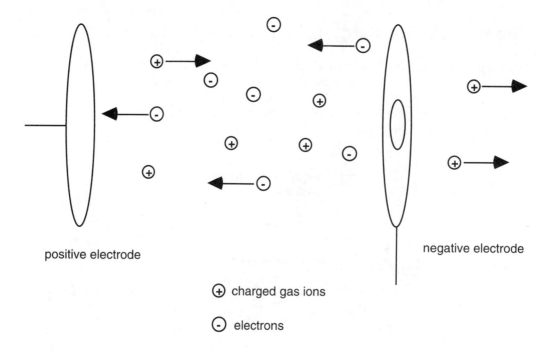

Figure 9.8   High Voltage Electrode Region of Cathode Ray Tube

A modern laboratory instrument using this approach to atomic mass measurement is the mass spectrometer. A modern mass spectrometer can accurately measure not only atomic masses, but also the masses of molecules and molecular fragments. It has become a powerful tool used for the characterization and identification of chemical substances. This instrument can be found in most modern chemical laboratories, including a tiny laboratory sent to the planet Mars!

EXAMPLE III

The Electric Discharge Tube "Mass Spectrometer"

The figure shown below represents the schematic of an electric discharge tube. In this schematic, what is the electric charge on the particles striking the detection screen? If the charges on the detected particles are equal in magnitude, which particle is more massive?

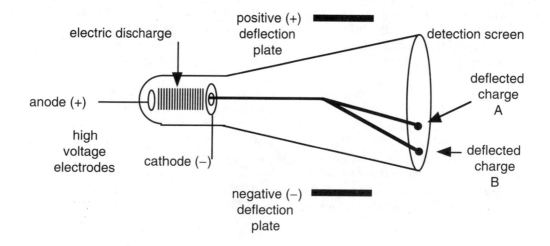

SOLUTION:
  Positive
  Particle A is more massive.

---□---

The chemical element that produces the least massive positive ion is the element hydrogen. It is, therefore, appropriate to view the positive ion of hydrogen as being a fundamental subatomic particle. This particle is called the **proton**. Its electric charge is equal in magnitude but opposite in sign compared to the electric charge on the electron. The proton has an atomic mass of 1.000 amu. Hydrogen, the simplest atom, is evidently made up of two subatomic particles—an electron and a proton.

9.6 A New Look at the Periodic Table

The discovery of the electron and the proton led inexorably to the question of a particular atom's subatomic composition. Simple mass bookkeeping dictated that a hydrogen atom was composed of one electron and one proton, but the mass bookkeeping was not so simple for the other atoms. The first step in solving this problem was taken in 1913, but this step was preceded by an important series of events involving the discovery of the periodic table (Chapter 6). This series of events was initiated by the English chemist John Newlands (Section 6.2). Since this series of events resulted in a new way of interpreting the periodic table, a brief review of the historic details is appropriate.

In February of 1863, John Newlands published the first in a series of papers in which he enunciated the essential aspects of the periodic law. In the third paper of this series, Newlands assigned a unique integer to each element having a unique atomic mass. The integers were used to order the elements according to increasing atomic mass. Since this ordering often resulted in a repetition of elemental characteristics in the eighth element from a given element, Newlands suggested a musical octave analogy. Newlands subsequently proposed that the periodicity of the chemical elements be referred to as the "Law of Octaves."

The number ordering of a list is not a mystical or illogical thing to do. But in the seventh paper of the series, Newlands revealed an almost mystical attitude. Newlands revealed in this paper that the numbers were not simply an ordering convenience. The number became "the number of the element," and it was through its number that an element manifested its exact identity. Newlands obviously felt that the element's ordering number had some fundamental significance.

This seventh paper anticipated the discovery of the importance of the atomic integers or atomic numbers. Newlands' atomic numbers, however, were not based on an empirical pattern, nor were they deduced from theory. Newlands introduced the integer atomic numbers following an almost Pythagorean instinct that a set of ordering integers must have fundamental significance. To be sure, Newlands reported real patterns related to these integers, but these relationships were the result of numerologic games played after the integer assumption. It is not surprising, therefore, that the Chemical Society of England refused to publish Newlands' paper "The Law of Octaves, and the Causes of Numerical Relations among the Atomic Weights" which was read before the Society on March 1, 1866. While the refusal to publish is understandable, it is sad that the only record of this paper is a brief note in the British journal *Chemical News* and Newlands' defensive response in the same journal. The latter response represented Newlands' final publication on the subject prior to the periodic table of Mendeleev.

The periodic law is an empirical relationship that stands independent of any integer assumptions. In March of 1869, Mendeleev presented to the Russian Chemical Society his paper on "The Relation of the Properties to the Atomic Weights of the Elements." Mendeleev's periodic table, published in 1869, was based on sound experimental observation and involved no integer assumptions. Following the publication of Mendeleev's periodic table, Newlands entered into a public priority debate regarding the discovery of the periodic law. In 1884, Newlands published a small book that contained a reprint of all of his papers relating to the periodic law. The book was published because "As a matter of simple justice, and in the interest of all true workers in science, both theoretical and practical, it is right that the originator of any proposal or discovery should have the credit of his labour." Although the Chemical Society did not publish Newlands' 1866 paper, he was awarded the Davy Medal by the Royal Society of England in 1887.

Newlands asked for, and was ultimately given, credit for the priority of his discovery of the periodic law. But his periodic table did not present the chemical community with a useful tool. Since Mendeleev's periodic table was based totally on sound empirical evidence, it had great utility. The integers, however, were gone, and they would not reappear in the periodic table until after 1913.

In January of 1913, the Dutch physicist Antonius van den Broek put forth a theory that the atomic numbers of Newlands really represented the number of fundamental positive charges in the atom. In that same year, the English physicist Henry Moseley showed experimentally that an element's integer position in the periodic table (atomic number) was related to some fundamental principle of atomic structure. Chemists soon came to realize that the **atomic number** represented the number of protons in an atom and also the number of electrons for a neutral atom.

9.7 The Elusive Neutron

Figure 9.9 represents the helium block on a modern periodic table. The block contains three bits of information—the symbol, the atomic mass, and the atomic number. According to the previous discussion, the atomic number 2 indicates that the helium atom contains two protons and two electrons. This subatomic composition, however, only accounts for slightly more than one-half of helium's atomic mass. This subatomic mass bookkeeping suggests a third subatomic particle. This particle is the **neutron**. The neutron carries no electric charge, and it has an atomic mass of 1.00 amu.

Since the neutron carries no electric charge, the cathode ray tube approach could not be used to detect this subatomic particle directly. Although its existence was postulated earlier, the neutron was not detected directly until 1932. It was detected in an experiment devised by the English physicist James Chadwick (problem 12 at the end of the chapter).

The helium block on the periodic table can now be interpreted more completely. The atomic number 2 indicates that helium contains two protons and two electrons. The difference between the atomic mass and the atomic number indicates that helium also contains two neutrons. In this case, the atomic composition bookkeeping is very tidy:

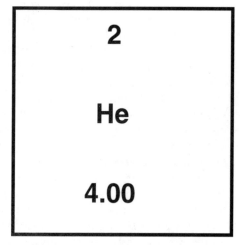

Figure 9.9 Periodic Table Block for Helium

2 electrons	= 2 × 0.0005	= 0.001 amu
2 protons	= 2 × 1.00	= 2.00 amu
2 neutrons	= 2 × 1.00	= 2.00 amu
Total Mass		4.00 amu[4]
Atomic Mass		4.00 amu

For the periodic table to become useful as a tool that codes subatomic structure, however, one more bookkeeping problem has to be considered.

---

4. This answer is rounded off according to the rule mentioned in Chapter 2. The number of decimal places is limited by the accuracy of the measurements of proton and neutron mass.

9.8 Isotopes

The atomic mass crisis discussed in Section 5.6 actually had two phases. The theoretical phase involved the determination of correct formulas independent of atomic mass (Section 7.15). The technical phase involved the development of very accurate laboratory techniques. In this latter phase, the American chemist Theodore Richards was awarded the Nobel Prize in chemistry (1914) for particularly accurate chemical atomic mass determinations. In one of his studies, he made the following type of observation:

Experiment I

One gram (1.0000 g) of lead sulfide made from the lead obtained in the ore Katangan curite gives the analysis 0.8654 g of lead.

Experiment II

One gram (1.0000 g) of lead sulfide made from the lead obtained in the ore Norwegian thorite gives the analysis 0.8664 g of lead.

If we assume that the atomic mass of sulfur is accurately known, this very accurate data allows for the calculation of two slightly different atomic masses for the element lead (Section 5.6). One way of explaining this apparent violation of the law of constant composition is to assume that while the identity of an element is fixed by its atomic number or number of protons, the atomic mass is not fixed. Accordingly, since the atomic number of lead is 82, a lead atom contains 82 protons and 82 electrons. But since the number of neutrons is not fixed, lead exists in two or more different forms of varying neutron content. Katangan curite contains one mixture of these various forms of lead and Norwegian thorite contains a slightly different mixture. The various forms of an element which differ only in neutron content are called **isotopes**.

In 1913, the English physicist Frederick Soddy predicted the existence of these various forms of the elements. This prediction was made prior to the discovery of the neutron and Richards' experimental results were regarded as proof of this prediction. With the discovery of the neutron, the existence of isotopes could be understood in terms of subatomic structure.

In a modern chemical laboratory, there is more direct evidence for the existence of isotopic forms of the chemical elements. Modern mass spectrometers (Section 9.5) can be used to detect isotopes and to measure their mass. Figure 9.10 shows a cathode ray tube "mass spectrometer" which illustrates this point for the two naturally occurring isotopes of hydrogen.

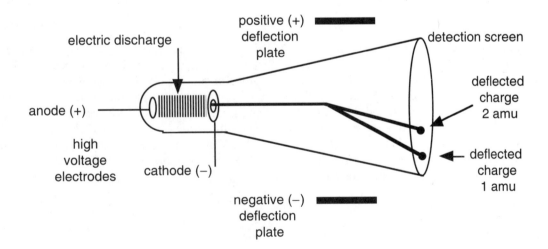

Figure 9.10   The Isotopes of Hydrogen

The intensity of the light spot created by the 1 amu isotope is greater than the intensity of the light spot created by the 2 amu isotope. In an actual mass spectrometer, these intensities are measured electrically, and they reflect the relative abundance of the isotopes. The chemical atomic mass that appears on the periodic table is a weighted average of the masses of the naturally occurring isotopes. For hydrogen, a chemical atomic weight of 1.00797 reflects that the "heavy" isotope of hydrogen is a minor component.

Because of the existence of isotopes, it is convenient to define a special symbol to represent the subatomic composition of an isotope. This new symbol uses the following definitions:

DEFINITION:	The **atomic number** of an element is the element's integer position on the periodic table. The atomic number is equal to the number of protons *and* electrons in the atom of the element.
EXAMPLE:	The atomic number of carbon is 6. Carbon, therefore, contains 6 protons and 6 electrons.
DEFINITION:	The **atomic mass number** of an isotope is equal to the integer atomic mass of the isotope. The atomic mass number is equal to the number of protons *plus* the number of neutrons in the isotope.
EXAMPLE:	A mass spectrometer indicates that the mass of a certain isotope of carbon is 13.001. The atomic mass number of this isotope is 13. Since the atomic number of carbon is 6, this isotope contains 6 protons, 6 electrons, and 7 neutrons.

Definition Summary

Atomic Number = # of protons

Atomic Number = # of electrons

Atomic Mass Number = # of protons + # of neutrons

Atomic Mass Number – Atomic Number = # of neutrons

The subatomic composition of a given isotope can be written by using the traditional symbol for the element preceded by a subscript notation of the atomic number and followed by a superscript notation of the atomic mass number:

atomic mass number

SYMBOL

atomic number

For example, the isotope of carbon with an integer mass of 13 is represented as $_6C^{13}$. Since carbon must always have an atomic number of 6, this isotopic symbol is sometimes abbreviated as C-13.

The isotopic symbol reflects the integer charge (protons or electrons) and integer mass characteristics of an isotope. In a similar way, the integer charge and integer mass characteristics of a subatomic particle can be represented with an appropriate particle symbol. With the use of modern particle accelerators, the atom has revealed a multitude of subatomic components with some very exotic characteristics. Figure 9.11 is a table listing the integer charge and integer mass characteristics of the subatomic particles discussed in this chapter. The particle symbol, which reflects the integer charge and integer mass, is also listed for each particle.

Particle	Symbol	Integer Charge	Integer Mass
electron	$_{-1}e^0$	-1	0 amu
proton	$_1P^1$	+1	1 amu
neutron	$_0n^1$	0	1 amu

Figure 9.11   Table of Subatomic Particles

EXAMPLE IV

Isotopic Composition

How many electrons, protons, and neutrons are contained in an atom of the isotope $_{15}P^{31}$?

SOLUTION:
15 electrons, 15 protons, and 16 neutrons

EXAMPLE V

Isotopic Symbols

The natural abundance of the most common isotope of nitrogen is so high the chemical atomic mass of nitrogen is almost equal to the mass of this most abundant isotope. Using the periodic table, write the isotopic symbol for the most abundant isotope of nitrogen.

SOLUTION: $_7N^{14}$

9.9 Radioactivity

As indicated in Section 9.5, the cathode ray tube is a primitive particle accelerator and "atom smasher." In 1896, the French scientist Henri Becquerel discovered that the disintegration of atoms into subatomic fragments also occurred as a natural process. He found that a mineral containing the element uranium could be used to expose photographic film in the absence of light. He concluded that the element uranium was spontaneously emitting a type of radiation which he called **radioactivity**. Subsequent studies in the early twentieth century revealed three distinct types of radioactive emissions—**alpha** (α), **beta** (β), and **gamma** (γ) radiation. Alpha and beta emissions were recognized as actual subatomic units, while gamma rays were found to be similar to X-rays (Section 9.4).

Although moving subatomic particles are usually investigated with strong magnets, the following schematic representing radioactive emissions under the influence of opposite electric charges conveys the essence of these investigations (Fig. 9.12).

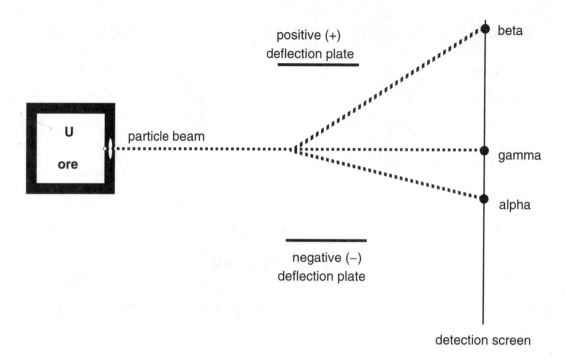

positive (+)
deflection plate

particle beam

beta

gamma

alpha

negative (−)
deflection plate

detection screen

Figure 9.12 The Effect of an Electric Charge on Radiation from Impure Uranium, Idealized Radioactivity Detector

Using methods similar to those used to evaluate the charge and mass of the electron and proton, the beta particle is found to be identical to the electron, while the alpha particle is found to be a cluster of two protons and two neutrons. The following revised table (Fig. 9.13) of subatomic particles includes the alpha particle and the beta particle (Fig. 9.12). The symbol for the alpha particle reflects that the alpha particle is really a helium atom stripped of its electrons. In a similar manner, the symbol for the proton in this revised table reflects that it is simply a hydrogen atom stripped of its electron. The positron, a particle that is introduced in problem 16 at the end of this chapter, is also included in this table.

Particle	Symbol	Integer Charge	Integer Mass
electron	$_{-1}e^0$	-1	0 amu
beta	$_{-1}e^0$	-1	0 amu
proton	$_1H^1$	+1	1 amu
neutron	$_0n^1$	0	1 amu
alpha	$_2He^4$	+1	4 amu
positron	$_1e^0$	+1	0 amu

Figure 9.13 Revised Table of Subatomic Particles

Of the elements that occur in nature, all elements with an atomic number of 84 and higher are naturally radioactive. In addition, certain naturally occurring isotopes of elements lower in atomic number than 84 are also radioactive. For example, the element carbon exists naturally in three isotopic forms: $_6C^{12}$, $_6C^{13}$, and $_6C^{14}$. Carbon-12 is the major component of natural carbon samples, and it is not radioactive. Carbon-13 is a minor non-radioactive component, but carbon-14 is a minor radioactive component. In general, for elements of lower atomic number than 84, a surplus or deficiency of neutrons in an isotope (when compared to the most abundant isotope of the element) tends to indicate radioactivity. The elements technetium (atomic number 43) and promethium (atomic number 61) are exceptions to the above generalizations; both exist only as radioactive isotopes. Promethium, technetium, and all elements higher in atomic number than 92 are synthetic elements. In all probability, they do not exist in the Earth's crust.

Each radioactive isotope emits a characteristic radiation. Both the radiation type (alpha, beta, and/or gamma) and radiation energy are characteristic of a given isotope and can be used to identify the isotope. The energy associated with radioactive disintegration is considered in the next section (Section 9.10).

In addition to a characteristic mode of disintegration, each isotope has a characteristic rate of disintegration. All radioactive isotopes disintegrate by a single disintegration rate law. For each isotope, there is a characteristic time interval required for half of the isotope to disintegrate. As a given sample of an isotope disintegrates, this time interval remains constant. Logically, this time interval is called the isotope's **half-life**. Because the disintegration of a radioactive isotope occurs at a predictable rate, known disintegration rates can be used to date objects containing radioactive isotopes. The concept of half-life is illustrated in Example VI.

EXAMPLE VI

Radioactive Half-Life

The element radon is an alpha emitter, and it disintegrates with a half-life of 3.8 days. In a sample of radon with an original mass of 100 milligrams, how many milligrams of radon will have disintegrated by emitting alpha radiation after 11.4 days?

SOLUTION: 12.5 milligrams

The time interval of 11.4 days represents exactly 3 half-lives:

$$\frac{11.4 \text{ days}}{3.80 \text{ days/half - life}} = 3.00 \text{ half - lives}$$

After one half-life, the original 100 milligrams is reduced to 50 milligrams. After the second half-life, the 50 milligram sample is reduced to 25 milligram. In the final half-life, the 25 milligram sample is reduced to 12.5 milligrams.

9.10 The Nuclear Atom

In 1911, the New Zealand scientist Ernest Rutherford used radioactivity as a tool to probe the actual substructure of the atom. In his experiment, alpha particles emitted by the element radium were allowed to penetrate the atoms of a thin piece of gold foil. Rutherford reasoned that the alpha particles would act as subatomic probes, and that their exit from the gold foil would reveal what they had experienced while they were inside of the atoms of gold. The alpha particles emerging from the gold foil could be detected by a fluorescent detection screen similar to the detection screen of a cathode ray tube (Section 9.4). A schematic of Rutherford's apparatus is shown in Figure 9.14.

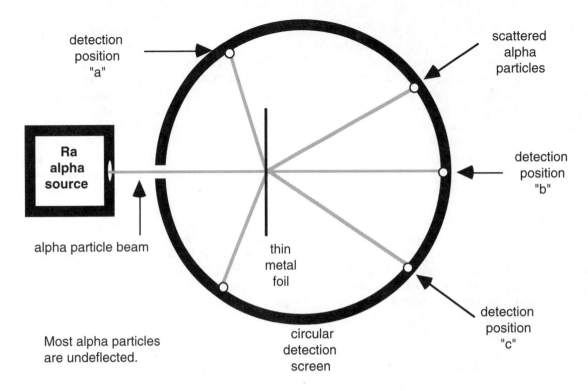

detection position "a"

scattered alpha particles

Ra alpha source

detection position "b"

alpha particle beam

thin metal foil

detection position "c"

Most alpha particles are undeflected.

circular detection screen

Figure 9.14 The Rutherford Experiment

Rutherford discovered that the alpha particles were scattered by what they experienced inside of the atoms of gold. Fluorescent detection indicated scattered alpha particles at positions like (a) and (c) in Figure 9.14, but most of the alpha particles were detected at position (b). Since these alpha particles were not scattered, Rutherford reasoned that they had experienced empty space. The atoms of gold evidently has a substructure that was essentially empty space, with subatomic particles arranged in such a way that occasionally an alpha particle probe experienced a gentle nudge (c) or an act of violence (a).

The results of the Rutherford experiment seem to be consistent with the nuclear model of the atom (Fig. 9.15). According to this model, the protons and neutrons of an atom are clustered in a tiny central portion of the atom called the **atomic nucleus**. The electrons, on the other hand, form the outer boundary (roughly spherical) of the atom, and they are located at a great distance from the nucleus. This leaves a tremendous volume of empty space between the electrons and the nucleus as demanded by the Rutherford experiment. The atom is held together by the mutual electric attraction of the electrons and the protons. The collapse of the atom due to this attractive force is evidently prevented by the inherent dynamic movement of electrons. The nature of this dynamic electron region is quite unique, and it is discussed further in the final section for this chapter (Section 9.11). Before this discussion, however, some aspects of the nuclear region of the atom should be explored.

The first thing to note about the nuclear model is that it suggests a natural force other than the gravitational and electric forces. The close proximity of mutually repelling protons in the nucleus suggests a force in nature that is stronger than the electric force. This force, called the nuclear **strong force**, is believed to act as a type of nuclear glue (attractive force) holding the nucleus together. The nuclear strong force, however, only acts over the very small distances realized in the atomic nucleus. Hence, under non-nuclear conditions, two protons would repel under the influence of the electric force (Coulombic force). If these two protons, however, are pushed very close together, to within atomic nuclear dimensions, they would attract with a vengeance under the influence of the nuclear strong force.

Recalling from Chapter 7 that stored energy is a function of force and distance (W=FS), the atomic nucleus is recognized as a storehouse of energy with both the electric and the nuclear strong forces playing a part. Consider, for example, the origins of radioactivity.

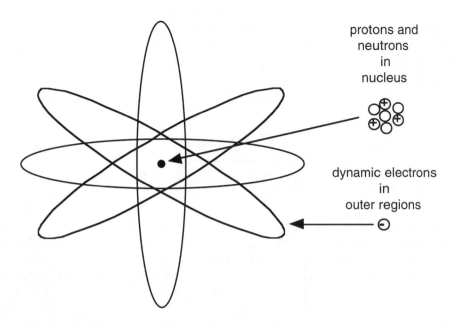

Figure 9.15   The Nuclear Model of the Atom

Since the element radium has an atomic number of 88, it occurs only in radioactive, isotopic forms. A simplistic, yet effective, way to understand the spontaneous radioactivity of elements with an atomic number of 84 or greater is to recall that the attractive nuclear strong force is capable of acting only over tiny distances. As the atomic nuclei get larger, evidently a limit of stability is reached at atomic number 83. Larger atomic nuclei attempt to reduce their dimensions by ejecting nuclear particles that carry away kinetic energy. The resultant repositioning of protons and neutrons remaining in the nucleus emits gamma radiation (kinetic energy) under the influence of the nuclear strong force. Since some isotopes with atomic numbers less that 84 are also radioactive, nuclear proton/neutron ratio is also known to be a stability factor.

Using the isotopic symbols introduced in Section 9.8 as nuclear composition symbols or nuclear symbols, the transformations that occur during radioactive emission can be represented by a nuclear bookkeeping statement called a **nuclear equation**. The ejection of alpha particles from a nucleus of radium-226 is shown below (Nuclear Equation 9.A):

Nuclear Equation 9.A
The Radioactive Decay of Ra-226

$$_{88}\text{Ra}^{226} \rightarrow {}_{86}\text{Rn}^{222} + {}_{2}\text{He}^{4}$$

Total Charge = 88   Total Charge = 86 + 2 = 88
Total Mass = 226   Total Mass = 222 + 4 = 226

Note that the nuclear equation simply reflects the conservation of nuclear mass and nuclear charge during the decay process. In this process, the alpha particle carries away nuclear energy as kinetic energy. Although Nuclear Equation 9.A looks something like a chemical equation, there is a significant difference. When the radium nucleus loses 2 protons and 2 neutrons (*i.e.* an alpha particle), its identity is altered. A nucleus with 86 protons is no longer a radium nucleus. It is a nucleus of the element radon. The nuclear equation reveals radioactivity as an elemental transmutation! The alchemists were right!! (See Chapter 5, Section 5.5.)

Since the nuclear equation represents only nuclear transformations, it does not reveal the fate of the dynamic electron region of the atoms involved. Nuclear transformations disrupt the electron proton equity in the resulting fragments. Although these imbalances are eventually corrected, nuclear transformations initially produce charged atoms or ions. Further, the subatomic particles and radiation emitted during nuclear transformations can interact with electrically neutral matter to form ions. For this reason the term **ionizing radiation** is sometimes used to describe these emissions.

Radioactivity is just one way in which nuclear energy is revealed. Examples VIII and IX explore two other nuclear processes, **fission** and **fusion**. Example VII illustrates radioactive beta decay.

EXAMPLE VII

Nuclear Energy: Radioactivity—Beta Emission

A carbon-14 nucleus is a beta emitter. Since a beta particle is actually an electron, this may seem to be an impossible situation given the nuclear atomic model. Actually, the beta particle is created in the carbon-14 nucleus by proton/neutron transformations. It is then ejected from the nucleus under the influence of nature's fourth fundamental force, the nuclear **weak force**. Although this esoteric force which directs beta decay is not as easy to characterize as the Coulombic force, the results of beta decay can be easily recorded by a nuclear equation.

Write the nuclear equation for the loss of a beta particle from $_6C^{14}$. Identify the isotope formed by this loss.

SOLUTION: $_7N^{14}$

The isotope can be identified by the following nuclear bookkeeping:

$$_6C^{14} \rightarrow \, _7N^{14} + \, _{-1}e^0$$

| Total Charge = 6 | Total Charge = 7 + (−1) = 6 |
| Total Mass = 14 | Total Mass = 14 + 0 = 14 |

EXAMPLE VIII

Nuclear Energy: Fission

In a nuclear reactor or an atomic bomb, huge amounts of energy are obtained by the fission process. A typical fission process is shown below:

$$_0n^1 + \, _{92}U^{235} \rightarrow \, _{55}Cs^{144} + \, —?— \, + 2 \, _0n^1$$

In this process, an unstable nucleus of uranium-235 is split into two daughter nuclei by neutron bombardment. In addition to the two daughter nuclei, additional neutrons are produced. The resulting nuclear reshuffling releases energy in the form of particle kinetic energy and radiation. Under the proper conditions, the neutrons produced can initiate secondary fission reactions. When this occurs, the fission is sustained as a chain reaction (Fig. 19.16).

The nuclear equation represented above is one of many possible fissions that can occur during a chain reaction. Use the principles of mass and charge conservation to deduce the identity of the missing isotope in this fission equation.

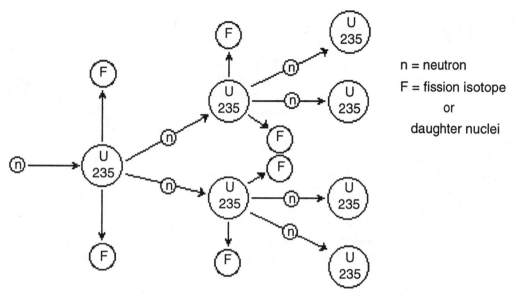

Figure 19.16 The Fission Chain Reaction of U-235

n = neutron
F = fission isotope
or
daughter nuclei

SOLUTION:[5] $_{37}Rb^{90}$

$$_0n^1 + {}_{92}U^{235} \rightarrow {}_{55}Cs^{144} + \text{—?—} + 2\,{}_0n^1$$

Total Charge = 0 + 92 = 92   Total Charge = 55 + (37) + 0 = 92
Total Mass = 1 + 235 = 236   Total Mass = 144 + (90) + 2 × 1 = 236

EXAMPLE IX

Nuclear Energy: Fusion

Another approach to exploiting nuclear energy is that of nuclear fusion. In this process, two small nuclei are forced close enough together so that the nuclear strong force pulls them into one atomic nucleus. The resulting nuclear reshuffling releases energy in the form of particle kinetic energy and radiation. One problem with this approach is the high temperatures required by the process. These high temperatures are required to provide the fusing nuclei with sufficient kinetic energy to overcome the initial electric repulsion of positive charges. In a thermonuclear or fusion bomb ("hydrogen bomb"), a fission bomb ("atomic bomb") is detonated to provide the energy for fusion! The production of controlled fusion is currently a subject of intensive study.

In the fusion process represented below, use the principles of mass and charge conservation to deduce the identity of the missing isotope:

$$_1H^2 + {}_1H^2 \rightarrow \text{—?—}$$

SOLUTION: $_2He^4$

---

5.   One major problem associated with fission energy produced in this manner is the radioactive nature of the daughter isotopes. Notice that both daughter isotopes produced in this example have proton/neutron ratios drastically different than their respective most abundant isotopes. In the operation of a nuclear reactor, dealing with such radioactive waste is a considerable challenge.

9.11 Electrons in Atoms

The nuclear model of the atom places the electrons at the outer periphery of the atom. Since the chemical bond is an attraction between atoms, it is logical to assume that the electron region of the atom plays an integral role in chemical bond formation. Following Ernest Rutherford's presentation of the nuclear model, a considerable research effort was devoted to elucidating the detailed structure of the electron region of the atom.

In 1914, the Danish physicist Niels Bohr suggested a model for the electron structure of the atom. This Bohr model of the atom was designed to explain the energy absorption and emission of gaseous elements. When gaseous elements absorb energy, they re-emit the absorbed energy as characteristic colors or energies of light. The characteristic red color of a neon advertising sign and the characteristic yellow color of a sodium vapor lamp are illustrations of this phenomenon.[6] Bohr visualized the positive atomic nucleus and orbiting negative electrons as an energy storage and emission system. This system could use the electric force to store and release energy in a manner analogous to the storage of energy by the gravitational force (Fig. 9.17 and 9.18):

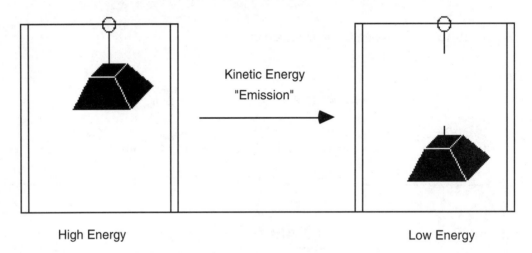

Figure 9.17 Storage of Energy by Gravitational Force

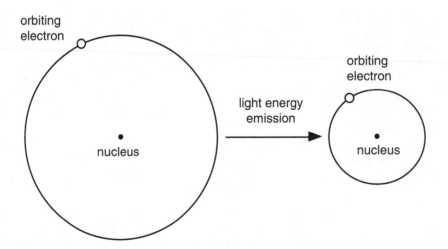

Figure 9.18 Storage of Energy by the Electric Force in the Bohr Atom

6. In general, the characteristic light emitted and absorbed by chemical substances can be used for the purposes of chemical identification. The study of the light emitted and absorbed by substances is called spectroscopy, and spectroscopy represents a powerful modern analytical tool.

In the Bohr model of the atom, the gravitational analogy broke down on one major point. In Figure 9.17, the forces within the rope can hold the weight at any intermediate position above the earth, thus the weight/earth system can store any intermediate energy. Since gaseous elements only store and emit specific characteristic energies, Bohr had to restrict the possible orbit positions of the electrons. The centrifugal force that holds electrons in orbits must hold the electrons to a specific set of orbits that correspond to the characteristic energy emissions of the elements. Since this position restriction has no easily detectable counterpart in the world of macroscopic objects that humans see with their eyes, the Bohr model of the atom represents a revolutionary point of view. Since the origin of chemical bonding is explained by a model related to the Bohr model, it is important to become familiar with some of the language used to describe this revolutionary point of view.

The Bohr model restricts the movement of electrons in an atom to specific orbits called **energy levels**. As the energy of an atom changes and electrons move from one energy level to another, they are not allowed to occupy any position between the two levels! This is an exceedingly awkward situation to describe. How does an electron get from point A to point B without occupying intermediate positions? In the next section, a rational explanation of this type of restriction will be offered. For the time being, it would be useful to have the language to express this concept. As electrons move from one energy level to another, they are said to make a **quantum transition**. The amount of energy absorbed or emitted during a quantum transition is called an **energy quanta**. Figure 9.19 is an attempt to show possible quantum transitions involving the first three energy levels of the hydrogen atom. In the next section, this **quantum theory** will be used to explain the origin of the chemical bond.

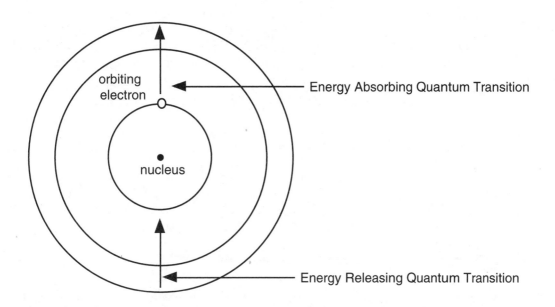

Figure 9.19   Quantum Transitions in the Bohr Hydrogen Atom

9.12 . Understanding the Chemical Bond

The nuclear model of the atom helped establish a new age of civilization—the nuclear age. For the chemist, it also brought new understanding to an old problem. That problem was understanding the basic nature of the chemical bond. Although a detailed discussion of chemical bonding theory is beyond the scope of this book, this important modern topic merits some brief discussion.

The normal chemical reaction process involves the transformation of chemical bonds as reactant molecules are transformed into product molecules. Since a chemical bond is a force of attraction between atoms, it follows from the nuclear model of the atom that the dynamic electron region of the atom must play an important role in chemical bond formation.

In 1923, the French physicist Louis de Broglie suggested a startling view of the dynamic electrons in an atom. In order to explain the quantum transitions of the Bohr model, de Broglie chose to regard electrons at the periphery of an atom, not as moving particles, but rather as a three dimensional smear that behaved as a standing wave—analogous to waves on water or vibrations on a string. The attractive feature associated with this point of view is that macroscopic standing waves do, indeed, manifest a kind of quantum behavior. For example, a string of a given length can only be made to vibrate with a specific set of standing wave frequencies determined by the tension on the string. This relationship between pitch (frequency) and length was studied by the Pythagoreans as early as the sixth century B.C.

The de Broglie wave model of the atom offered a real world analogy for understanding the quantum transition. It also provided the mathematical formalism for describing the quantum transition in atoms. This view was further developed in a more chemically useful form by the German physicist Erwin Schrödinger in 1926. Although this wave theory (also called quantum theory) of chemical bonding is highly mathematical, some insight can be obtained by using the water wave analogy.

A common characteristic of waves (sound, light, water, etc.) is that they are capable of merging with one another to form new waves. For example, when two pebbles are thrown into a pond, the water waves or ripples that meet create a combined disturbance. As the crest of one water wave meets the crest of the other water wave, a new larger crest is formed. The modern view of the chemical bond uses this point of view to explain how molecules are held together by the electric charges inside of the atoms of the molecule.

Although two negative electron particles would be expected to repel, the wave view of dynamic electrons allows for two single electron waves to merge and become one new double electron wave. To do this, the two single electron waves must drag along their respective nuclei, and the result is two atoms joined together by a double electron wave bond. Figure 9.20 represents a schematic of this process for two hydrogen atoms of equal size. The bond formed is called a **covalent bond**.

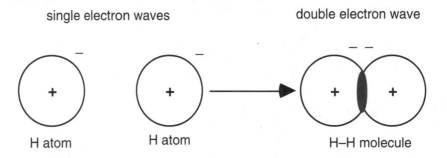

Figure 9.20   The Formation of a Covalent Bond

When the two electron waves overlap, the region of overlap becomes more intensely negative. Since the two atoms of hydrogen are of equal size, this overlap region is at the midpoint between the two nuclei. This is a particularly stable arrangement of electrical charge, hence the two electrons merged into one form the chemical bond.[7]

Figure 9.21 represents a schematic of the bonding process between two atoms that differ in size. If the size difference between the bonding atoms and/or the nuclear charge difference between the bonding atoms is appropriate, the chemical bond takes on characteristics that reflect these differences. The bond is called an **ionic bond**.

In the chemical bond represented in Figure 9.21, the electron overlap region is closer to the smaller hydrogen atom. This creates an unequal distribution of charge in the resulting molecule. A bond of this type is said to possess ionic character. If the ionic character of a chemical bond is sufficient, then the corresponding chemical compound will have properties that differ from covalently bonded molecules, including the ability to form charged atoms or ions in water solution. For example,

---

7.   Actually, the concept of a two electron chemical bond predates wave theory. In 1916, the American chemist Gilbert Newton Lewis proposed the concept of a two electron bond—hence Lewis structures.

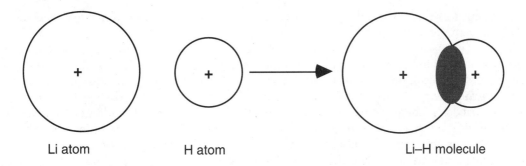

Li atom	H atom		Li–H molecule

Figure 9.21 The Formation of a Bond with Ionic Character

since sodium chloride is an ionic compound, a water solution of sodium chloride actually contains positive sodium ions (Na$^+$) and negative chloride ions (Cl$^-$). In the crystalline solid state, ionic compounds like sodium chloride exist as an ordered arrangement of opposite charged ions (Fig. 9.22).

The wave model of the chemical bonding process has given new meaning to the structural formula of a molecule (Lewis structure). This new view has allowed the modern chemist to correlate vast amounts of factual information with the structural formula of a compound. The result of this new atomic/molecular view is a theoretical tool that makes the modern chemist a true molecular architect.

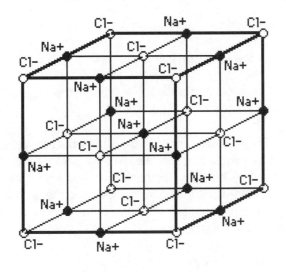

Ionic Solid

Figure 9.22 Sodium Chloride, A Crystalline Ionic Solid

9.13 The Electron Dot Lewis Structure

The wave model of the atom indicates that the electrons in a multi-electron atom are located in a series of energy levels that can be predicted by studying the energy emission and absorption of the gaseous element. As the distance of an energy level from the nucleus increases, its size and maximum electron content also increase. Experimental evidence involving the magnetic properties of elements indicates that electrons tend to be grouped as pairs within an energy level, and the bonding characteristics of many elements indicate that electron pairing within atoms prior to bonding begins after an energy level is half filled. In order to gain an understanding of how these electron energy level rules explain the bonding characteristics of specific elements, it is convenient to use a notation that represents electrons as dots. Although these dots may seem to be a return to particle electrons, the dots really represent electron waves as discussed in the previous section.

The bonding of the elements in the first two periods of the periodic table involves the first two wave model energy levels which can contain a maximum of 2 and 8 electrons respectively. It is no accident that the maximum number of electrons in the first two energy levels corresponds to the number of elements in the first two periods of the periodic table. All of the electron energy levels correspond to periods on the periodic table, and this aspect of the theory should not come as too much of a surprise. Any theory developed to explain the chemistry of the elements must ultimately correspond to the periodic table! If dots are used to represent electrons, and symbols are used to represent atomic nuclei, then the building up of the atoms of the first two periods of the periodic table can be represented as shown in Figure 9.23. Chemists refer to this building up of the periodic table according to the rules of quantum theory as the ***aufbau*** **principle**, from the German word for "build up."

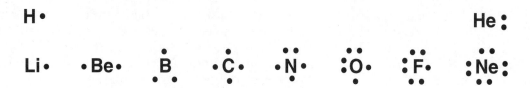

Aufbau Rules for Periods One and Two

1. Energy level one can contain a maximum of two electrons.
2. Energy level two can contain a maximum of eight electrons.
3. In preparation for bonding, electron pairing begins after an energy level is half filled.

Note: To avoid "dot clutter" in this notation, the filled two electron inner energy level of the period two elements is not shown. The symbols for the second period elements, therefore, represent the nuclei and the two electrons of the filled first energy level. The combination of the nucleus and the filled inner energy levels of an atom is called the kernel of the atom. For the period two elements above, the symbol represents the kernel.

Figure 9.23 *Aufbau for Period One and Two Elements*

Although the building up rules for the entire periodic table are much more complex than the rules presented for the first two periods, Figure 9.23 does illustrate the basic concept of the *aufbau*. The unpaired electrons in Figure 9.23 correspond to the directed valences of the atoms discussed in Chapter 8. During the process of bond formation, a single electron wave of one atom interacts with a single electron wave on another atom to form a double electron wave chemical bond as described in the previous section. Since the electrons in the incomplete outer energy level of the atom are responsible for bonding, all of these outer electrons are called **valence electrons**. Note that the valence of an atom is not necessarily equal to the number of valence electrons. For example, although the valence of nitrogen in NH_3 is three, nitrogen has five valence electrons. Note also that the completely paired valence electron energy levels of helium and neon explain the zero valence of these inert gases. Electron pair chemical bond formation can be conveniently represented using electron dot Lewis structures. The electron dot Lewis structures for NH_3, BF_3, and C_2H_4 are shown in Figure 9.24. In these structures note that the symbol of each atom represents the kernel of that atom—the nucleus and the filled inner energy level.

Figure 9.24 Electron Dot Lewis Structures

The shared electron pairs (shared dot pairs) in these electron dot structures correspond to the chemical bonds in the structural formulas introduced in Chapter 8. The unshared pairs of electrons that are shown in these electron dot Lewis structures can have a profound influence on the chemical and physical properties of compounds. For example, the unshared pair of electrons on the nitrogen atom in the NH_3 molecule can complete the eight electron valence energy level of the boron atom in BF_3 by forming a bond with the boron atom. The electron dot Lewis structure of the resulting NH_3BF_3 molecule is shown in Figure 9.25. This type of bonding involving unshared pairs of electrons is responsible for the secondary valences in the coordination compounds of the transition elements that was discussed in Chapter 6.

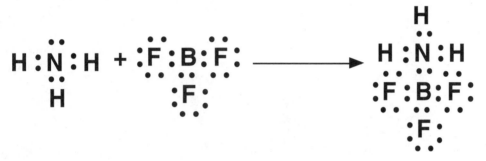

Figure 9.25 Bonding Involving Unshared Pairs of Electrons

EXAMPLE X

Drawing an Electron Dot Lewis Structure

Draw the electron dot Lewis structure of formaldehyde (CH_2O).

SOLUTION:
The procedure for drawing this structure is similar to the ball-and-stick model procedure described in Chapter 8 (Example II). Visualize the electron dot atomic pieces, and then use unpaired electrons to form bonds. Since some of the electron dot atomic pieces contain pairs of electron dots, the final structure will also contain unshared electron dot pairs.

Since the geometric arrangement of the valence electrons is identical to the geometric arrangement of the directed valences presented in Chapter 8, the molecular geometry of the molecule is identical to the molecular geometry predicted by the directed valence theory.

9.14 A Final Note

The investigation of the theoretical origins of valence concludes this book's discussion of chemical atomic/molecular theory. As indicated in the preface, this book has attempted to convey the essence of chemistry. Using the history of chemistry as a teaching vehicle, you have been asked to experience actual chemical thought and to use the modern chemist's basic intellectual tool—atomic/molecular theory. Most one semester courses in chemistry, including the course at Illinois State University, continue beyond this essence of chemistry into the realm of contemporary relevant problems. Hopefully, this book has provided a foundation for this continuing study. In this regard, your chemical journey is just beginning.

Bon Voyage
Otis S. Rothenberger
James W. Webb
American Chemists

Chapter Nine
Performance Objectives

P.O. 9.0

Review all of the boldfaced terminology in this chapter, and make certain that you understand the use of each term.

alpha particle	atomic mass number	atomic nucleus
atomic number	*aufbau* principle	beta particle
Coulomb's law	covalent bond	electrical force
electrolysis	electron	energy level
fission	fusion	gamma ray
gravitational force	half-life	ionic bond
ionizing radiation	isotopes	neutron
nuclear equation	nuclear strong force	nuclear weak force
proton	quanta, energy	quantum theory
quantum transition	radioactivity	valence electrons

P.O. 9.1

You must demonstrate a qualitative understanding of the electric charge laws and Coulomb's law.

EXAMPLE:
The diagrams below represent electrically charged systems. If all of the electric charges are equal in magnitude, which system contains the most stored energy?

a. ⊖ ⊕

b. ⊕ ⊖

c. ⊖   ⊕

d. ⊖   ⊕

e. Systems a & b contain the most energy.

SOLUTION: c

Textbook Reference: Section 9.3

ADDITIONAL EXAMPLE:
The diagrams below represent two electrically charged systems. If all of the electric charges are equal in magnitude, which system contains the most stored energy?

a. (+)   (+)

b. (+)   (+)

ANSWER: a

P.O. 9.2

You must demonstrate an understanding of electric discharge tube experiments.

EXAMPLE:
The figure shown below represents the schematic of an electric discharge tube. In this schematic, what is the electric charge on the particles striking the detection screen?

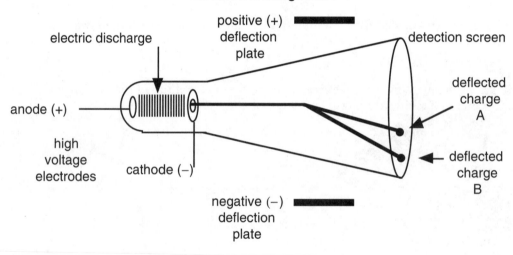

Electric Discharge Tube

SOLUTION: positive

Textbook Reference: Section 9.4

ADDITIONAL EXAMPLE:
If the charge on the detected particles (A & B above) are equal in magnitude, which particle is more massive?

ANSWER: A

P.O. 9.3

You must be able to use the periodic table to determine the subatomic composition of an isotope.

EXAMPLE:

Complete the following table:

Element	Electrons	Protons	Neutrons	Mass Number
Se	A	B	45	C
(1)	D	E	20	40
(2)	6	G	H	13
(3)	M	J	7	14
(4)	6	K	L	14

SOLUTION:

A=34 B=34 C=79 (1)=Ca D=20 E=20 (2)=C G=6 H=7 (3)=N M=7 J=7 (4)=C K=6 L=8

Textbook Reference: Section 9.8

ADDITIONAL EXAMPLE:

A certain isotope of oxygen has a mass number of 18. How many neutrons are contained in the nucleus of this isotope?

ANSWER: 10

P.O. 9.4

You must demonstrate an understanding of the detection of radioactivity by answering questions about an idealized radioactivity detector.

EXAMPLE:

The figure below represents an idealized radioactivity detector that is being used to measure the radiation being emitted by impure uranium:

Idealized Radioactivity Detector

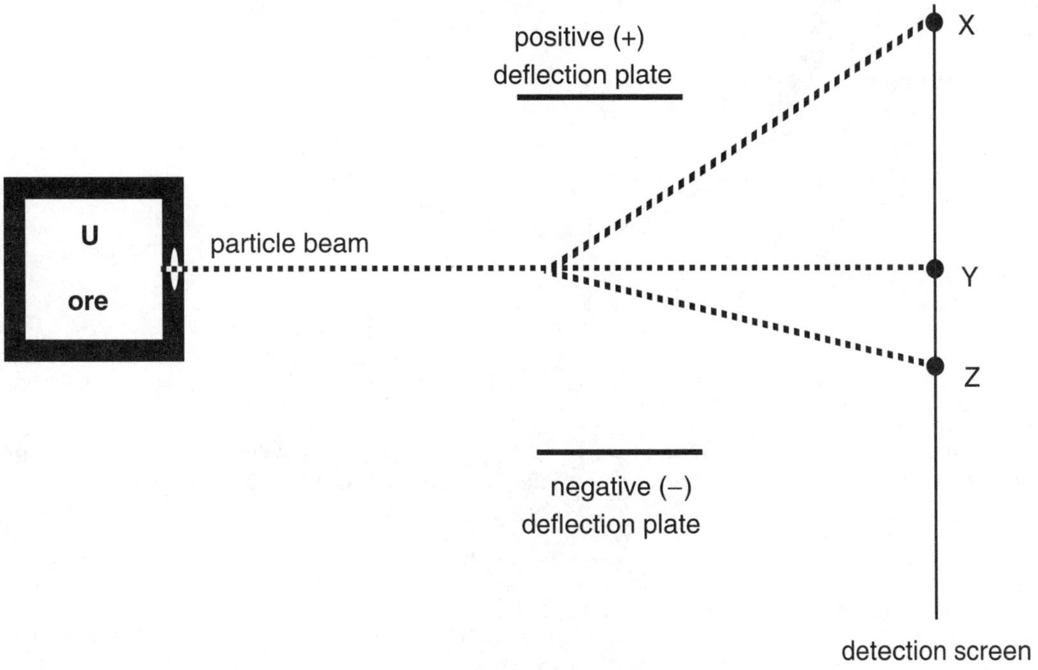

What type of radiation is being detected at position X?

SOLUTION: beta particles

Textbook Reference: Section 9.9

ADDITIONAL EXAMPLE:
At which position on the detection screen of the above idealized radioactivity detector are alpha particles being detected?

ANSWER: position Z

P.O. 9.5

You must demonstrate an understanding of the experimental basis of the nuclear model of the atom—the Rutherford alpha particle scattering experiment.

EXAMPLE:

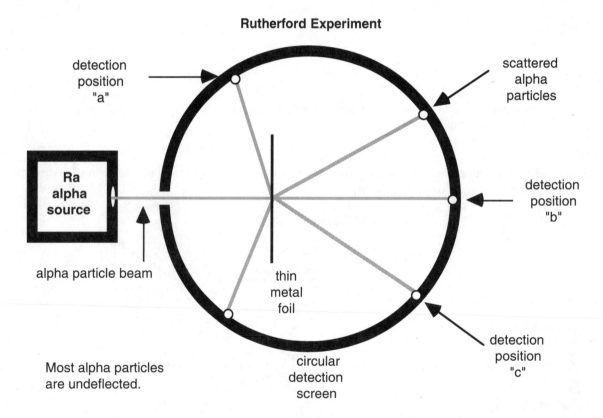

The figure above represents the schematic of an apparatus used by Ernest Rutherford to probe the substructure of the atom. Are more alpha particles detected at location b or c?

SOLUTION: location b

Textbook Reference: Section 9.10

ADDITIONAL EXAMPLE:
The figure above represents the schematic of an apparatus used by Ernest Rutherford to probe the substructure of the atom. Are any alpha particles detected at location c?

ANSWER: yes

P.O. 9.6

You must be able to read a nuclear symbol.

EXAMPLE:
What is the integer mass of the isotope represented by the symbol $_5B^{11}$?

SOLUTION: 11

Textbook Reference: Sections 9.8 and 9.10

ADDITIONAL EXAMPLE:
How many neutrons are contained in the nucleus of the isotope represented by the symbol $_7N^{15}$?

ANSWER: 8

P.O. 9.7

You must be able to identify a radioactive isotope given the isotope's nuclear symbol.

EXAMPLE:
Which of the following isotopes is most likely to be radioactive?

$$_2He^4 \qquad _6C^{12} \qquad _{13}Al^{27} \qquad _6C^{14} \qquad _{20}Ca^{40}$$

SOLUTION: $_6C^{14}$

Textbook Reference: Section 9.9

ADDITIONAL EXAMPLE:
Which of the following isotopes is most likely to be radioactive?

a) $_{19}K^{40}$ b) $_2He^4$ c) $_{13}Al^{27}$ d) $_{20}Ca^{40}$ e) $_6C^{12}$

ANSWER: a

P.O. 9.8

You must be able to complete a partial nuclear equation.

EXAMPLE:
What is the identity of the missing particle in the following nuclear equation?

$$_{38}Sr^{90} \rightarrow -?- + _{-1}e^0$$

SOLUTION: $_{39}Y^{90}$

Textbook Reference: Section 9.10

ADDITIONAL EXAMPLE:
What is the identity of the missing particle in the following nuclear equation?

$$_{1}H^{2} + _{1}H^{3} \rightarrow _{0}n^{1} + \text{—?—}$$

ANSWER: alpha particle

P.O. 9.9

You must demonstrate an understanding of radioactivity.

EXAMPLE:
The half-life of hydrogen-3 ($_{1}H^{3}$) is 12 years. How much of a 100 gram sample of hydrogen-3 will remain after 24 years?

SOLUTION: 25 grams

Textbook Reference: Section 9.9

ADDITIONAL EXAMPLE:
Write a balanced nuclear equation for the loss of a beta particle by $_{39}Y^{90}$. What is the identity of the new nucleus formed by this radioactive decay process?

ANSWER: $_{40}Zr^{90}$

P.O. 9.10

You must demonstrate an understanding of the nuclear fission and fusion processes.

EXAMPLE:
The nuclear reaction shown below represents the fission of uranium-235:

$$_{0}n^{1} + _{92}U^{235} \rightarrow _{61}Pm^{158} + \text{—?—} + 4\,_{0}n^{1}$$

What is the identity of the unknown nucleus?

SOLUTION: Ga

Textbook Reference: Section 9.10

ADDITIONAL EXAMPLE:
Is nuclear fission an exothermic or endothermic process?

ANSWER: exothermic

P.O. 9.11

You must be able to draw an electron dot Lewis structure for a molecule containing period one and/or period two elements.

EXAMPLE:
Draw the electron dot Lewis structure for a molecule of water.

SOLUTION:

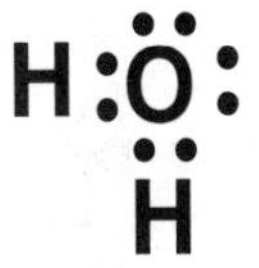

H:Ö:
H

Textbook Reference: Section 9.13

ADDITIONAL EXAMPLE:
Draw the electron dot Lewis structure for a molecule of acetylene (C_2H_2).

ANSWER:

H:C:::C:H

Laws of Electric Charge

1. The diagrams below represent electrically charged systems. Rank these electrically charged systems in order of increasing stored energy. Do the charges in each of these systems attract or repel one another? All of the electric charges are equal in charge magnitude.

    a. (−) (+)

    b. (+) (−)

    c. (−)   (+)

    d. (−)   (+)

STUDENT SOLUTION:

2. How would the answers to question 1 be changed if all of the charges were positive?

    a. (+) (+)

    b. (+) (+)

    c. (+)   (+)

    d. (+)   (+)

STUDENT SOLUTION:

3.   The diagram below represents the movement of two positive charges that are under the influence of the Coulombic (electrostatic) force. In this diagram, the two particles are moving toward each other. Does the process represented by this diagram liberate energy or consume energy?

Two Positive Charges Moving Toward Each Other

STUDENT SOLUTION:

4.   If the particles in the previous question were oppositely charged, would the process liberate energy or consume energy?

STUDENT SOLUTION:

Problems Involving Classic Experiments

5.   A Faraday electrolysis experiment is performed by passing the same amount electricity through molten NaCl and molten $ZnCl_2$. The masses of metals produced are 0.552 g of sodium and 0.785 g of zinc. Are these quantities consistent with the connection between the quantitative laws of electrolysis and valence? What mass of Al would be produced from molten $AlCl_3$ with the same amount of electricity?

STUDENT SOLUTION:

6. The following diagram represents a schematic of an electric discharge tube:

Electric Discharge Tube

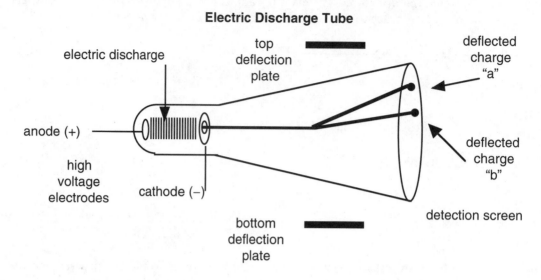

a. What is the sign of the charge on the detected particles a and b?
b. What is the sign of the charge on the top deflection plate?
c. If the charges on particles a and b are equal in magnitude, which particle is more massive?

STUDENT SOLUTION:

7. The diagram below represents the schematic of an ideal apparatus designed to detect alpha, beta, and gamma radiation:

Idealized Radioactivity Detector

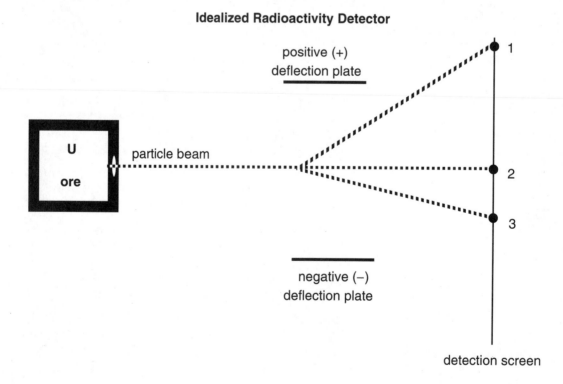

a. Which location on the detection screen corresponds to the detection of beta particles?

b. Which location on the detection screen corresponds to the detection of alpha particles?

STUDENT SOLUTION:

8. The diagram below represents the schematic of an apparatus used by Ernest Rutherford to probe atomic substructure:

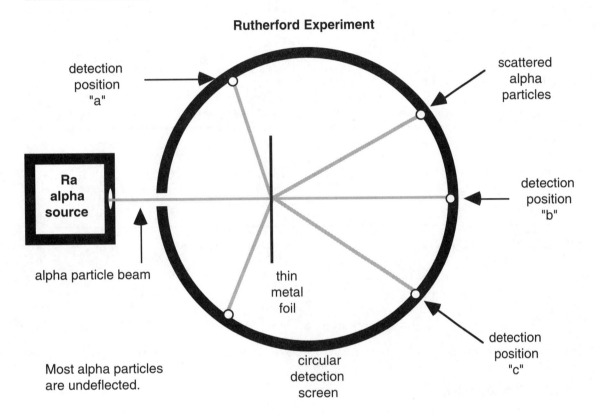

Give an atomic explanation for the results of this experiment.

STUDENT SOLUTION:

Problems Involving Subatomic Composition and Structure

9. The charge and integer mass of several subatomic particles is listed below. Identify each particle and write the correct symbol for the particle.

    a. +1 and 1 amu
    b. −1 and 0 amu
    c. 0 and 1 amu
    d. +2 and 4 amu

    STUDENT SOLUTION:

10. Complete the following table:

Element	Electrons	Protons	Neutrons	Mass Number
F	A	B	10	C
(1)	D	18	E	40
(2)	2	G	H	4
(3)	M	J	15	31
(4)	19	K	L	40

    STUDENT SOLUTION:

11. How many electrons does each atom in the previous problem have in its outer valence electron energy level?

    STUDENT SOLUTION:

Problems Involving Nuclear Processes

12. Write a balanced nuclear equation for the loss of an alpha particle by $_{84}Po^{218}$.

    STUDENT SOLUTION:

13. Write a balanced nuclear equation for the loss of a beta particle by $_{38}Sr^{90}$.

   STUDENT SOLUTION:

14. Write a balanced nuclear equation for the fission of uranium-235 to give $_{36}Kr^{97}$, another nucleus, and an excess of 3 neutrons.

   STUDENT SOLUTION:

15. Complete the following nuclear equation:

$$_0n^1 + {}_{92}U^{235} \rightarrow 2\,_0n^1 + {}_{40}Zr^{97} + \text{---?---}$$

   STUDENT SOLUTION:

16. The radioactive decay of certain synthetic isotopes does not proceed by alpha or beta emission. For example, nitrogen-12 is a radioactive isotope that emits a strange subatomic fragment called a positron. The positron, one of many fundamental subatomic particles discovered during this century, is really a positive electron, $_1e^o$. Write a balanced nuclear equation for the loss of a positron by $_7N^{12}$.

   STUDENT SOLUTION:

17. Positron emitters such as carbon-11 are used in a medical diagnostic technique call positron emission tomography (PET). What is the element produced when carbon-11 decays by emitting a positron?

$$_6C^{11} \rightarrow {}_1e^o + \text{---?---}$$

   STUDENT SOLUTION:

18. In 1932 James Chadwick discovered the neutron by bombarding an element with alpha particles. Based on the following reaction, what element did he bombard?

$$-?- + {}_2He^4 \rightarrow {}_6C^{12} + {}_0n^1$$

STUDENT SOLUTION:

19. Plutonium-239 is synthetic fissionable isotope that can be used in a nuclear reactor or an atomic bomb. Uranium-238 is the most abundant natural isotope of uranium, but it is not fissionable. Pu-239 can be produced by bombarding U-238 in a breeder nuclear reactor. The series of nuclear reactions is as follows:

$$_{92}U^{238} + {}_0n^1 \rightarrow A$$
$$A \rightarrow B + {}_{-1}e^0$$
$$B \rightarrow {}_{94}Pu^{239} + C$$

Identify A, B, and C.

STUDENT SOLUTION:

20. Complete the following thermonuclear (fusion) reaction:

$$_3Li^7 + {}_1H^2 \rightarrow 2 \, {}_2He^4 + -?-$$

STUDENT SOLUTION:

21. The half-life of strontium-90 ($_{38}Sr^{90}$) is 28 years. How much of a 16 milligram sample of strontium-90 will remain after 56 years?

STUDENT SOLUTION:

22. Carbon dating is useful for estimating the age of archaeological artifacts. The measurement is based on carbon-14, a naturally occurring radioactive isotope of carbon. The concentration of carbon-14 is constant in the atmosphere, and living plants and animals incorporate this isotope via the biological carbon dioxide cycle. When the organism dies, the incorporation of carbon-14 ceases, and the carbon-14 contained in the organism decays. The half-life of carbon-14 is 5730 years.

An organic sample taken from an archaeological dig is measured for carbon-14 activity. Its activity decreased to $\frac{1}{4}$ of the carbon-14 activity found in living tissue. Estimate the age of the organic sample.

STUDENT SOLUTION:

23. Which of the following isotopes is most likely to be radioactive:

a) $_7N^{14}$ b) $_7N^{12}$ c) $_{11}Na^{23}$ d) $_{84}Po^{210}$ e) $_6C^{14}$

STUDENT SOLUTION:

24. In 1919, Ernest Rutherford performed the first artificial transmutation by bombarding nitrogen-14 with high velocity alpha particles. The nuclear equation representing this experiment is shown below. Complete this nuclear equation:

$$_7N^{14} + \text{alpha particle} \rightarrow {}_1H^1 + \text{—?—}$$

STUDENT SOLUTION:

25. In the nuclear fusion reaction described in Example IX, 1 gram of deuterium releases about the same energy as six tons of coal. In sea water, one of every 6500 hydrogen atoms is deuterium. How many liters of sea water would contain enough deuterium atoms with sufficient potential energy equivalent to six tons of coal?

STUDENT SOLUTION:

Problems Involving Electron Bonding Theory

26. Draw the electron dot Lewis structure for each of the following molecules:

$$C_2H_6 \quad H_2O_3 \quad N_2 \quad BeH_2 \quad H_2S \quad HCN$$

STUDENT SOLUTION:

27. How many unshared pairs of electrons does each of the molecules in the previous question contain?

STUDENT SOLUTION:

Problems Involving Household Chemistry and Science

28. In the article "Apparatus for Demonstrating Electrolysis on the Overhead Projector," Doris and Kenneth Kolb describe a simple apparatus that can be used to investigate electrolysis (*J. Chem. Educ.* **1986**, 63, 517). It is very easy to construct an apparatus similar to the one described in this article. Sharpen two new number 2 pencils. Cut a notch about 2 cm down from the eraser end of each pencil exposing the graphite. Place a small pencil between the two pencils as a spacer. Place this spacer pencil about 2 cm below the notches. Tape a 9 volt battery onto the pencils so that the snap terminals of the battery are just below the notches. Wrap a short piece of copper wire onto each battery terminal then around the exposed graphite on each pencil. This apparatus can be used to demonstrate electrolysis by immersing the pencil tips into a cup containing a dilute solution of Epsom salts ($MgSO_4 \bullet 7H_2O$). The purpose of the magnesium sulfate is to provide a solution that will conduct electricity. The bubbles that appear at each pencil tip indicate that the electrolysis of water is taking place:

$$2\,H_2O \rightarrow 2\,H_2 + O_2$$

Look at the electrolysis of water diagram in Section 7.15 (Figure 7.15). Is the volume of gas that appears to be forming at each of the pencil electrodes consistent with Figure 7.15? The terminal polarity should be marked on the battery.

During the electrolysis process, different chemical reactions are taking place at each electrode. Because these reactions are different, they affect the pH of the solution differently. Add a few drops of red cabbage indicator (problem 35, Chapter 6) to the solution during electrolysis. Note the color changes at each electrode. Which electrode is producing an acid pH, and which electrode is producing a basic pH?

STUDENT SOLUTION:

29. Static electricity has intrigued natural scientists for years. Even the ancient Greeks were aware of it. There is an old parlor trick that can be used to illustrate some interesting aspects of static electricity. Balance a nickel on a table, and then balance a paper match on the nickel. Place a small glass over the two objects. The objective of the parlor trick is to cause the match to fall from the nickel without moving the glass or the table. The solution to this trick is to run a comb through your hair to establish a static charge on the comb. Bring the comb close to the glass, and the static charge built up on the comb will cause the match to fall from the nickel.

A few experiments with this trick can be very instructive. Watch the motion of the match very carefully. What is the nature of the motion of the match as it falls off the coin? Does this motion suggest that the match has the same charge as the comb or the opposite charge? Is there anything to suggest that the match has two electrical charges? Try the trick with a thick-walled glass and a thin-walled glass of about equal size. Which glass makes it easier to topple the match? Rinse the glass that seems to work best with water. Without drying this glass, try the trick again. Is it harder or easier to topple the match? See how much fun a chemist can have with a comb, a nickel, a match, and a glass?

STUDENT SOLUTION:

Library Problems

30. Nobel Laureate Linus Pauling spent much of his career studying the nature of the chemical bond. He won the Nobel Prize in Chemistry in 1964 for his work on protein structure and the Nobel Peace Prize in 1963 for his fight against nuclear weapons testing. Read the articles "Interview with Linus Pauling" (Ridgeway, David *J. Chem. Educ.* **1976**, 53, 471) and "G.N. Lewis and The Chemical Bond" (Pauling, Linus *J. Chem. Educ.* **1984**, 61, 201). Compare the contributions that Pauling and Lewis made toward understanding the chemical bond. How were each influenced by contemporaries in their field? What was unique about Pauling's acceptance into Oregon Agriculture College? What advice does Pauling give to young people entering the field of science?

STUDENT SOLUTION:

31. The word "radioactivity" has a negative connotation. People fear radioactivity because it is a consequence of nuclear fallout, and it can result from contamination from nuclear reactors. However, radioisotopes are used in medical science as important diagnostic and therapy tools. Read the articles "Radioactivity: A Natural Phenomenon" (Ronneau, C. *J. Chem. Educ.* **1990**, 67, 736) and "Radioactivity in the Service of Man" (Yalow, Rosalyn *J. Chem. Educ.* **1982**, 59, 735). (Rosalyn Yalow received the Nobel Prize in Medicine and Physiology in 1977 for her work in developing the analytical technique known as radioimmunoassay.) What are the consequences of high doses of radiation on human beings? Differentiate between artificial and natural sources of radiation. How much do natural sources contribute to the average dose of radiation that humans receive? Compare the risks and benefits of using low level doses of radioactive isotopes in medical science.

STUDENT SOLUTION:

32. The synthetic elements 104, 106, 107, and 109 have been named for scientists who have made significant contributions to the development of atomic theory. The names of these elements are Rutherfordium (104), Seaborgium (106), Bohrium (107), and Meitnerium (109). Identify these scientists and describe their contributions to the development of atomic theory.

STUDENT SOLUTION:

Formal Laboratory Exercise

The Electrodeposition of Zinc on Copper
Having Some Fun with Electrolysis

Introduction

In Section 9.2 of this chapter, the process of electrolysis and its importance to subatomic structural theory were discussed. Electrolysis also has important industrial applications. For example, electrolysis is used in the production of hydrogen, chlorine, sodium hydroxide, and numerous other pure elements including metallic aluminum. Electrolysis is used to purify some metals, and it is used for electroplating—the deposition of a thin film of metal onto another surface.

Electroplating is done to protect metals against corrosion and wear. It used to improve the appearance of metals as in production of jewelry. Chromium, silver, and gold are metals commonly used in electroplating. Electroplating involves the application of a voltage across two electrodes immersed in salt solution. The metal being plated is connected to the negative terminal (cathode) of the power source, and the salt solution contains ions of the metal to be deposited (e.g. silver ions). The electrolysis conditions are important to produce a smooth, uniform coating of the metal and to prevent side reactions. Often other ions such as cyanide ions (CN^-) are added to the salt solution. These ions do not participate in the overall stoichiometry of the electrolysis reaction; however, they do influence the uniformity of the deposition by forming metal complexes with the metal ions in the salt solution. These coordination complexes were discussed in textbook Section 6.7.

The purpose of this formal laboratory exercise is to study the process of electrolysis and electrodeposition. Since this is the final formal laboratory exercise in this textbook, the electrolysis system to be studied was selected so as to maximize the "fun" of the exercise and to produce a lasting memento of the study of chemistry.

Procedural Overview

In this experiment, a coating of zinc metal will be electrodeposited on the surface of a penny using a nine volt transistor battery. The construction of the electrolysis apparatus used in this procedure is described in problem 28 (Problems Involving Household Chemistry and Science).

The penny that is to be electroplated with zinc is first placed on a small piece of aluminum foil (Fig. 9.26).

The aluminum foil acts as a connector from the penny to the negative battery terminal. By placing the penny on a piece of aluminum foil and touching the foil with the graphite tip of the pencil that is connected to the negative terminal of the battery, the penny is made the cathode in the electrolysis reaction. The electrolyte solution containing the zinc ions is carefully placed on the surface of the penny. When the graphite tip of the pencil connected to the positive terminal of the battery contacts this solution, the circuit is complete and the electroplating begins.

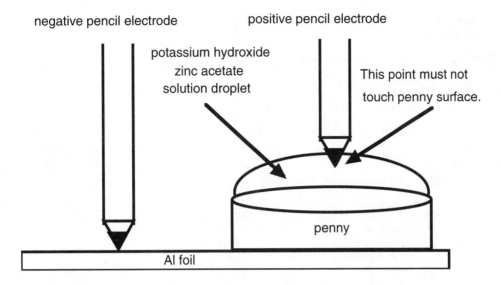

negative pencil electrode   positive pencil electrode

potassium hydroxide
zinc acetate
solution droplet

This point must not
touch penny surface.

penny

Al foil

Figure 9.26   Storage of Energy by the Electric Force in the Bohr Atom

This experiment will illustrate the importance in the electrolyte composition in obtaining a good metal deposition. An interesting reaction between the zinc deposit and penny surface will also be illustrated. For this experiment to be successful, it is important to prevent the electrolyte solution on the surface of the penny from flowing onto the aluminum foil. Also, do not touch the penny with the positive electrode; this will short out the electrolytic cell and prevent deposition. Although only small quantities of a potassium hydroxide solution are used, it is important to be careful when handling this caustic solution. Potassium hydroxide solutions can cause skin burns. In case of accidental skin contact, wash the affected skin area with a steady flow of cold water. When handling the pennies, use forceps. Immediately clean up any small (less than 1 mL) spills with a water wash. Inform the laboratory instructor immediately if larger spills occur. As in all formal laboratory exercises, eye protection is essential.

Materials

new pennies, pencil electrolysis apparatus (see problem 28—Problems Involving Household Chemistry and Science), hotplate or Bunsen burner, forceps, dropper, aluminum foil (6 x 6 cm)

Chemicals

potassium hydroxide solution (1 M), zinc acetate solution (0.1 M)

Procedure

1. Place two pennies on the aluminum foil.

2. Add 10 drops of the zinc acetate solution to the surface of one penny.

3. Add 5 drops of the zinc acetate solution and 5 drops of the potassium hydroxide solution to the second penny. Mix the solutions on this penny with a toothpick.

4. Using the pencil electrolysis apparatus, contact the negative terminal of the battery to the aluminum foil and the positive terminal of the battery to the solution on the penny surface containing just the zinc acetate solution. Do not allow the positive terminal to touch the penny (Fig. 9.26).

5. Allow the electrolysis to occur for one minute.

6. Repeat the electrolysis on the penny containing the zinc acetate and potassium hydroxide solutions.

7. Carefully remove the pennies with the forceps and wash the pennies with tap water.

8. Dry the pennies on a paper towel.

9. Describe the surfaces of the two pennies.

10. Using a forceps to hold the penny from the zinc acetate/potassium hydroxide electrolysis, gently heat the penny in the Bunsen burner flame. The laboratory instructor will demonstrate this technique. Alternatively, place this penny on a hotplate at a medium setting. Again, the laboratory instructor will demonstrate this technique.

11. Allow the penny to cool.

12. Describe the change that occurred on the surface of the penny.

Data Table

Description of Penny Surfaces after Electrolysis

Penny from Zinc Acetate Solution Electrolysis:

Penny from Zinc Acetate/Potassium Hydroxide Solution Electrolysis:

Description of Penny Surface after Heating:

Conclusion and Discussion

1. Which electrolysis solution composition resulted in the best deposition of zinc?

2. What was the role of the potassium hydroxide solution in the mixed solution electrolysis? Reread the introduction for a hint to the answer of this question.

3. What is the reason for the rather startling color change on the surface of the heated penny? Ask your laboratory instructor for help on the question.

Chapter Nine Notes

Chapter Nine Notes

Appendix I—Sample Exams

Sample Exam I: Chapters 1–5

Multiple Choice

1. There are a number of formal arguments that are used during various stages of the scientific reasoning process. The following argument forms represent two of these formal arguments. Which of these argument forms is invalid?

   FORM A
   If H, then E
   H
   Therefore E

   FORM B
   If H, then E
   Not E
   Therefore not H

   a) FORM A   b) FORM B   c) Both argument forms are valid.
   d) An argument's validity depends on statements H and E.

2. Estimate the length of this exam page.

   a) 0.0230 m   b) 500 cm   c) 23.0 cm   d) 2300 mm   e) 0.0500 m

3. If oxygen with an atomic mass of 16.0 amu is used as the standard for the atomic mass system, then carbon will have an atomic mass of 12.0 amu. During the early part of the nineteenth century, several different atomic mass standards were used. What would the atomic mass of carbon be under a system that assigned the oxygen atomic mass standard as 100 amu?

   a) 192   b) 19.2   c) 133   d) 75.0   e) none of these

4. The following chemical equation represents the combustion of butane (C_4H_{10}):

   $$2\ C_4H_{10} + 13\ O_2 \rightarrow 8\ CO_2 + 10\ H_2O$$

   What is the gravimetric ratio that describes the quantity of oxygen gas that is required to react with butane according to this reaction?

   a) $\dfrac{O_2}{C_4H_{10}}$ b) $\dfrac{13\ O}{C_4H_{10}}$ c) $\dfrac{13\ O_2}{2\ C_4H_{10}}$ d) $\dfrac{O}{2\ C_4H_{10}}$ e) none of these

5. The schematic representations of three chemical reactions involving seven substances (A through G) are shown below:

   SUBSTANCE A + SUBSTANCE B → SUBSTANCE C
   SUBSTANCE A → SUBSTANCE D + SUBSTANCE E
   SUBSTANCE F + SUBSTANCE G → SUBSTANCE B

The study of the reactant and product masses for each of these reactions indicates that each reaction schematic is complete as shown. There are no hidden substances involved in any of the reactions. Based on these reactions only, how many of the seven substances can definitely be identified as chemical compounds?

a) 1   b) 2   c)   3 d)   4 e) none of these

6. Which of the following compounds contains the element iron?

a) $IrCl_4$ b) nCl_3 c) $FeCl_2$ d) FrCl   e) none of these

7. If a 1.00 gram iron nail rusts completely, will the mass of the rust be greater than or less than the mass of the iron nail?

a) less than   b) greater than   c) The mass of the rust will equal the mass of the iron nail.

8. How many atoms of oxygen are represented by the formula $CH_2N(COOH)_2CH_2N(COOH)_2$?

a) 6   b) 5   c) 10   d) 8   e) none of these

9. The compound sodium cyclamate can be synthesized from petroleum. The formula of sodium cyclamate is $C_6H_{12}NSO_3Na$. What is the molecular mass of this compound in amu?
Atomic Masses: C = 12.0   H = 1.01   N = 14.0   S = 32.1   O = 16.0   Na = 23.0 amu
NOTE: Significant figure rules allow 4 figures in this case.

a) 185.1   b) 217.1   c) 201.2   d) 189.1   e) none of these

10. The density of palladium is known to be 12.0 g/mL. The density of platinum is known to be 21.5 g/mL. In an attempt to discover the identity of an unknown metal, a chemist determines the volume of a 64.9 gram sample of the metal. If the unknown metal is palladium, what should this volume be in milliliters?

a) 3.02   b) 5.41   c) 779   d) 0.185   e) none of these

11. A chemical compound is known to contain only the elements aluminum and oxygen. A quantitative analysis of this compound indicates that it contains 52.9% aluminum. What mass of this compound in grams would need to be decomposed in order to produce 72.4 grams of the aluminum?

a) 38.3   b) 154   c) 0.731   d) 137   e) none of these

12. An unknown element, X, forms a compound with the element oxygen. The formula of this compound is X_2O_7. The compound has a definite composition of 38.8% X. If the atomic mass of oxygen is known to be 16.0 amu, what is the atomic mass of X in amu?

a) 71.0   b) 88.3   c) 35.5   d) 10.1   e) none of these

13. An economically important chemical compound of the element aluminum has the formula $Al_2(SO_4)_3$. In theory, chemists should be able to synthesize this compound from its elements. What mass of this compound in grams could be produced from 93.5 grams of aluminum and an excess of the other elements? Atomic Masses: Al = 27.0   S = 32.1   O = 16.0 amu

a) 14.8   b) 1190   c) 7.38   d) 593   e) none of these

14. How many moles of sulfur are contained in a 10.2 gram sample of sulfur? The atomic mass of sulfur is 32.1 amu.

a) 3.15   b) 32.1   c) 0.318   d) 327   e) none of these

15. A binary compound of the elements calcium and chlorine has the formula $CaCl_2$. If 24.1 grams of calcium is reacted with 76.3 grams of chlorine to form this compound, which element is present in stoichiometric excess? (*i.e.* Which element will have some mass left unreacted?)
    Atomic Masses: Ca = 40.1   Cl = 35.5 amu

a) calcium   b) chlorine   c) Neither element is present in stoichiometric excess.
d) insufficient information

16. Qualitative analysis indicates that a certain chemical compound contains only the elements potassium and sulfur. A quantitative analysis of the compound reveals that a 90.6 gram sample of the compound contains 29.7 grams of potassium. What is the empirical formula of this compound? Atomic Masses: K = 39.1   S = 32.1 amu

a) K_3S_4 b) K_2S_5 c) K_5S_2 d) K_4S_3 e) none of these

17. Balance the following chemical equation:

$$C_5H_{10}S + O_2 \rightarrow CO_2 + H_2O + SO_2$$

When this equation is balanced with integer coefficients, what is the stoichiometric coefficient of carbon dioxide (CO_2) on the right hand side of the equation?

a) 8   b) 10   c) 2   d) 6   e) none of these

18. At high temperatures, the gas methane (CH_4) reacts with copper oxide (CuO) to produce carbon dioxide (CO_2), water, and the element copper. Write a chemical equation for this reaction. When this equation is balanced with integer coefficients, what is the stoichiometric coefficient of copper oxide on the left hand side of the equation?

a) 1   b) 2   c) 3   d) 4   e) none of these

19. The following chemical equation represents the combustion of the octane component of gasoline:

$$2\ C_8H_{18} + 25\ O_2 \rightarrow 16\ CO_2 + 18\ H_2O$$

According to this equation, how many grams of carbon dioxide (CO_2) are formed by the complete combustion of 25.6 grams of octane (C_8H_{18})? Atomic Masses: C = 12.0   O = 16.0   H = 1.01 amu

a) 8.29   b) 6270   c) 78.9   d) 9.88   e) none of these

20. The propulsion system on the first lunar landing craft involved the chemical reaction of nitrogen tetroxide (N_2O_4) with hydrazine (N_2H_4) to form nitrogen (N_2) and water. According to this description of the landing craft propulsion system, how many grams of nitrogen tetroxide (N_2O_4) are required to exactly react with 22.2 grams of hydrazine (N_2H_4)? Atomic Masses: N = 14.0   O = 16.0   H = 1.01 amu

a) 265   b) 31.9   c) 63.8   d) 15.4   e) none of these

Sample Exam II: Chapters 6–7

Multiple Choice

1. Which of the following is the correct formula for potassium carbonate?

   a) $K_2(CO_3)_3$ b) K_2CO_3 c) KCO_3 d) $K(CO_3)_2$ e) none of these

2. Which of the following is the correct formula for lithium sulfide?

   a) Li_2S_3 b) Li_2S c) LiS d) LiS_2 e) none of these

3. The element aluminum reacts with iron (III) oxide to produce the element iron and aluminum oxide. When this equation is balanced with integer coefficients, what is the stoichiometric coefficient of aluminum on the left hand side of the equation?

   a) 3   b) 4   c) 2   d) 6   e) none of these

4. Two chemical formulas for compounds of the unknown elements X and Y are shown below:

$$Al_2X_3 \qquad\qquad YX$$

   If the valence of aluminum is known to be three, what is the valence of the unknown element Y that is consistent with the given formulas?

   a) 1   b) 2   c) 3   d) 4   e) none of these

5. If calcium oxide is dissolved in water, will the resulting solution be acidic or basic?

   a) acidic   b) basic

6. In forming chemical compounds, the element nitrogen can assume every integer oxidation state between –3 and +5. What is the oxidation state of nitrogen in laughing gas (N_2O)?

   a) +1   b) +2   c) +3   d) +4   e) +5

7. Consider the following chemical equation:

$$NO_2 + H_2O \rightarrow HNO_2 + HNO_3$$

Is the chemical reaction represented by this equation an oxidation/reduction reaction?

a) yes   b) no   c) insufficient information

8. For an average person, the metabolic rate during a vigorous shiver is about 467 watts. How many kilocalories does this metabolic rate correspond to during 19.0 minutes of shivering? Given: 4.18 J/cal

    a) 8870   b) 127   c) 2230   d) 532000   e) none of these

9. A chemical reaction takes place inside of a metal container that is immersed in 1520 grams of a liquid. The initial temperature of the liquid is 17.0°C. The chemical reaction causes the temperature of the liquid to change. After the chemical reaction, the final temperature of the liquid is 28.0°C. Is the chemical reaction exothermic or endothermic, and how many calories of heat are exchanged as a result of the chemical reaction? The specific heat of the liquid is 0.461 cal/g°C.

    NOTE: Answer must reflect correct sign and magnitude.

    a) +16700   b) –7710   c) +7710   d) –16700   e) none of these

10. The addition of ten calories of heat changes the temperature of one gram of metal element X from 20.0°C to 21.0°C. The addition of ten calories of heat changes the temperature of one gram of metal element Y from 20.0°C to 22.4°C. Which element has the higher atomic mass?

    a) element X   b) element Y
    c) An element's atomic mass is not related to temperature change.

11. If the following chemical reactions take place at 25°C, which reaction(s) will be exothermic?

    i. $HCl\ (g) \rightarrow H\ (g) + Cl\ (g)$ Q = ?
    ii. $H_2\ (g) \rightarrow 2H\ (g)$ Q = ?
    iii. $H_2O\ (g) \rightarrow 2H\ (g) + O\ (g)$ Q = ?

    a) i   b) ii   c) iii   d) i and iii   e) All of these reactions are endothermic.

12. Consider the following bond dissociation energies:

    O_2 118 kcal/mole
    N_2 225 kcal/mole
    NO 150 kcal/mole

    Use these bond dissociation energies to determine the value for Q in kilocalories that is required to complete the following thermochemical equation:

    $$O_2\ (g) + N_2\ (g) \rightarrow 2NO\ (g) \qquad\qquad Q = ?\ kcal$$

    a) 493   b) 193   c) 43   d) 257   e) none of these

13. According to the kinetic molecular theory, an increase in the temperature of a substance corresponds to an increase in the average kinetic energy of the molecules of the substance.

    a) true   b) false

14. If the following physical change takes place at constant temperature, will the change liberate heat or consume heat?

$$Liquid\ Substance \rightarrow Gaseous\ Substance$$

a) consume   b) liberate   c) Energy is not liberated or consumed by this process.

15. The combustion of 1.12 grams of a hydrocarbon results in the formation of 3.33 grams of carbon dioxide and the release of 13.2 kilocalories of heat energy. What quantity of the hydrocarbon in grams would have to be burned in order to release 338 kilocalories of heat energy?

a) 85.3   b) 3980   c) 28.7   d) 0.0437   e) none of these

16. The following thermochemical equation represents the combustion of butane:

$$2\ C_4H_{10}\ (g) + 13\ O_2\ (g) \rightarrow 8\ CO_2\ (g) + 10\ H_2O\ (l) \qquad Q = -1380\ kcal$$

Using this thermochemical equation, calculate the mass of carbon dioxide in grams that would be released when 485 kilocalories is produced by this combustion. Atomic Masses: C = 12.0 O = 16.0   H = 1.01 amu

a) 15.5   b) 1900   c) 1000   d) 124   e) none of these

17. Consider the following incomplete thermochemical equation:

$$2\ C\ (s) + 3\ F_2\ (g) + Cl_2\ (g) \rightarrow 2\ CF_3Cl\ (l) \qquad Q = -?\ kcal$$

If 5.03 grams of carbon is found to release 69.6 kilocalories when it reacts with an excess of fluorine and chlorine according to this equation, what absolute value for Q in kilocalories is required to complete the thermochemical equation? Atomic Masses: C = 12.0   F = 19.0 Cl = 35.5 amu

a) 1.73   b) 332   c) 166   d) 14.6   e) none of these

18. The following equation represents the process of anaerobic glycolysis:

$$C_6H_{12}O_6 \rightarrow 2\ C_3H_6O_3 \qquad Q = -32\ kcal$$

Does the chemical reaction represented by this chemical equation liberate energy or consume energy?

a) liberate   b) consume   c) insufficient information

19. The following chemical equations were discussed in class. Which of the these chemical equations represents the most common net process by which living organisms produce energy for life activity?

    (i) $6 CO_2 + 6 H_2O \rightarrow C_6H_{12}O_6 + 6 O_2$

    (ii) $C_6H_{12}O_6 + 6 O_2 \rightarrow 6 CO_2 + 6 H_2O$

    (iii) $C_6H_{12}O_6 \rightarrow 2 CO_2 + 2 C_2H_6O$

    (iv) $C_6H_{12}O_6 \rightarrow 2 C_3H_6O_3$

a) i   b) ii   c) iii   d) iv

20. Consider the following balanced themochemical equation:

$$CH_4 (g) + 2 O_2 (g) \rightarrow CO_2 (g) + 2 H_2O (l) \qquad\qquad Q = -213 \text{ kcal}$$

According to this equation, what mass of CH_4 in grams would have to be reacted in order to provide sufficient heat to vaporize 816 grams of water under conditions where the phase change constant for the water is 540 cal/g? Atomic Masses: C = 12.0   H = 1.01   O = 16.0 amu

a) 5870   b) 441   c) 7.73   d) 33.1   e) none of these

Sample Exam III: Chapters 8–9

Multiple Choice

1. According to the directed valence theory presented in the textbook, which of the following Lewis structures are not correct?

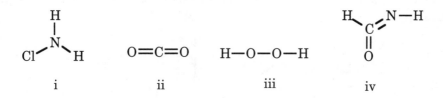

            i   ii   iii   iv

    a) i   b) ii   c) iii   d) iv   e) all correct

2. After drawing Lewis structures, decide which of the following molecules could contain a double bond: i) CH_5N ii) C_2H_6 iii) CCl_2O

    a) i   b) ii   c) iii   d) i and ii   e) i and iii

3. After drawing Lewis structures, decide which of the following molecules could contain an oxygen atom bonded to a hydrogen atom (*i.e.* –O–H):

    (i) CH_4O
    (ii) NH_3O
    (iii) $CHFO$

    a) i   b) ii   c) iii   d) i and ii   e) ii and iii

4. After drawing Lewis structures, decide which of the following molecules could contain a triple bond:

    (i)   HCN
    (ii) C_2H_3Cl
    (iii) N_2H_4

    a) i   b) ii   c) iii   d) i and ii   e) i and iii

5. How many isomers are represented by the molecular formula C_2H_7N?

    a) 1   b) 2   c) 3   d) 4   e) 5

6. Which of the following molecular structures represents a pair of chiral molecules?

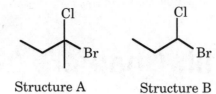

Structure A   Structure B

a) Structure A   b) Structure B
c) Both molecular structures represent a pair of chiral molecules.

7. Which of the following Lewis line structures is the correct structure for 2,4-dimethyl-4-propyloctane?

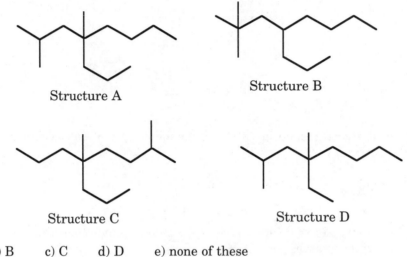

Structure A

Structure B

Structure C

Structure D

a) A   b) B   c) C   d) D   e) none of these

8. What is the molecular formula of the compound represented by the following structural formula?

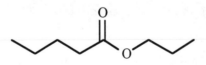

a) $C_8H_{18}O_2$ b) $C_8H_{16}O_2$ c) $C_9H_{18}O_2$ d) $C_9H_{20}O_2$ e) none of these

9. Which of the following functional group combinations best describes the molecule represented by the structural formula in the previous question?

a) alkane/alcohol   b) aromatic/amine   c) aromatic/ester
d) alkane/ester   e) alkane/carboxylic acid

10. A chemist is working with a radioactive isotope that has a half-life of 31 days. If a 370 mg sample of this isotope decays for 124 days, how many milligrams of the original isotope will remain in the sample?

a) 92.5   b) 185   c) 11.9   d) 23.1   e) 46.3

11. What is the identity of the missing atomic nucleus in the following nuclear equation:

$$_0n^1 + _{92}U^{235} \rightarrow _{54}Xe^{144} + ---?--- + 2\ _0n^1$$

a) Rb   b) Sr   c) Y   d) Zr   e) Kr

12. The most abundant isotope of iron makes up 91.8% of the natural element. How many neutrons are contained in the nucleus of the most abundant isotope of iron?

a) 29   b) 26   c) 30   d) 115   e) none of these

13. Two analogous chemical reactions are represented by the equations shown below. In one of these equations, the structure of one of the products is missing. By analogy to the complete equation, which of the suggested structures represents the missing product?

TWO ANALOGOUS CHEMICAL REACTIONS

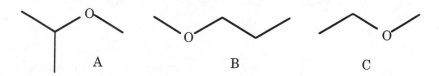

SUGGESTED STRUCTURES FOR MISSING PRODUCT

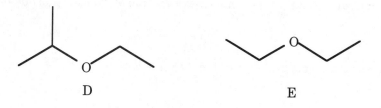

a) A   b) B   c) C   d) D   e) E

14. The figure below represents the cross-sectional schematic of a low pressure discharge tube. The legend next to the schematic identifies the essential components of the tube.

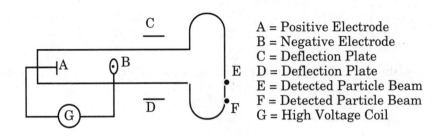

A = Positive Electrode
B = Negative Electrode
C = Deflection Plate
D = Deflection Plate
E = Detected Particle Beam
F = Detected Particle Beam
G = High Voltage Coil

What is the electric charge on the particles striking the detection screen at the positions indicated in the diagram?

a) positive   b) negative   c) insufficient information

15. In the low pressure discharge tube represented in the previous question, what is the charge on the deflection plate labeled with a "D"?

a) positive   b) negative   c) insufficient information

16. The diagram below represents the movement of two protons that are under the influence of the Coulombic (electrostatic) force only. In this diagram, the two particles are moving toward each other. Does the process represented by this diagram liberate energy or consume energy?

a) liberate   b) consume   c) neither

17. What is the identity of the missing particle in the following nuclear equation?

$$_{92}U^{239} \rightarrow \ _{93}Np^{239} + \text{—?—}$$

a) electron   b) alpha particle   c) neutron   d) proton   e) positron

18. Does the following nuclear equation represent an exothermic or an endothermic process?

$$_{92}U^{238} \rightarrow \ _{2}He^4 + \ _{90}Th^{234}$$

a) exothermic   b) endothermic   c) insufficient information

19. The figure below represents an idealized radioactive emission detector. The opposite charges on the detector's deflection plates are as indicated. What type of radioactive emission is striking the detection screen at the position marked "D"?

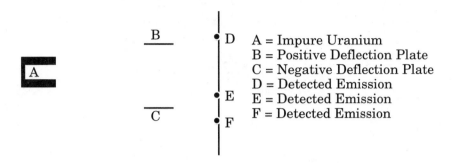

A = Impure Uranium
B = Positive Deflection Plate
C = Negative Deflection Plate
D = Detected Emission
E = Detected Emission
F = Detected Emission

a) neutrons   b) alpha particles   c) beta particles   d) gamma rays

20. Write a balanced nuclear equation for the loss of a positron by $_{17}Cl^{33}$. What is the identity of the new nucleus formed by this radioactive decay process?

a) $_{16}S^{33}$ b) $_{18}Ar^{33}$ c) $_{17}Cl^{32}$ d) $_{17}Cl^{34}$ e) none of these

Appendix II

Answers to Problems

Chapter One

1. The nature of an academic discipline is characterized by two things: subject matter and method of study.

4. The effects of war and depression are obvious, but what happened in 1960? HINT: It had something to do with a chemist named Russell Marker.

5. II and IV are valid, and I and III are invalid.

6. Scientific theories are "proved" by an invalid argument , but they are disproved by a valid argument.

7. If we could get inside Holmes' mind, we would see all four arguments in use. The conclusion he articulates, however, is arrived at by argument I.

9. invalid; invalid.

10. The hypotheses are now disproved by a valid argument.

11. Although the hypotheses become more "certain," the argument by which they are proved is still invalid (argument I).

Chapter Two

1. 28 cm, 0.28 m

2. 0.18 L, 180 mL, 180 cc

3. 370 g, 0.370 kg

4. 6 miles

5.

6.

7.

8.

9.

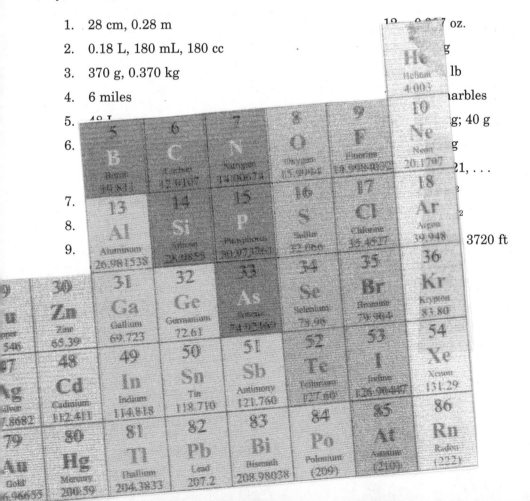

Chapter Three

1. a) yes b) insufficient information
   c) compound

2. Substance A is a compound; there is
   insufficient information to identify
   substance B.

5. 0.935 g/mL

6. 13600 g; 2720 nickels

7. 93.3 mL

8. density = 0.962 g/mL; thus the liquid
   is cyclohexanol.

9. 1720 g

10. 5.67 mL

11. density = 2.70 g/mL; thus, the metal
    may be Al.

12. Ice floats on liquid water, *i.e.* ice is
    less dense than liquid water.

13. 5.88% hydrogen; 94.1% oxygen

14. 40.1% sulfur; 59.9% oxygen

15. 294 g

16. 72.9 g

17. 131 g

18. 4905 kg

19. 1.92 g

20. 271 g

21. 170 g

22. 107 g

Chapter Four

2.

3.

   Li - O

   See Chapter 5 for a discussion of this
   problem.

4. 72.7% O; 3.00 g

8. For carbon monoxide and carbon
   dioxide 133 g and 266 g, respectively,
   of oxygen are necessary to combine
   with 100 g carbon.

9. For methane, ethylene, and acetylene,
   33.3 g, 16.7 g, and 8.34 g, respectively,
   of carbon are necessary to combine
   with 100 g hydrogen.

10. 75.1

Chapter Five

1. b

2. c

3. d

4. a

5. $C_3H_8O_3$

6. $C_8H_5NO_6$

7. a) 10 b) 10 c) 9 d) 12 e) 8

8. 92.1

9. 211.0

10. 106.0

11. a) 58.1 b) 100.1 c) 234.0
    d) 106.1 e) 58.1 f) 262.9
    g) 96.1 h) 195.1

12. Chemical Equation

13. $2H_2O_2 \rightarrow 2H_2O + O_2$

14. $2K + 2H_2O \rightarrow 2KOH + H_2$

15. $2C_2H_6N + 9O_2 \rightarrow 4CO_2 + 6H_2O + 2NO_2$

16. $2C_5H_{10} + 15O_2 \rightarrow 10CO_2 + 10H_2O$

17. $2ZnS + 3O_2 \rightarrow 2ZnO + 2SO_2$

18. $2Cu + O_2 \rightarrow 2CuO$

19. 32.0

20. 27.0

21. 204.0

22. 24.2

23. 27.0

24. a) 20.6   b) 3.89   c) 5.50

25. a) 7.67   b) 0.677   c) 1.35
    d) 0.0833

26. a) i   b) i   c) i   d) i

27. Chlorine is in excess.

28. CF_3

29. a) CH_4 b) SO_3 c) Na_2O

30. A: CH_2 B: CS_2

31. N_2O

32. N_2O_3

33. $CuFeS_2$

34. $CaSiO_3$

35. $MgSO_4 \cdot 7H_2O$

36. a) 62.0%   b) 12.0%   c) 15.4%
    d) 90.5%   e) 82.6%   g) 12.5%
    h) 61.5%

37. 35.9%

38. 79.8%

39. 61.3%

40. 38.5%

41. 8.01 metric tons

42. a) 15.5 metric tons
    b) 3.00 metric tons
    c) 3.85 metric tons
    d) 22.6 metric tons
    e) 20.6 metric tons
    g) 3.12 metric tons
    h) 15.4 metric tons

43. 15.0 tons

44. 171 g

45. 97.9 g

46. 1977

47. 23.5 g

48. 1.47 kg

49. 19.7 g

50. a) 47.5%   b) 63.9 g

51. 3.23 g

52. 2.52 g

53. a) 411 g   b) 84.4 g

54. a) 130 g   b) 102 g

55. 85.0%

56. 75.2%

57. 76.1%

58. 66.7%

59. 77.3%

Chapter Six

1. 2, 3

2. 1, 2

3. a

4. d

5. 5

6. $Mg + 2H_2O \rightarrow Mg(OH)_2 + H_2$

7. a) Ba   b) P

8. a

9. $MgO + SO_2 \rightarrow MgSO_3$

10. Liquid

11. Metals: Cr, Ni, Ga

12. Mg-basic c-acidic

13. B

14. S<7 Ca>7

15. VA, metal, solid

16. sodium phosphate, calcium
    hypochlorite, zinc chloride, sodium
    chlorate, hydrogen sulfide, calcium
    sulfate, silicon dioxide, carbon
    tetrachloride

17. $Al_2(SO_4)_3$, Li_2O, Na_2SO_4, $CuCl_2$, NaOH,
    ZnO, $SiCl_4$, CS_2

18. c

19. b

20. b

21. d

22. b

23. e

24. b

25. a

26. $2KNO_3 \rightarrow 2KNO_2 + O_2$

27. $2NH_4Cl + CaO \rightarrow 2NH_3 + CaCl_2 + H_2O$

28. $CO_2 + 2NaOH \rightarrow Na_2CO_3 + H_2O$

29. $Na_2O + H_2O \rightarrow 2NaOH$

30. $4Ag + 2H_2S + O_2 \rightarrow 2Ag_2S + 2H_2O$

31. $2KClO_3 \rightarrow 3O_2 + 2KCl$

32. $2Cu + O_2 \rightarrow 2CuO$

33. $N_2 + O_2 \rightarrow 2NO$

    $2NO + O_2 \rightarrow 2NO_2$

34. LiF: Li = +1, F = –1

    CaC_2: Ca = +2, C = –1

    $KMnO_4$: K = +1, Mn = +7, O = –2

    H_2SO_4: H = +1, S = +6, O = –2

    CuO: Cu = +2, O = –2

    CuO_2: C = +2, O = –1

    N_2O: N = +1, O = –2

    NO: N = +2, O = –2

35. b

36. e

37. a) displacement   b) redox
    c) addition   d) displacement
    e) acid/base   f) acid/base   g) redox
    h) addition   i) displacement
    j) redox   k) acid/base

38. a) Na_2S b) $CO_2 + H_2O$
    c) $Pb + CO_2 + H_2O$ d) H_2SO_4
    e) $Ca(OH)_2$ f) $I_2 + KCl$

Chapter Seven

1. 1460000 J

2. 1125 J, 0.269 kcal

3. 19.2 m

4. 697 m

5. 76.5 kg

6. 23900 J (rounded off)

7. 254 cal

8. 159°C

9. 8370 cal

10. 21600 cal

11. 3675 J

12. 9.90 m/s

13. 646000 cal

14. $0.09

15. 21.4%

16. 294 J

17. 6.97 m

18. ii

19. Both will have the same K.E.

20. endothermic

21. a) and c) liberate   b) and d) consume

22. endothermic

23. exothermic, 1730 cal (rounded off)

24. exothermic

25. a) exothermic   b) endothermic

26. 83.9 kcal

27. 499 kcal

28. 155 g

29. 33.7 kcal

30. 140 g

31. 64.0 g

32. 560 kcal

33. 30.7 kcal

34. a) 16.3 g   b) 0.620 g

35. b

36. a) and c) endothermic   b) exothermic

37. hot day

38. less than 136 kcal

39. 1.50 kcal/mole

40. –128.4 kcal/mole

41. 52.9 kcal/mole

42. −7.9 kcal

43. −235.4 kcal

44. None of the engines violates the first law. Engine one violates the second law. Engine three might violate the second law.

45. no

46. $N_2H_3I_3$

47. element X

48. element Y

Chapter Eight

1. See section 8.3

2. "Bravais" arrangement I

3. 2 isomers

4. 2 isomers

5.

$$N \equiv N \qquad H-C \equiv C-H$$

6. double: CS_2 C_2H_4

   triple: HCN N_2

7. c

8. b

9.

10.

11.

12.

   *Because of this chiral carbon atom, there are two enantiomers of this molecule.

13.

14.

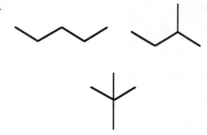

15.

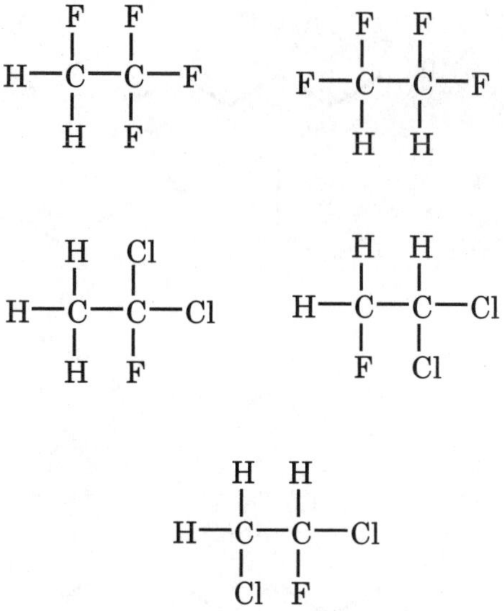

$$H-\underset{\underset{H}{|}}{\overset{\overset{F}{|}}{C}}-\underset{\underset{F}{|}}{\overset{\overset{F}{|}}{C}}-F \qquad F-\underset{\underset{H}{|}}{\overset{\overset{F}{|}}{C}}-\underset{\underset{H}{|}}{\overset{\overset{F}{|}}{C}}-F$$

$$H-\underset{\underset{H}{|}}{\overset{\overset{H}{|}}{C}}-\underset{\underset{F}{|}}{\overset{\overset{Cl}{|}}{C}}-Cl \qquad H-\underset{\underset{F}{|}}{\overset{\overset{H}{|}}{C}}-\underset{\underset{Cl}{|}}{\overset{\overset{H}{|}}{C}}-Cl$$

$$H-\underset{\underset{Cl}{|}}{\overset{\overset{H}{|}}{C}}-\underset{\underset{F}{|}}{\overset{\overset{H}{|}}{C}}-Cl$$

16. The chiral carbon is marked with an asterisk.

17. Caffeine = $C_8H_{12}N_4O_2$
Ibuprofen = $C_{13}H_{18}O_2$
Cholesterol = $C_{27}H_{46}O$

18. $C_{10}H_{14}O$

19. A = 2,4-dimethyl-4-propyloctane
B = 2,2-dimethyl-4-propyloctane
C = 2,5-dimethyl-5-propyloctane
D = 2,4-dimethyl-4-ethyloctane

20.

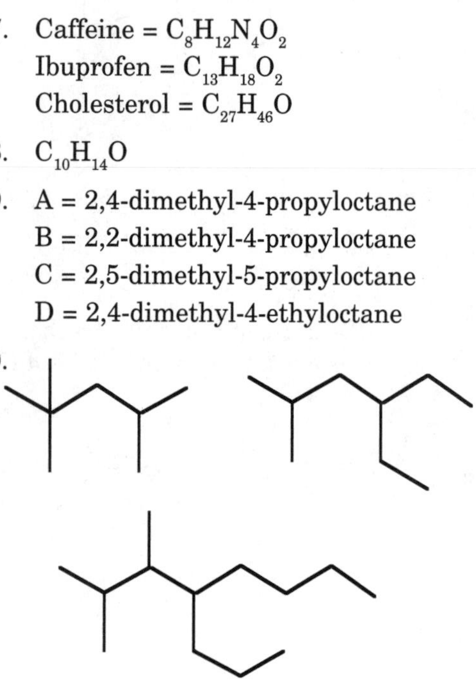

21. 4 Alcohols

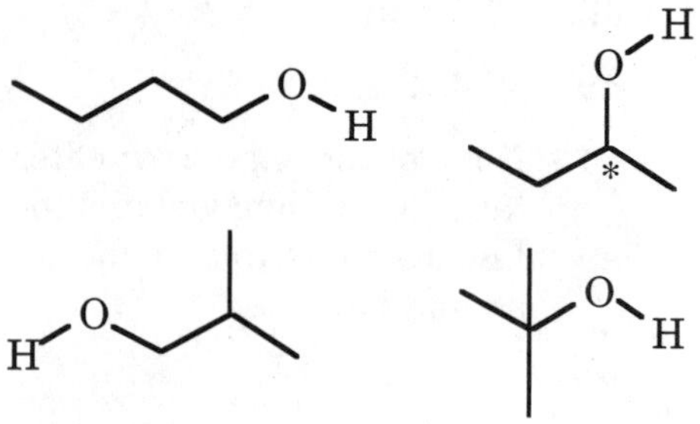

3 Ethers

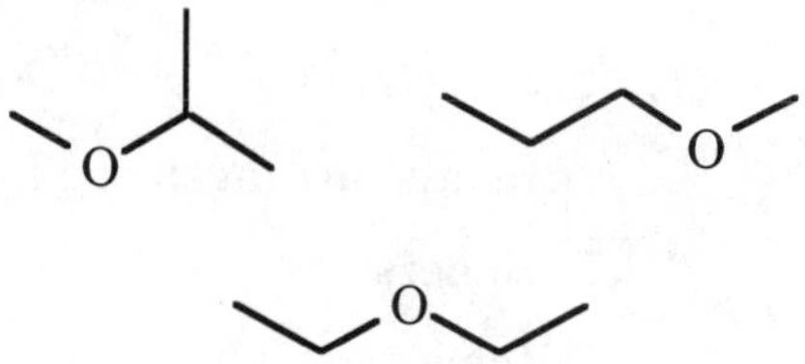

22. A = alkane, alkene, aromatic
B = alkane, ketone
C = alkane, carboxylic acid
D = alkane, ketone, alcohol

23. $C_7H_6O_2$ (d)

24. $C_{10}H_{22}$

25.

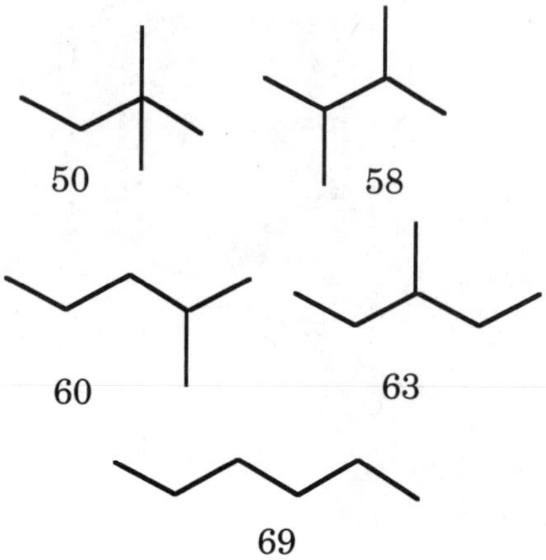

50

58

60

63

69

404

26. A < C < D < B
27. methanol
28.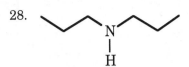

29. structure B

30.

A:

B:

C:

D:

E:

F:

Chapter Nine

1. A = B < C < D; Attract
2. D < C < A = B; Repel
3. Consume
4. Liberate
5. Yes; 0.216 g Al
6. A: positive; B: negative; C: particle b
7. A: 1; B: 3
8. See section 9.10
9. A: proton; B: electron; C: neutron;
   D: alpha particle
10. A = 9; B = 9; C = 19
    1) = Ar; D = 18; E = 22
    2) = He; G = 2; H = 2
    3) = S; M = 16; J = 16
    4) = K; K = 19; L = 21
11. F = 7; Ar = 8; He = 2; S = 6; K = 1
12. $_{84}Po^{218} \rightarrow _{2}He^4 + _{82}Pb^{214}$
13. $_{38}Sr^{90} \rightarrow _{-1}e^o + 39Y^{90}$
14. $_{o}n^1 + _{92}U^{235} \rightarrow _{36}Kr^{97} + _{56}Ba^{136} + 3 \, _{o}n^1$
15. $_{52}Te^{137}$

16. $_{7}N^{12} \rightarrow _{1}e^o + _{6}C^{12}$
17. $_{5}B^{11}$
18. $_{4}Be^9$
19. A = $_{92}U^{239}$; B = $_{93}Np^{239}$; C = $_{-1}e^o$
20. $_{o}n^1$
21. 4 mg
22. 11460 years
23. b, d, e
24. $_{8}O^{17}$
25. 29.2 L
26 – 27.

H H
H :C:C: H   H :Be: H
H H

H :O:O:O: H   H :S: H

:N:::N: H:C:::N:

405

Sample Exam I

c, c, d, c, c, c, b, d, c, b, d, c, d, c, b, b, b, d, c, b

Sample Exam II

b, b, c, b, b, a, a, b, b, b, e, c, a, a, c, d, b, a, b, d

Sample Exam III

d, c, d, a, b, c, a, b, d, d, b, c, e, a, b, b, a, a, c, a

Appendix III

Chemicals & Materials Required for Formal Laboratory Exercises

Each of the textbook chapters concludes with a formal laboratory exercise. These exercises explore a laboratory oriented problem related to the associated chapter. The problems are stated in a more formal manner than the household chemistry problems, and the experiments associated with these problems are designed to be done in a supervised chemical laboratory. Safety issues associated with the equipment and chemicals used in these experiments require proper pre-laboratory instruction. An essential prerequisite to this pre-laboratory instruction is the complete reading of the laboratory exercise.

The formal laboratory exercises are designed to supplement the content of the chapter and to give students an opportunity to experience what chemists actually do. The exercises do not represent an attempt at discovery learning. In many of the experiments, the "unknown" identities are revealed in the introduction. At Illinois State University, students are allowed to work on an experimental procedure until satisfactory results are obtained.

This appendix lists all of the materials and chemicals required for the formal laboratory exercises. All of the solution concentrations are also listed. Solution concentrations are listed as either moles/liter (M) or % wt/vol. Estimated quantities for 100 students are indicated in parentheses.

Chemicals, Solutions

acetic acid, 5% (2 L)
aluminum sulfate, 0.1 M (50 mL)
ammonium acetate, 1 M (50 mL)
ammonium chloride, 2 M (50 mL)
ammonium hydroxide, 6 M (300 mL)
boric acid, 0.3 M (50 mL)
copper sulfate, 0.05 M (100 mL)
ethanol (2 L)
hydrochloric acid, 1 M (10 L)
hydrogen peroxide, 3% (2 L)
iodine solution, 0.25 g I2 + 0.75 g KI in 100 mL water, stir well (100 mL)
litmus indicator solution, 0.25 g of powder/L, heat, filter (1.5 L)
nickel sulfate, 0.1 M (100 mL)
phenolphthalein solution, 1% in ethanol (50 mL)
potassium hydroxide, 1 M (5 L)
sodium carbonate, 0.2 M (50 mL)
sodium hydroxide, 0.1 M (6 L)
sodium hydroxide, 1 M (5 L)
sodium phosphate (dibasic), 0.2 M (50 mL)
Yamada's universal indicator solution, dilute solution 1:4 with water (1.5 L)
zinc acetate, 0.1 M (100 mL)

Chemicals, Solids

aluminum foil (1 roll)
aspirin standard (100 g)
aspirin tablets (200)
calcium carbonate (10 g)
copper carbonate (200 g)
magnesium metal, ribbon (20 g)
povidone (10 g)
starch (10 g)
U.S. five cent coins (100)
U.S. pennies, 1983 or later (200)

Materials

6x4 depression well plate
9V battery pencil battery
beaker, 150 mL
Bunsen burner
buret, 25 or 50 mL
clay triangle
crucible tongs
droppers
Erlenmeyer flasks, 125 and 250 mL
forceps
funnel
graduated cylinder, 100 mL
hotplate
laboratory balance, 1 mg
metric ruler
nickel alloy spatula
porcelain crucible and lid
ring stand, ring, and utility clamp
styrofoam cup
thermometer
wash bottle
wire gauze

Index

Student Note Card

Exam II

The Periodic Table

IA	IIA	IIIB	IVB	VB	VIB	VIIB	VIIIB			IB	IIB	IIIA	IVA	VA	VIA	VIIA	VIIIA
1 H 1.01																1 H 1.01	2 He 4.00
3 Li 6.94	4 Be 9.01											5 B 10.8	6 C 12.0	7 N 14.0	8 O 16.0	9 F 19.0	10 Ne 20.2
11 Na 23.0	12 Mg 24.3											13 Al 27.0	14 Si 28.1	15 P 31.0	16 S 32.1	17 Cl 35.5	18 Ar 40.0
19 K 39.1	20 Ca 40.1	21 Sc 45.0	22 Ti 47.9	23 V 50.9	24 Cr 52.0	25 Mn 54.9	26 Fe 55.9	27 Co 58.9	28 Ni 58.7	29 Cu 63.6	30 Zn 65.4	31 Ga 69.7	32 Ge 72.6	33 As 74.9	34 Se 79.0	35 Br 79.9	36 Kr 83.8
37 Rb 85.5	38 Sr 87.6	39 Y 88.9	40 Zr 91.2	41 Nb 92.9	42 Mo 95.9	43 Tc (98)	44 Ru 101	45 Rh 103	46 Pd 106	47 Ag 108	48 Cd 112	49 In 115	50 Sn 119	51 Sb 122	52 Te 128	53 I 127	54 Xe 131
55 Cs 133	56 Ba 137	57 La * 139	72 Hf 179	73 Ta 181	74 W 184	75 Re 186	76 Os 190	77 Ir 192	78 Pt 195	79 Au 197	80 Hg 201	81 Tl 204	82 Pb 207	83 Bi 209	84 Po 209	85 At 210	86 Rn 222
87 Fr (223)	88 Ra 226	89 Ac † (227)	104 Rf (261)	105 Db (262)	106 Sg (263)	107 Bh (262)	108 Hs (265)	109 Mt (266)	110 Uun	111 Uuu	112 Uub						

*	58 Ce 140	59 Pr 141	60 Nd 144	61 Pm (145)	62 Sm 150	63 Eu 152	64 Gd 157	65 Tb 159	66 Dy 163	67 Ho 165	68 Er 167	69 Tm 169	70 Yb 173	71 Lu 175
†	90 Th 232	91 Pa (231)	92 U 238	93 Np (237)	94 Pu (244)	95 Am (243)	96 Cm (247)	97 Bk (247)	98 Cf (251)	99 Es (252)	100 Fm (257)	101 Md (258)	102 No (259)	103 Lr (260)

Notes:

Student Note Card

Exam III

The Periodic Table

IA	IIA	IIIB	IVB	VB	VIB	VIIB	VIIIB			IB	IIB	IIIA	IVA	VA	VIA	VIIA	VIIIA
1 H 1.01																1 H 1.01	2 He 4.00
3 Li 6.94	4 Be 9.01											5 B 10.8	6 C 12.0	7 N 14.0	8 O 16.0	9 F 19.0	10 Ne 20.2
11 Na 23.0	12 Mg 24.3											13 Al 27.0	14 Si 28.1	15 P 31.0	16 S 32.1	17 Cl 35.5	18 Ar 40.0
19 K 39.1	20 Ca 40.1	21 Sc 45.0	22 Ti 47.9	23 V 50.9	24 Cr 52.0	25 Mn 54.9	26 Fe 55.9	27 Co 58.9	28 Ni 58.7	29 Cu 63.6	30 Zn 65.4	31 Ga 69.7	32 Ge 72.6	33 As 74.9	34 Se 79.0	35 Br 79.9	36 Kr 83.8
37 Rb 85.5	38 Sr 87.6	39 Y 88.9	40 Zr 91.2	41 Nb 92.9	42 Mo 95.9	43 Tc (98)	44 Ru 101	45 Rh 103	46 Pd 106	47 Ag 108	48 Cd 112	49 In 115	50 Sn 119	51 Sb 122	52 Te 128	53 I 127	54 Xe 131
55 Cs 133	56 Ba 137	57 La * 139	72 Hf 179	73 Ta 181	74 W 184	75 Re 186	76 Os 190	77 Ir 192	78 Pt 195	79 Au 197	80 Hg 201	81 Tl 204	82 Pb 207	83 Bi 209	84 Po 209	85 At 210	86 Rn 222
87 Fr (223)	88 Ra 226	89 Ac † (227)	104 Rf (261)	105 Db (262)	106 Sg (263)	107 Bh (262)	108 Hs (265)	109 Mt (266)	110 Uun	111 Uuu	112 Uub						

*	58 Ce 140	59 Pr 141	60 Nd 144	61 Pm (145)	62 Sm 150	63 Eu 152	64 Gd 157	65 Tb 159	66 Dy 163	67 Ho 165	68 Er 167	69 Tm 169	70 Yb 173	71 Lu 175
†	90 Th 232	91 Pa (231)	92 U 238	93 Np (237)	94 Pu (244)	95 Am (243)	96 Cm (247)	97 Bk (247)	98 Cf (251)	99 Es (252)	100 Fm (257)	101 Md (258)	102 No (259)	103 Lr (260)

Notes: